Foundations of
Ultracentrifugal Analysis

CHEMICAL ANALYSIS

A SERIES OF MONOGRAPHS ON ANALYTICAL CHEMISTRY AND ITS APPLICATIONS

VOLUME 42

A WILEY-INTERSCIENCE PUBLICATION

JOHN WILEY & SONS

New York / London / Sydney / Toronto

Foundations of Ultracentrifugal Analysis

HIROSHI FUJITA

Department of Polymer Science
Osaka University
Toyonaka, Japan

A WILEY-INTERSCIENCE PUBLICATION

JOHN WILEY & SONS
New York / London / Sydney / Toronto

Library of Congress Cataloging in Publication Data:

Fujita, Hiroshi, 1922–
    Foundations of ultracentrifugal analysis.

    (Chemical analysis; v. 42)
    "A Wiley-Interscience publication."
    Includes bibliographical references and index.
    1. Centrifuges. 2. Sedimentation analysis.
I. Title. II. Series.

QD54.C4F83   544'.1   74-20899
ISBN 0-471-28582-X

Printed in the United States of America

10 9 8 7 6 5 4 3 2 1

Dedicated to Dr. John Warren Williams,
Professor Emeritus of the University of Wisconsin,
with esteem and affection

PREFACE

The ultracentrifuge has become an almost indispensable instrument for macromolecular research in modern laboratories. Its models of current design are indeed versatile and useful, enabling the laboratory worker to perform both sedimentation velocity and equilibrium experiments with accuracy, reliability, and ease under a variety of precisely controlled external conditions. The interpretation of the experimental results, however, requires great caution. It can only be achieved successfully with adequate attention to our current knowledge on the theory of ultracentrifugal analysis. In addition, a theory when understood may provide the guiding principle in the design of a more advantageous or novel type of experiment.

Because of such facts I attempt in this book to describe in a logical and coherent fashion those theoretical contributions which seem to me to have played a fundamental role in the developments of ultracentrifugal methods. The emphasis is mathematical, but the discussion deals, in many cases, with problems of primary interest to the experimentalist, such as the ways and conditions in which a derived theoretical relation is compared with experimental data to evaluate the parameters contained in it. It is hoped that the material summarized herein will afford the reader with deeper understanding and appreciation of the theory of ultracentrifugal analysis in its present form.

The arrangement of subjects follows essentially the same pattern as in my previously published monograph of a similar kind, *Mathematical Theory of Sedimentation Analysis* (Academic Press, New York and London, 1962), except that a new chapter dealing with sedimentation equilibrium of chemically reacting solutes is added. The discussion covers a much wider range of topics, including those which have appeared recently or have received renewed interest in the past decade. Certain problems treated in the 1962 book are omitted or given less space when considered to be of less importance. Solutions to integral or differential equations are presented as given, except for cases where the derivations are simple or substantial for the understanding of the theory concerned. Instead, every effort is made to acquaint the reader with the assumptions and approximations to which a given solution is subjected. Though useful and presumably relevant on the elementary level, the kinetic-theoretical or mechanistic approaches to sedimentation phenomena are not described here because they have proved inadequate for the general and rigorous formulation. As is now well recognized, such formulation must be done on the basis of nonequilibrium thermodynamics governing flow processes in liquid systems. Chapter 1 may be regarded as an introduction to this new thermodynamics, and I hope the reader will find it to be informative.

As a general statement one can say that it is the fields of biochemistry and molecular biology where the ultracentrifuge has found its greatest success and where incentives to the recent developments in experimental and theoretical aspects of the ultracentrifugal methods have originated. Thus, in writing the present monograph, I have thought it appropriate to make its general context more attractive to the technical workers in these fields, and have made some efforts in this direction. For example, whenever possible, illustrative examples for the theoretical results obtained are taken from journals with which these technicians are familiar. Problems are often formulated or stated with special reference to systems in which macromolecules of biological interest are studied. Because of their greater significance in protein chemistry, methods of investigating self-associating solutes by sedimentation transport and equilibrium experiments are discussed in detail. Considerable space is given to an exposition of the density-gradient sedimentation equilibrium in view of its importance for studies on DNA and other nucleic acids.

I wish to express my gratitude to Dr. J. W. Williams, Professor Emeritus of the University of Wisconsin and my teacher, and Dr. R. C. Deonier of the University of Southern California who read an initial draft of this book with care and provided many valuable comments, and to Professor E. T. Adams, Jr., of Texas A & M University who recommended me to the editors of the *Chemical Analysis* series. I am much indebted to Professor T. Kotaka of the Institute for Chemical Research, Kyoto University, for a draft from which the final manuscript for Section VI in Chapter 5 has been prepared. Thanks are rendered to my associates Dr. T. Norisuye and Mr. I. Omura for checking equations and drawing figures. Mrs. Y. Norisuye and Miss S. Kubo have displayed their skill in the preparation of the manuscript, which I acknowledge with thanks. Without the encouragement, advice, and assistance of these persons among several others, this monograph would not have been completed.

HIROSHI FUJITA

Toyonaka, Japan
June 23, 1973

CONTENTS

CHAPTER 1 FUNDAMENTAL EQUATIONS FOR FLOW PROCESSES IN THE ULTRACENTRIFUGE

CHAPTER 2   SEDIMENTATION TRANSPORT IN TWO-COMPONENT SYSTEMS

CHAPTER 5 SEDIMENTATION EQUILIBRIUM IN NONREACTING SYSTEMS

CHAPTER 7 APPROACH TO SEDIMENTATION EQUILIBRIUM

Foundations of
Ultracentrifugal Analysis

1

FUNDAMENTAL EQUATIONS FOR FLOW PROCESSES IN THE ULTRACENTRIFUGE

A. INTRODUCTION

The starting point for the phenomenological theory of sedimentation processes in the ultracentrifuge is the derivation of flow (or flux) equations which describe the isothermal mass transport of thermodynamic components in a centrifugal field. For a binary solution, that is, a system which consists of a homogeneous solute and a solvent, a useful flow equation for the solute may be derived by resorting to a kinetic theory approach, in which transport of solute molecules through the solvent is considered to be the result of a centrifugal force, a buoyant force, and a diffusion force. This approach prevailed in the early days of the development of sedimentation theory; even now some textbooks of physical chemistry or biophysical chemistry adopt it in an account of sedimentation phenomena, probably because of its simplicity.

It has become clear in recent years that the kinetic theory must be replaced by nonequilibrium thermodynamics (or thermodynamics of irreversible processes) in order to achieve the general and rigorous derivation of flow equations for the ultracentrifuge. Although many others have contributed, the chief credit for this important recognition must be given to Hooyman.[1] In this chapter, we show how the flow equations basic to the whole subject of this monograph are deduced from fundamental principles of nonequilibrium thermodynamics. In doing this, we shall often cite basic assumptions (or postulates) and equations of this new thermodynamics without going into their details. The reader will find necessary information on them in any of the recently published textbooks[2-6] or review articles.[7,8]

B. THE COORDINATE SYSTEM

The ordinary ultracentrifuge cell is a truncated sector of a cylinder as shown in Fig. 1.1. It is necessary to use a cell of this type because the force causing the motion of components in it acts radially from the axis of the rotor; use of a rectangular cell would produce convection arising from accumulation of matter at the side walls of the cell. Strictly speaking, even a sector-shaped cell is unable to prevent the components from convective

1

flows. The Coriolis force, which is necessarily involved in a rotating system as considered here, produces a transverse effect. Hooyman et al.[9] showed that the Coriolis effect is almost completely negligible under the ordinary working conditions with the current ultracentrifuges. Therefore, we neglect it in the theoretical treatments given in the present monograph.

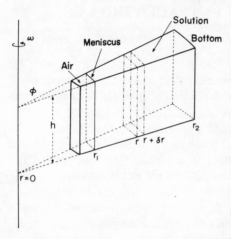

Fig. 1.1. The sector-shaped ultracentrifuge cell. A thin slice of volume indicated by chain lines is for the convenience of discussions given later. $r_2 - r_1$ is the length of solution column, h is the depth of solution column, and ϕ is the sector angle.

The flow equations assume simple forms if use is made of the system of cylindrical coordinates as shown in Fig. 1.1. Here r is the radial distance measured from the rotor axis, h is the distance parallel to the rotor axis, and ϕ is the sector angle. For cells designed for the Beckman-Spinco Model E ultracentrifuge, ϕ is 2, 2.5, or 4° and h ranges from 1.5 to 30 mm. The cell is filled with a solution which extends from $r = r_1$ to $r = r_2$. At the cylindrical boundary at $r = r_1$, the solution is in contact with an air bubble so that we have at this position a liquid-air interface, which is called the *meniscus*. Because the air bubble is admitted into the cell at atmospheric pressure while filling and before sealing the cell, the pressure of the solution at the meniscus is maintained at that of the atmosphere even though the rotor may spin under vacuum. The rigid cylindrical boundary at $r = r_2$ is called the *cell bottom*. Sometimes a thin layer of immiscible liquid which is denser than a given solution is inserted between the cell bottom and the solution in order to facilitate the measurement of the distribution of refractive indices or refractive index gradients in the region near the cell bottom. Such a liquid is usually called *bottom liquid*. When this is used, the liquid-liquid interface between it and the solution must be taken as the position for r_2. The distance between r_1 and r_2 is referred to as the length of the solution column, or simpy the solution depth or height. This length may be adjusted by changing the volume of a solution admitted into the cell. In conventional sedimentation velocity experiments it is about 1 cm, whereas in current sedimentation equilibrium experiments it may be adjusted to less than about 3 mm in order to speed up the attainment of the equilibrium (for the reason see Chapter 7).

Physically, it is obvious that under the geometrical conditions described above (with the neglect of the Coriolis effect), each component in the solution column moves only in the radial direction (either centrifugal or centripetal) when it is subjected to a centrifugal force, and that the physical situation at any given time during the centrifugation is identical along a circular arc of fixed radius. This suggests that for a given speed of rotation only r and time t appear as independent variables in the final forms of equations which describe flow processes in the ultracentrifuge cell.

C.   DEFINITIONS OF FLOWS (OR FLUXES)

Consider a system which consists of $q+1$ *nonreacting* components. For convenience in presentation, the component 0 is specified as the solvent, and all others (labeled $1, 2, \ldots, q$) are called solutes. For the moment, all solutes are assumed to be nonelectrolytes. Introduction of electrolyte components is deferred to Section K.

Suppose for simplicity that this system is contained in a straight tube of uniform cross section and is subject to some "forces" which may produce a unidirectional motion of each component along the tube. Such forces may be gradients of the total potential [the chemical potential plus the potentials due to external forces (mechanical, electric, magnetic, etc.)] and of temperature along the tube. Because any uniform motion of the tube as a whole does not give rise to irreversible processes in the system, we omit it from the subsequent considerations.

We set a coordinate origin 0 at some point on the tube and express the distance in the direction of the tube in terms of a variable x. Suppose there is at a certain point x a plane frame which is perpendicular and fixed to the tube. Each component passes through this frame either in the positive or the negative direction of the x-axis. We define the *flow* or *flux* of component k as the number of grams of component k per second which crosses 1 cm^2 of this frame, and we designate it by the symbol $(J_k)_c$. The unit of $(J_k)_c$ is g/cm^2-sec. Here the subscript c implies that the frame considered is fixed to the tube. At a given time $(J_k)_c$ may vary with x, and at a given x it may change with time t. Let the c-scale concentration (grams of a component per cubic centimeter of solution) be denoted by c_k and let the average velocity of molecules (or particles) of component k at the position x and a given time t be denoted by $(u_k)_c$. Then we have

$$(J_k)_c = c_k(u_k)_c \tag{1.1}$$

It should be noted that $(u_k)_c$ does not mean the local velocity of a particular molecule of component k but the average of velocities of all molecules of component k in a volume element at the position considered, where the volume element is taken very small macroscopically but large enough microscopically to accommodate a great number of molecules. Also we should notice that $(u_k)_c$ is the velocity *relative to the tube* fixed in space. Both c_k and $(u_k)_c$ are generally functions of x and t.

Next we allow the frame to move at each point on the x-axis with the velocity of solvent, $(u_0)_c$, at that point, and we denote the flow of component k relative to this solvent-fixed frame by $(J_k)_0$. Then we have obviously

$$(J_0)_0 = 0 \tag{1.2}$$

The value of $(J_k)_0$ for component $k(k \neq 0)$ is the product of c_k and the velocity of component k relative to the solvent-fixed frame. Hence

$$(J_k)_0 = c_k[(u_k)_c - (u_0)_c] \tag{1.3}$$

We refer to $(J_k)_0$ as the *solvent-fixed flow* of component k, whereas $(J_k)_c$ defined by equation 1.1 is called the *cell-fixed flow* of that component. The combination of equations 1.1 and 1.3 yields

$$(J_k)_0 = (J_k)_c - c_k(u_0)_c \tag{1.4}$$

Use of equation 1.2 gives

$$(u_0)_c = \frac{(J_0)_c}{c_0} \tag{1.5}$$

Hence we have the relation

$$(J_k)_0 = (J_k)_c - \left(\frac{c_k}{c_0}\right)(J_0)_c \tag{1.6}$$

We define a velocity $(u)_M$ by

$$(u)_M = \frac{\sum_{k=0}^{q} c_k(u_k)_c}{\sum_{k=0}^{q} c_k} \tag{1.7}$$

and call it the velocity of the local center of mass (relative to the tube) or the local barycentric velocity. The value of this velocity depends on the position x and the time t considered. Equation 1.7 may be put in the form

$$(u)_M = \frac{1}{\rho} \sum_{k=0}^{q} (J_k)_c \tag{1.8}$$

where ρ, the local density of the solution, is expressed by the sum of c_k over all components, that is,

$$\rho = \sum_{k=0}^{q} c_k \tag{1.9}$$

Obviously, ρ is a function of x and t. We define still another flow of component k relative to the frame moving with the velocity $(u)_M$, denote it by $(J_k)_M$, and term it the *mass-fixed flow* of component k. The frame to which this quantity is referred is called the mass-fixed frame. By definition we have

$$(J_k)_M = c_k[(u_k)_c - (u)_M] \tag{1.10}$$

After summation of equation 1.10 over all components and use of equation 1.7, we obtain

$$\sum_{k=0}^{q} (J_k)_M = 0 \tag{1.11}$$

This relation is basic for the mass-fixed flows, and it may be compared to equation 1.2, which applies for solvent-fixed flows. It is seen that the mass-fixed flows of $q+1$ components are not independent of each other, but are related linearly. Thus if q of them are known, the remaining one is automatically determined. By combining equations 1.1, 1.8, and 1.10 we obtain

$$(J_k)_M = (J_k)_c - \frac{c_k}{\rho} \sum_{k=0}^{q} (J_k)_c \tag{1.12}$$

which is to be compared to equation 1.6 for $(J_k)_0$. Obviously, $(J_k)_M$ is a function of x and t.

Finally, we define a velocity $(u)_V$ by

$$(u)_V = \frac{\sum_{k=0}^{q} \bar{v}_k c_k (u_k)_c}{\sum_{k=0}^{q} \bar{v}_k c_k} \tag{1.13}$$

where $\bar{v}_k$ is the partial specific volume of component k in the solution at the position x and the time t considered. By using the well-known relation[10]

$$\sum_{k=0}^{q} \bar{v}_k c_k = 1 \qquad (1.14)$$

equation 1.13 is reduced to

$$(u)_V = \sum_{k=0}^{q} \bar{v}_k c_k (u_k)_c \qquad (1.15)$$

We call $(u)_v$ the velocity of the local center of volume; it generally varies with x and t. The *volume-fixed flow* of component k is then defined as the flow relative to that frame which moves with the velocity $(u)_V$, and is denoted here by $(J_k)_V$. It is a function of x and t and can be represented by

$$(J_k)_V = c_k[(u_k)_c - (u)_V] \qquad (1.16)$$

or

$$(J_k)_V = (J_k)_c - c_k(u)_V \qquad (1.17)$$

Introduction of equation 1.1 into equation 1.15 gives

$$(u)_V = \sum_{k=0}^{q} \bar{v}_k (J_k)_c \qquad (1.18)$$

Therefore, equation 1.17 may be written

$$(J_k)_V = (J_k)_c - c_k \sum_{i=0}^{q} \bar{v}_i (J_i)_c \qquad (1.19)$$

If both sides are multiplied by $\bar{v}_k$ and summed over all components, there results

$$\sum_{k=0}^{q} \bar{v}_k (J_k)_V = 0 \qquad (1.20)$$

where equation 1.14 has been used. Equation 1.20 indicates that the volume-fixed flows of $q+1$ components are not independent of each other, but are related linearly. It may be compared to equation 1.11 for mass-fixed flows.

The solvent-fixed, mass-fixed, and volume-fixed flows defined above all may be regarded as special cases of the flows $(J_k)_R$ defined by a general relation:

$$\sum_{k=0}^{q} (a_k)_R (J_k)_R = 0 \qquad (1.21)$$

where the $(a_k)_R (k = 0, 1, \ldots, q)$ are the set of coefficients characterizing the frame to which the $(J_k)_R$ refer.

For

$$(a_0)_R = 1, \qquad (a_k)_R = 0 \qquad (k = 1, 2, \ldots, q) \qquad (1.22)$$

equation 1.21 reduces to equation 1.2 for solvent-fixed flows; for

$$(a_k)_R = 1 \qquad (k = 0, 1, \ldots, q) \qquad (1.23)$$

equation 1.21 agrees with equation 1.11 for mass-fixed flows; and for

$$(a_k)_R = \bar{v}_k \qquad (k = 0, 1, \ldots, q) \qquad (1.24)$$

equation 1.21 is identical to equation 1.20.

By assigning other values to $(a_k)_R$ we may define various frames of reference. The group of such frames is hereafter called the R-*group of reference frames* for flows. It should be noted that the cell-fixed frame is in general not contained in this group: No set of $(a_k)_R$ exists that satisfies equation 1.21 when the $(J_k)_R$ are taken to be $(J_k)_c$, except for a special case which is discussed in Section G.3. In other words, except for such a case, the cell-fixed flows of all components in the solution are linearly independent of each other.

It can be shown that flows referred to any frame of the R-group are linearly related to those corresponding to any other frame of this group. Let two frames of reference belonging to the R-group be denoted by R′ and R″. By definition

$$\sum_{k=0}^{q} (a_k)_{R'} (J_k)_{R'} = 0, \qquad \sum_{k=0}^{q} (a_k)_{R''} (J_k)_{R''} = 0 \qquad (1.25)$$

If the local velocity of frame R″ relative to frame R′ is represented by $(u)_{R''R'}$, we have the relation

$$(J_k)_{R''} = (J_k)_{R'} - c_k (u)_{R''R'} \qquad (1.26)$$

Multiplication by $(a)_{R''}$, summation over all components, and use of the second equation of 1.25 allows equation 1.26 to be solved for $(u)_{R''R'}$ to give

$$(u)_{R''R'} = \frac{\sum_{k=0}^{q}(a_k)_{R''}(J_k)_{R'}}{\sum_{k=0}^{q}(a_k)_{R''}c_k} \tag{1.27}$$

provided

$$\sum_{k=0}^{q}(a_k)_{R''}c_k \neq 0 \tag{1.28}$$

Substitution of equation 1.27 back into equation 1.26 yields

$$(J_k)_{R''} = (J_k)_{R'} - \frac{c_k\sum_{i=0}^{q}(a_i)_{R''}(J_i)_{R'}}{\sum_{i=0}^{q}(a_i)_{R''}c_i} \tag{1.29}$$

which shows that $(J_k)_{R''}$ and $(J_k)_{R'}$ are linearly related. Since the derivation given above did not use the first equation of 1.25, equation 1.29 holds even if the frame R' does not belong to the R-group. Thus we may employ $(J_k)_c$ for $(J_k)_{R'}$. Then, with $(a_0)_{R''} = 1$ and $(a_k)_{R''} = 0$ $(k = 1, 2, \ldots, q)$, equation 1.29 reduces to equation 1.6. Similarly, equation 1.12 is derived as a special case of equation 1.29. Of course, equation 1.29 provides explicit relations between any pairs of $(J_k)_0$, $(J_k)_M$, and $(J_k)_V$. For example, we can deduce the following relation between $(J_k)_V$ and $(J_k)_M$:

$$(J_k)_V = (J_k)_M - c_k\sum_{i=0}^{q}\bar{v}_i(J_i)_M \tag{1.30}$$

In this case,

$$\sum_{k=0}^{q}(a_k)_{R''}c_k = \sum_{k=0}^{q}\bar{v}_k c_k = 1$$

so that condition 1.28 is satisfied.

For the ultracentrifuge cell, in which each component moves only in the radial direction, we define the cell-fixed flow of component k, $(J_k)_c$, by taking a cylindrical surface concentric with the cell bottom as the frame of reference. That is, it is defined as the number of grams of component k per second which crosses 1 cm^2 of such a cylindrical surface fixed at a particular radial position r. By allowing this surface to move with the velocities $(u_0)_c$, $(u)_M$, and $(u)_V$, we can define the solvent-fixed flows, the mass-fixed flows, and the volume-fixed flows appropriate to the ultracentrifuge cell for all components. Here $(u)_M$ and $(u)_V$ are defined by equations 1.7 and 1.15, respectively, with $(u_k)_c$ being interpreted as the velocity in the radial direction of component k (relative to the cell) at the particular radial position r and time t considered. For the various flows and the velocities of components thus defined, all the relations presented on the preceding pages hold exactly as they stand.

D. HYDRODYNAMIC EQUATIONS

The first step toward the derivation of flow equations is the formulation of three hydrodynamic equations relevant to a given system. These are mathematical expressions for the conservation of mass, of momentum, and of energy, and they can be deduced as special cases of a general equation called the *continuity equation*. In this section, we derive the continuity equation appropriate to the ultracentrifuge cell.

Let G^* be any extensive quantity, and let G be its value per gram of solution. Consider a cylindrical slice as shown by chain lines in Fig. 1.1. The amount of G^* contained in this slice is represented by $\rho G \delta V$, where δV is the volume of the slice and ρ is the local density of the solution. δV is equal to $\phi h r\, \delta r$, where ϕ is the sector angle and h is the depth of the cell. The change in $\rho G \delta V$ for a short interval of time δt is therefore expressed by

$$\phi h r\, \frac{\partial(\rho G)}{\partial t}\, \delta r\, \delta t \tag{1.31}$$

Let us denote by $(J_G)_c$ the amount of G^* per second which flows in the radial direction through 1 cm^2 of a cell-fixed cylindrical surface. Then the amount of G^* flowing into the slice for time δt through the surface at r minus that flowing out for the same time through the surface at $r + \delta r$ may be represented by

$$-\phi h\, \frac{\partial}{\partial r}\left[(J_G)_c r\right] \delta r\, \delta t \tag{1.32}$$

Finally, we denote by Φ_G the rate at which G^* is produced per cubic centimeter within the slice. Then the amount of G^* produced for time δt in the slice is expressed by

$$\phi h r \Phi_G\, \delta r\, \delta t \tag{1.33}$$

The quantity Φ_G is often referred to as the source function. Conservation of G^* requires that the quantity given by equation 1.31 be equal to the sum of the two quantities represented by equations 1.32 and 1.33. Therefore, it follows that

$$\frac{\partial(\rho G)}{\partial t} + \frac{1}{r}\frac{\partial}{\partial r}[r(J_G)_c] = \Phi_G \tag{1.34}$$

This is the desired continuity equation for an extensive quantity G^*.

1. Equation for the Conservation of Mass

First, let us take the mass of component k as G^*. Then $G = c_k/\rho$ and $(J_G)_c = (u_k)_c c_k$. Furthermore, $\Phi_G = 0$ if it is assumed that no components in the solution react chemically with each other. Then equation 1.34 becomes

$$\frac{\partial c_k}{\partial t} + \frac{1}{r}\frac{\partial}{\partial r}[rc_k(u_k)_c] = 0 \tag{1.35}$$

This is called the continuity equation for component k. We note that equation 1.35 may be written

$$\frac{\partial c_k}{\partial t} + \frac{1}{r}\frac{\partial}{\partial r}[r(J_k)_c] = 0 \tag{1.36}$$

It will be seen later that this equation becomes important in deriving the "differential equations" for the ultracentrifuge.

Summing equation 1.35 over all components and using equations 1.7 and 1.9, we obtain

$$\frac{\partial \rho}{\partial t} + \frac{1}{r}\frac{\partial}{\partial r}[r\rho(u)_M] = 0 \tag{1.37}$$

This equation can be put in the form

$$\frac{D\rho}{Dt} + \frac{\rho}{r}\frac{\partial}{\partial r}[r(u)_M] = 0 \tag{1.38}$$

if we introduce a "substantial" derivative D/Dt defined by

$$\frac{D}{Dt} = \frac{\partial}{\partial t} + (u)_M\frac{\partial}{\partial r} \tag{1.39}$$

2. Equation of Motion

Next, let us take the radial component of momentum of the solution as G^*. Then G becomes equal to $(u)_M$, and $(J_G)_c$ is given by $(u)_M^2\rho$. Thus equation 1.34 takes the form

$$\frac{\partial[(u)_M\rho]}{\partial t} + \frac{1}{r}\frac{\partial}{\partial r}[(u)_M^2\rho r] = \Phi_G \tag{1.40}$$

Using equations 1.37 and 1.39, we can transform this equation into

$$\rho\frac{D(u)_M}{Dt} = \Phi_G \tag{1.41}$$

By Newton's law of mechanics, Φ_G for the G^* considered here is equal to the radial component of the force acting per unit volume of solution. If we assume that the motion of each component is not affected by the side walls of the cell except in a very thin layer near each wall, the contribution from shearing stresses may be ignored, and it can be shown that Φ_G is given by

$$\Phi_G = \rho F - \frac{\partial P}{\partial r} \tag{1.42}$$

Here F is the centrifugal force acting on 1 g of the solution and P is the hydrostatic pressure. Let F_k be the centrifugal force acting on 1 g of component k. Then we have

$$\rho F = \sum_{k=0}^{q} c_k F_k \tag{1.43}$$

Thus equation 1.41 becomes

$$\rho\frac{D(u)_M}{Dt} = \sum_{k=0}^{q} c_k F_k - \frac{\partial P}{\partial r} \tag{1.44}$$

which is called the *equation of motion* of the solution.

3. Equation for the Conservation of Energy

Finally, let G^* be the total energy of the solution. Then

$$G = \frac{1}{2}(u)_M^2 + E = \frac{1}{2}(u)_M^2 + \frac{\sum_{k=0}^{q}\bar{E}_k c_k}{\rho} \tag{1.45}$$

where $\bar{E}_k$ denotes the partial *specific* energy of component k and E is the energy of 1 g of the solution with exclusion of its barycentric kinetic energy. Next, $(J_G)_c$ for the present G^* is represented by

$$(J_G)_c = \tfrac{1}{2}(u)_M^2 \rho(u)_M + \sum_{k=0}^{q} \bar{E}_k (u_k)_c c_k$$

which may be rewritten

$$(J_G)_c = [\tfrac{1}{2}(u)_M^2 + E]\rho(u)_M + (J_E)_M \tag{1.46}$$

with

$$(J_E)_M = \sum_{k=0}^{q} \bar{E}_k (J_k)_M \tag{1.47}$$

Introduction of equation 1.46 into equation 1.34, followed by rearrangement of terms by use of equations 1.37 and 1.39, yields

$$\rho \frac{D}{Dt}\left[E + \frac{1}{2}(u)_M^2\right] = \Phi_G - \frac{1}{r}\frac{\partial}{\partial r}[r(J_E)_M] \tag{1.48}$$

It can be shown easily that Φ_G for the present G^* is given by

$$\Phi_G = \sum_{k=0}^{q}(u_k)_c c_k F_k - \frac{1}{r}\frac{\partial}{\partial r}[rP(u)_M] - \sum_{k=0}^{q}\frac{1}{r}\frac{\partial}{\partial r}[rP\bar{v}_k(J_k)_M] \tag{1.49}$$

Note that the third term on the right-hand side relates to the compression work done by pressure. Substitution of equation 1.49 into equation 1.48 and simplification by use of equation 1.44 leads to the desired result:

$$\rho \frac{DE}{Dt} = \sum_{k=0}^{q} F_k(J_k)_M - \frac{P}{r}\frac{\partial}{\partial r}[r(u)_M] - \sum_{k=0}^{q}\frac{1}{r}\frac{\partial}{\partial r}[rP\bar{v}_k(J_k)_M]$$

$$- \frac{1}{r}\frac{\partial}{\partial r}[r(J_E)_M] \tag{1.50}$$

E. GENERALIZATION OF THE SECOND LAW OF THERMODYNAMICS

According to the second law of thermodynamics, we have the following

inequality for a small irreversible change of a *closed* system:

$$d\mathfrak{S} > \frac{dQ}{T_{surr}} \qquad (1.51)$$

where $d\mathfrak{S}$ is the change in entropy of the system, dQ is the heat absorbed by the system during the process, and T_{surr} is the absolute temperature of the surroundings. Thus if we denote by $d_e\mathfrak{S}$ the amount of entropy which flows into a system when it is subject to a small change, and write

$$d\mathfrak{S} = d_e\mathfrak{S} + d_i\mathfrak{S} \qquad (1.52)$$

the quantity $d_i\mathfrak{S}$ is always positive, provided that the change proceeds irreversibly and the system is closed. This leads to the following interpretation: When a closed system changes irreversibly, a positive amount of entropy is always produced inside the system.

For an *open* system $d_e\mathfrak{S}$ may not be equal to dQ/T_{surr}, but may contain additional contributions from substances which flow in or out of the system. Under such circumstances, both $d\mathfrak{S}$ and $d_e\mathfrak{S}$ could be either positive or negative, and we can say nothing general about the sign of the entropy production $d_i\mathfrak{S}$ from ordinary thermodynamics. So in nonequilibrium thermodynamics the second law of thermodynamics is generalized by the following postulate: *For any system, either closed or open, the entropy production accompanying its irreversible change is always positive.* Obviously, this generalized second law contains the second law of ordinary thermodynamics as one of its special cases. Being a postulate, it can be judged valid only if its theoretical consequences agree with experimental observations.

1. Equation for the Rate of Entropy Production

If G in equation 1.34 is taken as S, the entropy per gram of the solution, the quantity $(J_G)_c$ may then be written

$$(J_G)_c = \sum_{k=0}^{q} \bar{S}_k c_k (u_k)_c = \rho S(u)_M + (J_S)_M \qquad (1.53)$$

where

$$(J_S)_M = \sum_{k=0}^{q} \bar{S}_k (J_k)_M \qquad (1.54)$$

and $\bar{S}_k$ is the partial *specific* entropy of component k. Introducing equation

1.53 into equation 1.34, with G taken to be S, and using equation 1.37, we obtain

$$\rho \frac{DS}{Dt} = \Phi_S - \frac{1}{r} \frac{\partial}{\partial r}[r(J_S)_\mathrm{M}] \tag{1.55}$$

Here, for clarity, the subscript of Φ_G has been changed to S. The quantity $(J_S)_\mathrm{M}$ represents the flow of entropy arising from the motion of components relative to the local center of mass.

To proceed further toward a desired expression for the rate of entropy production Φ_S, we must introduce an important postulate, which is generally referred to as *the assumption of local equilibrium*. This is stated as follows: *In any particular volume element of a nonequilibrium system there exist all kinds of thermodynamic functions and they are related to the state variables at that volume element in exactly the same fashion as they are in the corresponding equilibrium system*. It will be noticed that the former half of this statement has been tacitly presumed throughout the preceding discussion on continuity equations. That is, we have already adapted various thermodynamic quantities such as energy, entropy, and density to nonequilibrium systems without specific mention of their significance. It must be kept in mind that these quantities have well-defined meanings only in equilibrium systems and that adaptation of them to the description of nonequilibrium systems is only a postulate. The latter half of the statement given above implies that, for example, the molar Gibbs free energy $G(r,t)$ at given position r and time t in a nonequilibrium system depends on pressure $p(r,t)$, temperature $T(r,t)$, and mole fractions $x_k(r,t)$ just as it does in the corresponding equilibrium system, and its functional form is not affected by the particular r and t considered. Acceptance of this statement allows us to write the following well-known equation of Gibbs for any small irreversible change of a system:

$$d\mathcircledS = T d\mathfrak{S} - P dV + \sum_{k=0}^{q} \mu_k dm_k \tag{1.56}$$

where $\mathcircledS$, $\mathfrak{S}$, V, and m_k represent the energy, entropy, volume, and mass of component k, and μ_k is the chemical potential *per gram* of component k. We will use a notation $\hat{\mu}_k$ for chemical potential *per mole* of component k. If Euler's relation

$$E = TS - PV + \sum_{k=0}^{q} \mu_i w_i \tag{1.57}$$

is inserted into equation 1.56 we obtain

$$dE = T\,dS - P\,d\left(\frac{1}{\rho}\right) + \sum_{k=0}^{q} \mu_k\,dw_k \tag{1.58}$$

where E and S are entropy and energy per gram of the system, as defined previously, and w_k is the weight fraction of component k. When equation 1.58 is valid, the following equation may be shown to hold:

$$\rho\,\frac{DS}{Dt} = \frac{\rho}{T}\,\frac{DE}{Dt} - \frac{P}{\rho T}\,\frac{D\rho}{Dt} - \frac{\rho}{T}\sum_{k=0}^{q} \mu_k\,\frac{Dw_k}{Dt} \tag{1.59}$$

Expressions for DS/Dt, DE/Dt, and $D\rho/Dt$ to be substituted into this equation are given, respectively, by equations 1.55, 1.50, and 1.38. The corresponding equation for Dw_k/Dt can be derived by substituting the relation $c_k = \rho w_k$ into equation 1.35 and using equations 1.37 and 1.39. The result is

$$\rho\,\frac{Dw_k}{Dt} = -\frac{1}{r}\,\frac{\partial}{\partial r}[r(J_k)_M] \tag{1.60}$$

Introduction of these quantities into equation 1.59, followed by proper rearrangement of terms, leads to the desired expression for the rate of entropy production. It reads

$$\Phi_S = \frac{1}{T}\sum_{k=0}^{q}(J_k)_M X_k \tag{1.61}$$

where X_k stands for a "force" acting on component k and is defined by

$$X_k = F_k - \frac{\partial}{\partial r}(\mu_k) \tag{1.62}$$

If the potential of centrifugal force is denoted by ψ, we have

$$\psi = -\tfrac{1}{2}r^2\omega^2 \tag{1.63}$$

and

$$F_k = -\frac{\partial \psi}{\partial r} = r\omega^2 \tag{1.64}$$

where ω is the angular speed of the rotor. Therefore, X_k may be written

$$X_k = r\omega^2 - \frac{\partial}{\partial r}(\mu_k) = -\frac{\partial}{\partial r}(\psi_k) \tag{1.65}$$

where ψ_k is the *total potential* of component k defined by

$$\psi_k = \mu_k - \tfrac{1}{2}r^2\omega^2 \tag{1.66}$$

Inasmuch as the entropy production is an effect associated with irreversible processes in a system, one would naturally expect that the quantity Φ_S ought to be expressed in terms of quantities which are descriptive of such processes. Equation 1.61 is consistent with this prediction, being expressed in terms of flows and of the forces causing them. Two important points should be noticed. One is that the expression given above for Φ_S is valid under three physical conditions imposed on its derivation: The system is at constant temperature, effects of shearing stresses on the fluid motion are negligible, and no components react chemically with each other. When any one of these conditions is not obeyed, one term corresponding to it is added to Φ_S. For example, when the temperature is nonuniform in the radial direction of the cell, the additional term is $-(Q/T)(\partial \ln T/\partial r)$, where Q is the flux of heat. For the details of these additional terms the reader is referred to a suitable textbook on nonequilibrium thermodynamics, for example, one by Fitts.[5] Two postulates were introduced in the derivation of equation 1.61: the generalization of the second law of thermodynamics and the assumption of local equilibrium. However, since no mathematical approximation has been introduced, equation 1.61 is exact as far as our physical conditions and postulates are obeyed. Another point to notice is that the mass-fixed flows of components appear in this "exact" expression for Φ_S. It implies that this type of material flow is the most basic to the phenomenological formulation of mass transport in nonequilibrium systems. Unless an additional condition is imposed, $(J_k)_M$ cannot be replaced by flows referred to any other frames, for example, by $(J_k)_c$.

F. MECHANICAL EQUILIBRIUM

A system is at *mechanical equilibrium* when $D(u)_M/Dt$ is zero everywhere in it, that is, when $(u)_M$ is independent of position and time. When the solution in the ultracentrifuge cell is at mechanical equilibrium, the equation of motion 1.44 therefore reduces to

$$\sum_{k=0}^{q} c_k F_k - \frac{\partial P}{\partial r} = 0 \tag{1.67}$$

Since by equation 1.64

$$\sum_{k=0}^{q} c_k F_k = r\omega^2 \rho \tag{1.68}$$

we have at mechanical equilibrium

$$\frac{\partial P}{\partial r} = \rho r \omega^2 \tag{1.69}$$

Integration of this equation, with the boundary condition that $P = 1$ atm at $r = r_1$, gives the pressure distribution in the cell. To effect this operation, however, it is necessary to know the local density ρ as a function of r, which in turn requires the determination of c_k of all components as functions of r.

Acceptance of the assumption of local equilibrium allows us to write the following Gibbs-Duhem relation for any volume element in a nonequilibrium system:

$$- S \, dT + \left(\frac{1}{\rho} \right) dP = \sum_{k=0}^{q} w_k \, d\mu_k \tag{1.70}$$

This gives at constant temperature

$$\frac{\partial P}{\partial r} = \sum_{k=0}^{q} c_k \frac{\partial \mu_k}{\partial r} \tag{1.71}$$

Substitution for $\partial P / \partial r$ from equation 1.67 and consideration of equation 1.62 leads to an important relation:

$$\sum_{k=0}^{q} c_k X_k = 0 \tag{1.72}$$

which indicates that forces X_k of $q + 1$ components are linearly related. Note that this relation *holds at thermal and mechanical equilibrium*.

If R'' is chosen as the mass-fixed frame M, equation 1.29 takes the form

$$(J_k)_M = (J_k)_{R'} - C \cdot c_k \tag{1.73}$$

where C is a constant independent of the subscript k. Introducing equation 1.73 into equation 1.61 and considering equation 1.72, we obtain

$$\Phi_S = \frac{1}{T} \sum_{k=0}^{q} (J_k)_{R'} X_k \tag{1.74}$$

Since, as has been noted in Section C, R' can be an arbitrary frame of reference, we obtain an important conclusion: When the system is at thermal and mechanical equilibrium, $(J_k)_M$ in equation 1.61 may be replaced by the flow of component k relative to any conceivable frame of reference, for example, the cell-fixed frame of reference.

G. PHENOMENOLOGICAL FLOW EQUATIONS AND COEFFICIENTS

In order to proceed further it is necessary to introduce two more postulates. For the system considered here, one of them may be stated as follows. Let a set of flows of $q+1$ components relative to a frame of reference A be denoted by $\{(J_k)_A\}$ and let a set of forces acting on $q+1$ components be denoted by $\{Y_k\}$. Then *each of these flows is a linear homogeneous function of all $q+1$ forces*. Mathematically,

$$(J_k)_A = \sum_{i=0}^{q} (L_{ki})_A Y_i \qquad (k=0,1,\ldots,q) \qquad (1.75)$$

with coefficients $(L_{ki})_A$ which are independent of flows and forces. Equation 1.75 is called the *phenomenological flow equation* for component k, and the $(L_{ki})_A$ are called *phenomenological coefficients* referred to the frame A. Specifically, the coefficients with $k=i$ are termed the *main coefficients* and those with $k \neq i$ are called the *cross-term coefficients*. The latter are concerned with effects which are exerted on the flow of a particular component by forces acting on other components. Such coupling effects are sometimes called flow interactions. It is important that the forces Y_k need not be the same as forces X_k defined by equation 1.62. Often Y_k is called a *generalized force* for component k.

In a sense equation 1.75 may be regarded as a generalization of the familiar Fick first law of diffusion for binary solutions to multicomponent systems. However, it is nothing but a postulate within the framework of nonequilibrium thermodynamics, and it is valid only if its theoretical consequences are consistent with experimental results.

1. Onsager's Reciprocal Relations

When $(J_k)_A$ and Y_k satisfy the condition

$$\Phi_S = \frac{1}{T} \sum_{k=0}^{q} (J_k)_A Y_k \qquad (1.76)$$

they are called *conjugate flows* and *forces*. Equation 1.61 indicates that $\{(J_k)_M\}$ and $\{X_k\}$ form a set of conjugate flows and forces. Our final postulate is that in the phenomenological flow equations for this set, that is,

$$(J_k)_M = \sum_{i=0}^{q} (L_{ki})_M X_i \qquad (k=0,1,\ldots,q) \qquad (1.77)$$

the matrix formed by coefficients $(L_{ki})_M$ (mass-fixed phenomenological coefficients) *is symmetric.* That is,

$$(L_{ki})_M = (L_{ik})_M \qquad (k \neq i) \qquad (1.78)$$

A relation like equation 1.78 between cross-term phenomenological coefficients is called *Onsager's reciprocal relation*. Some authors mention that for any set of conjugate flows and forces the matrix of the phenomenological coefficients is symmetric. However, as has been pointed out by Coleman and Truesdell,[11] this is an incorrect statement. The use of conjugate flows and forces is not a sufficient condition to ensure Onsager's reciprocal relation. However, this does not necessarily mean that no set of conjugate flows and forces other than $\{(J_k)_M\}$ and $\{X_k\}$ can have cross-term phenomenological coefficients which obey this relation (see Appendix A).

2. Transformation of Phenomenological Flow Equations

Equation 1.77 cannot be inverted to obtain X_k as a linear combination of $q+1$ flows $(J_k)_M$, since, according to equation 1.11, these flows are not linearly independent. However, if the system is not at mechanical equilibrium, it is possible, in principle, to evaluate all $(L_{ik})_M$, $q+1$ in total, from appropriate experiments, since $q+1$ forces are linearly independent for such a system. In this evaluation we may use the following relation:

$$\sum_{k=0}^{q} (L_{ki})_M = 0 \qquad (1.79)$$

This follows if equation 1.77 is introduced into equation 1.11 and use is made of the fact that forces X_k are allowed to vary independently of each other. When the system is at mechanical equilibrium, no unique evaluation of $(L_{ki})_M$ can be made, and equation 1.77 ceases to be a pertinent point of departure for the treatment of flow processes in such a system.

We now apply equation 1.77 to the solvent component, that is,

$$(J_0)_M = \sum_{i=0}^{q} (L_{0i})_M X_i \tag{1.80}$$

If we multiply both sides by c_k/c_0, subtract the resulting expression from equation 1.77, and use the relation

$$(J_k)_0 = (J_k)_M - \frac{c_k}{c_0} (J_0)_M \tag{1.81}$$

which is derived from equation 1.29 with $R'' = 0$ and $R' = M$, there results

$$(J_k)_0 = \sum_{i=0}^{q} \left[(L_{ki})_M - \frac{c_k}{c_0} (L_{0i})_M \right] X_i \tag{1.82}$$

Let us define a set of new coefficients $(L_{ki})_0$ by

$$(L_{ki})_0 = (L_{ki})_M - \frac{c_k}{c_0} (L_{0i})_M - \frac{c_i}{c_0} (L_{k0})_M + \frac{c_i c_k}{c_0^2} (L_{00})_M \tag{1.83}$$

$$(k, i = 0, 1, \ldots, q)$$

Then equation 1.82 may be written

$$(J_k)_0 = \sum_{i=1}^{q} (L_{ki})_0 X_i + \left[\frac{(L_{k0})_M}{c_0} - \frac{c_k (L_{00})_M}{c_0^2} \right] \sum_{i=0}^{q} c_i X_i \tag{1.84}$$

At mechanical equilibrium the last term on the right-hand side vanishes because of equation 1.72. Thus

$$(J_k)_0 = \sum_{i=1}^{q} (L_{ki})_0 X_i \qquad (k = 0, 1, \ldots, q) \tag{1.85}$$

which indicates that the solvent-fixed flow of each component can be expressed as a linear combination of forces X_k acting on *solute components only* if the condition of mechanical equilibrium is obeyed. With equation 1.78 it follows from equation 1.83 that

$$(L_{ki})_0 = (L_{ik})_0 \qquad (k \neq i) \tag{1.86}$$

That is, Onsager's reciprocal relations are obeyed by cross-term $(L_{ki})_0$'s.

Since X_k's for q solute components are linearly independent, that is, they are allowed to vary independently of each other, the q^2 coefficients $(L_{ki})_0$ can, in principle, be evaluated from adequate experiments. Furthermore, since q solvent-fixed flows are also linearly independent, equations 1.85 for q solute components can be solved for X_k to give

$$X_k = \sum_{i=1}^{q} (R_{ki})_0 (J_i)_0 \qquad (k=1,2,\ldots,q) \tag{1.87}$$

[Note that this equation for the solvent component yields only an obvious result $(J_0)_0 = 0$]. It is a simple matter to show that coefficients $(R_{ki})_0$ also satisfy Onsager's reciprocal relations.

In view of the above-mentioned properties, equations 1.85 may be taken as an adequate set of phenomenological equations for treating flow processes in nonequilibrium systems at mechanical equilibrium. In what follows, we confine our consideration to such systems, since, as is generally assumed, the condition of mechanical equilibrium usually applies very well for ordinary sedimentation and diffusion experiments.

Equation 1.85 is still inconvenient for practical purposes, because solvent-fixed flows may not be experimentally measurable. We seek instead the equations for cell-fixed flows of solute components. To obtain them we first transform equation 1.85 to the equations for volume-fixed flows. Equation 1.29 with $R'' = V$ and $R' = 0$ gives

$$(J_k)_V = (J_k)_0 - c_k \sum_{i=1}^{q} \bar{v}_i (J_i)_0 \tag{1.88}$$

Substitution of equation 1.85 yields

$$(J_k)_V = \sum_{i=1}^{q} (L_{ki})_V X_i \qquad (k=0,1,\ldots,q) \tag{1.89}$$

where

$$(L_{ki})_V = (L_{ki})_0 - c_k \sum_{j=1}^{q} \bar{v}_j (L_{ji})_0 \tag{1.90}$$

Thus we find that at mechanical equilibrium each of the volume-fixed flows of q solute components can be represented by a linear combination of forces X_k acting on q solute components. Values of the q^2 phenomenological coefficients $(L_{ki})_V$ can be determined uniquely from appropriate experiments. However, it follows from equation 1.90 that these coefficients do not satisfy Onsager's reciprocal relation, that is,

$$(L_{ki})_V \neq (L_{ik})_V \qquad (k \neq i) \tag{1.91}$$

If desired, the flow equation for $(J_0)_V$ may be obtained by substituting equation 1.89 into equation 1.20.

Finally, we introduce equation 1.85 into the relation

$$(J_k)_M = (J_k)_0 - \frac{c_k}{\rho} \sum_{i=1}^{q} (J_i)_0 \tag{1.92}$$

which is obtained by putting $R'' = M$ and $R' = 0$ in equation 1.29. The result reads

$$(J_k)_M = \sum_{i=1}^{q} (L_{ki}^*)_M X_i \qquad (k = 0, 1, \ldots, q) \tag{1.93}$$

where

$$(L_{ki}^*)_M = (L_{ki})_0 - \frac{c_k}{\rho} \sum_{j=1}^{q} (L_{ji})_0 \tag{1.94}$$

It is seen that each of the mass-fixed flows of the q solute components can be expressed as a linear homogeneous function of forces X_k acting on these q components when the system is at mechanical equilibrium. The q^2 coefficients $(L_{ki}^*)_M$ appearing in this set of q linear equations can, in principle, be evaluated uniquely from adequate experiments, but their cross-term coefficients do not satisfy Onsager's reciprocal relation, as is readily shown from equation 1.94, that is,

$$(L_{ki}^*)_M \neq (L_{ik}^*)_M \qquad (k \neq i) \tag{1.95}$$

We must observe that in equation 1.77 the summation extends over all $q + 1$ components, whereas the summation in equation 1.93 is taken over the solute components only.

In an analogous manner, we can derive similar phenomenological equations for flows of components referred to any other frame belonging to the R-group. However, since such flows may not be experimentally measurable, the equations for them are of little use for practical purposes. The desired equations for cell-fixed flows could be obtained if there existed a relation which expresses the cell-fixed flow of a particular component in terms of the flows of solute components referred to a certain frame of the R-group. Unfortunately, such a relation is as yet not known. Therefore, to proceed further we are forced to introduce an approximation.

3. Practical Flow Equations

We assume that the partial specific volumes $\bar{v}_k$ of all components may be considered to be independent of pressure and composition. This is equivalent to assuming the solution to be incompressible and the volume to be unchanged upon mixing. As shown in Appendix B, the velocity of the local center of volume, $(u)_V$, becomes zero when this assumption holds; in other words, the volume-fixed frame of reference coincides with the cell-fixed one. In fact, if $(u)_V = 0$, equation 1.17 yields

$$(J_k)_V = (J_k)_c \tag{1.96}$$

Due to this identity, equations 1.89 with $(L_{ki})_V$ given by equation 1.90 can be used as the phenomenological equations for cell-fixed flows of solute components if the $\bar{v}_k$'s of all components are independent of pressure and composition. Thus

$$(J_k)_c = \sum_{i=1}^{q} (L_{ki})_V X_i \qquad (k = 1, 2, \ldots, q) \tag{1.97}$$

Note that the $(L_{ki})_V$'s do not obey Onsager's reciprocal relation. When $(u)_V$ is zero, it follows from equation 1.18 that

$$\sum_{k=0}^{q} \bar{v}_k (J_k)_c = 0 \tag{1.98}$$

If desired, the value of $(J_0)_c$ may be obtained by substituting equations 1.97 for other components into this relation. It is important to note that when $(u)_V$ is zero, cell-fixed flows of components are no longer independent of each other but are related linearly by equation 1.98. That is, under this condition, the cell-fixed frame of reference belongs to the R-group.

Phenomenological equations for flows referred to a frame belonging to the R-group are often called *theoretical flow equations*, and those for cell-fixed flows are termed *practical flow equations*. Equation 1.97 is a special case of the latter, although, as can be understood from the argument given above, it also belongs to the former.

When the system is subject to a very strong centrifugal force, as is the case with ordinary sedimentation velocity experiments, the partial specific volumes of components may change due to a high pressure gradient set up in the cell. The effect may be particularly appreciable when organic liquids are used as solvents. Some systems may exhibit a measurable change of volume upon mixing as a result of appreciable dependence of the partial specific volumes on composition. If we consider these possibilities, it appears too restrictive to impose the condition that the partial specific volumes are independent of both pressure and composition. Kirkwood et al.[12] worked out practical flow equations for free diffusion in an isothermal multicomponent system in which the partial specific volumes vary with composition. No corresponding equations for the case of sedimentation are as yet available in the literature. Effects of the pressure dependence of partial specific volumes on the practical flow equations also remain unexplored. Probably, as is generally believed, these effects on the flow equations are minor except in some special circumstances, but it is highly desirable that they be elucidated as soon as possible. Until that time, the framework of sedimentation analysis must remain as lacking in rigor.

H. FLOW EQUATIONS FOR SEDIMENTATION IN THE ULTRACENTRIFUGE

We are now prepared to write the practical flow equations for sedimentation in a multicomponent solution. The forces X_k to be substituted into equations 1.97 are obtained from equation 1.66, that is,

$$X_k = r\omega^2 - \left(\frac{\partial \mu_k}{\partial r} \right)_T \tag{1.99}$$

where the subscript T has been attached to the gradient of μ_k in order to indicate that we are considering flow processes in an isothermal solution.

If we assume that the solution is incompressible, the c-scale concentrations of the $q+1$ components become dependent on each other,* and any one of them may be expressed as a function of all others. Let the q *solute* concentrations be chosen as independent variables. The gradient of μ_k in the direction of r at constant T may then be rewritten

$$\left(\frac{\partial \mu_k}{\partial r} \right)_T = \sum_{j=1}^{q} \left(\frac{\partial \mu_k}{\partial c_j} \right)_{T,P,c_m} \frac{\partial c_j}{\partial r} + \bar{v}_k \frac{\partial P}{\partial r} \tag{1.100}$$

$$(k = 0, 1, \ldots, q, \qquad m = 1, 2, \ldots, q)$$

The subscript c_m affixed to $\partial \mu_k / \partial c_j$ indicates that all solute concentrations on the c-scale other than that of solute j are held constant while the chemical potential μ_k is differentiated with respect to c_j; thus m runs from 1 to q. The partial specific volume $\bar{v}_k$ has been inserted for the equivalent

---

*Under the imposed condition, the partial specific volumes of the $q+1$ components become functions of the composition of the solution only. If this fact is combined with equation 1.14, this statement follows immediately.

quantity $(\partial\mu_k/\partial P)_{T,c_1,c_2,...,c_q}$, which describes the pressure dependence of μ_k at constant temperature and constant composition under the condition of negligible compressibility.

Substituting equation 1.100 into equation 1.99, together with equation 1.69 for $\partial P/\partial r$, we arrive at

$$X_k = (1 - \bar{v}_k\rho)\omega^2 r - \sum_{j=1}^{q} \left(\frac{\partial\mu_k}{\partial c_j}\right)_{T,P,c_m} \frac{\partial c_j}{\partial r} \qquad (1.101)$$

$$(k = 0, 1, ..., q, \qquad m = 1, 2, ..., q)$$

It should be noted that these expressions are valid at *thermal and mechanical equilibrium* of an *incompressible* solution. The first term on the right-hand side represents the *net* sedimentation force acting (per gram) on component k. The factor $1 - \bar{v}_k\rho$ is often termed *the buoyancy factor* or *the Archimedes factor* of component k. The second term involving a summation of the gradients of solute concentrations is referred to as the diffusion force acting (per gram) on component k. When $\omega = 0$, that is, the solution is not centrifuged, the first term vanishes, but the second term may not if there is a nonuniform concentration distribution in the solution. This is the case treated in ordinary diffusion experiments. In recent years, there have been many important contributions to theoretical and experimental aspects of diffusion phenomena in multicomponent solutions, especially in ternary solutions, and some of them have led to the substantial check of Onsager's reciprocal relation. We touch on this subject in Section L.

The result of substitution of equation 1.101 into equation 1.97 may be put in the form

$$(J_k)_c = (s_k)_v c_k \omega^2 r - \sum_{j=1}^{q} (D_{kj})_v \frac{\partial c_j}{\partial r} \qquad (k = 1, 2, ..., q) \qquad (1.102)$$

where

$$(s_k)_v = \frac{1}{c_k} \sum_{i=1}^{q} (L_{ki})_v (1 - \bar{v}_i\rho) \qquad (1.103)$$

$$(D_{kj})_v = \sum_{i=1}^{q} (L_{ki})_v \left(\frac{\partial\mu_i}{\partial c_j}\right)_{T,P,c_m} \qquad (1.104)$$

As has been noted above, equation 1.101 holds under the conditions of thermal and mechanical equilibrium and of negligible compressibility, whereas equation 1.102 requires for its validity the additional condition that the partial specific volumes of all components be independent of the composition of the solution. In actual systems, the partial specific volumes are strictly independent of neither pressure nor composition. Therefore, strictly speaking, equations 1.102 are of limited value for sedimentation processes in actual solutions. Yet a more general form of $(J_k)_c$ is not available in the literature. Thus, at the present stage of our knowledge, we cannot help making a compromise in order to proceed further toward statements which guide evaluation of sedimentation velocity experiments. This compromise is to assume that equations 1.102 may be applied as they stand, even if the partial specific volumes vary with the composition.* However, the assumption of negligible compressibility will be retained, since equation 1.101 for X_k requires this restriction for its validity. Thus, in what follows, we shall treat equations 1.102 as valid if the system is *incompressible* and at *thermal and mechanical equilibrium*.

1. Equations for Binary and Ternary Solutions

For binary solutions, that is, solvent $0 +$ solute 1, the set of equations 1.102 reduces to a single equation for solute 1:

$$(J_1)_c = (s_1)_V c_1 \omega^2 r - (D_{11})_V \frac{\partial c_1}{\partial r} \tag{1.105}$$

where

$$(s_1)_V = \frac{1}{c_1} (L_{11})_V (1 - \bar{v}_1 \rho) \tag{1.106}$$

$$(D_{11})_V = (L_{11})_V \left(\frac{\partial \mu_1}{\partial c_1} \right)_{T,P} \tag{1.107}$$

For ternary solutions, that is, solvent $0 +$ solute $1 +$ solute 2, the corresponding equations are those for solute 1 and solute 2 and have the following forms:

$$(J_1)_c = (s_1)_V c_1 \omega^2 r - (D_{11})_V \frac{\partial c_1}{\partial r} - (D_{12})_V \frac{\partial c_2}{\partial r} \tag{1.108}$$

---

*For the treatment of sedimentation equilibrium phenomena this compromise need not be made. The condition of mechanical equilibrium also becomes unnecessary (see Chapter 5, Section I.C).

$$(J_2)_c = (s_2)_V c_2 \omega^2 r - (D_{21})_V \frac{\partial c_1}{\partial r} - (D_{22})_V \frac{\partial c_2}{\partial r} \tag{1.109}$$

$$(s_1)_V = \frac{1}{c_1}(L_{11})_V(1 - \bar{v}_1\rho) + \frac{1}{c_1}(L_{12})_V(1 - \bar{v}_2\rho)$$
$$\tag{1.110}$$
$$(s_2)_V = \frac{1}{c_2}(L_{21})_V(1 - \bar{v}_1\rho) + \frac{1}{c_2}(L_{22})_V(1 - \bar{v}_2\rho)$$

$$(D_{11})_V = (L_{11})_V \left(\frac{\partial \mu_1}{\partial c_1}\right)_{T,P,c_2} + (L_{12})_V \left(\frac{\partial \mu_2}{\partial c_1}\right)_{T,P,c_2}$$

$$(D_{12})_V = (L_{11})_V \left(\frac{\partial \mu_1}{\partial c_2}\right)_{T,P,c_1} + (L_{12})_V \left(\frac{\partial \mu_2}{\partial c_2}\right)_{T,P,c_1}$$
$$\tag{1.111}$$
$$(D_{21})_V = (L_{21})_V \left(\frac{\partial \mu_1}{\partial c_1}\right)_{T,P,c_2} + (L_{22})_V \left(\frac{\partial \mu_2}{\partial c_1}\right)_{T,P,c_2}$$

$$(D_{22})_V = (L_{21})_V \left(\frac{\partial \mu_1}{\partial c_2}\right)_{T,P,c_1} + (L_{22})_V \left(\frac{\partial \mu_2}{\partial c_2}\right)_{T,P,c_1}$$

As early as 1929, Lamm[13] derived for a binary solution a flow equation which is of exactly the same form as equation 1.105. However, the frame of reference to which the coefficients in his equation corresponding to $(s_1)_V$ and $(D_{11})_V$ are related has long been a subject of controversy. Furthermore, there was a question as to whether the term ρ in his equation is the density of the solvent or that of the solution. These questions arose because Lamm had used a kinetic theory for the derivation of the equation. Here we can see the disadvantage or even the inadequacy of this type of approach to an unambiguous treatment of sedimentation processes in solution. From the above-mentioned development based on nonequilibrium thermodynamics it is now clear that the coefficients in question should be *referred to the volume-fixed frame* and that ρ is the density of the *solution*. It is also apparent that Lamm's equation is valid only for incompressible binary solutions at thermal and mechanical equilibrium.

Because the second term on the right-hand side of equation 1.105 has the familiar form of the Fick law for diffusion, we may call the quantity $(D_{11})_V$ the diffusion coefficient of component 1. However, it is more properly termed the mutual diffusion coefficient of a binary solution. To show this we substitute equation 1.105 into equation 1.98 for $q = 1$ and solve for $(J_0)_c$ to give

$$(J_0)_c = -\frac{\bar{v}_1}{\bar{v}_0}(s_1)_V c_1 \omega^2 r + \frac{\bar{v}_1}{\bar{v}_0}(D_{11})_V \frac{\partial c_1}{\partial r} \qquad (1.112)$$

The frequently used relation 1.14 gives for $q = 1$

$$\bar{v}_0 c_0 + \bar{v}_1 c_1 = 1 \qquad (1.113)$$

Differentiation of this equation with respect to r gives

$$\bar{v}_0 \frac{\partial c_0}{\partial r} + \bar{v}_1 \frac{\partial c_1}{\partial r} = 0 \qquad (1.114)$$

provided $\bar{v}_0$ and $\bar{v}_1$ may be treated as constant. With the help of these relations, equation 1.112 may be put in the form

$$(J_0)_c = (s_0)_V c_0 \omega^2 r - (D_{11})_V \frac{\partial c_0}{\partial r} \qquad (1.115)$$

where

$$(s_0)_V = -\frac{\bar{v}_1 c_1 (s_1)_V}{1 - \bar{v}_1 c_1} \qquad (1.116)$$

It is seen that $(D_{11})_V$ appears as the diffusion coefficient in the expression for $(J_0)_c$ corresponding in form to equation 1.105 for $(J_1)_c$. This implies that for description of isothermal diffusion in a binary solution in which $\bar{v}_0$ and $\bar{v}_1$ are independent of pressure and composition, only a single diffusion coefficient, $(D_{11})_V$, referred to the volume-fixed frame is sufficient. For this reason Hartley and Crank[14] proposed to call it the *mutual diffusion coefficient* of a binary solution, or simply the diffusion coefficient of the *system*. In contrast, both $(s_0)_V$ and $(s_1)_0$ generally have different values, and may be termed the sedimentation coefficients of components 0 and 1, respectively. However, both are linearly related by equation 1.116.

2. Definitions of Sedimentation Coefficients and Diffusion Coefficients

For systems containing more than one solute component we call the value of $(s_k)_V$ given by equation 1.103 the *sedimentation coefficient of component k* (referred to the volume-fixed frame). The set of $(D_{ij})_V$ ($i, j = 1, 2, ..., q$) defined by equation 1.104 is called the *set of volume-fixed diffusion coefficients* for a given multicomponent system. It is important to recognize that none of these diffusion coefficients can be assigned to any particular solute component in the system; instead, each should be regarded only as a member of the set of $(D_{ij})_V$. It is customary to call $(D_{ii})_V$ the *main diffusion coefficients* and $(D_{ij})_V$ $(i \neq j)$ the *cross-term diffusion coefficients*. When the cross-term diffusion coefficients are nonzero, there are said to be interactions between diffusion flows of different solute components.[15] Theory and methods for studying such flow interactions in ternary solutions have been worked out in great detail by Gosting and co-workers.[16-23] Recently, Kim[24] has extended them to quaternary solutions. The set of sedimentation coefficients does not contain any cross-term coefficient, but this does not mean that flows of solutes due to centrifugal forces do not interfere with each other.

I. THE SVEDBERG RELATION AND ITS EXTENSIONS

Since the same phenomenological coefficients $(L_{ki})_V$ enter into the expressions for both the sedimentation coefficients and the diffusion coefficients, we may correlate measurements of these coefficients. This correlation for a two-component system leads to the familiar Svedberg relation for molecular weight.

According to the assumption of local equilibrium, the chemical potential μ_1 per gram of a *nonelectrolyte* solute 1 in a binary solution may be written in the form

$$\mu_1 = (\mu_1{}^0)_c + \frac{RT}{M_1} \ln(y_1 c_1) \tag{1.117}$$

Here $(\mu_1{}^0)_c$ is the reference chemical potential of component 1 appropriate to the c-concentration scale, M_1 is its molecular weight, and y_1 is its practical activity coefficient on the c-scale. Thus y_1 has the following limiting property:

$$\lim_{c_1 \to 0} y_1 = 1 \tag{1.118}$$

Substitution of equation 1.117 into equation 1.107 yields

$$(D_{11})_V = (L_{11})_V \frac{RT}{M_1 c_1} \left[1 + c_1 \left(\frac{\partial \ln y_1}{\partial c_1} \right)_{T,P} \right] \tag{1.119}$$

An expression for M_1 is then obtained by eliminating the coefficient $(L_{11})_V$ from equations 1.106 and 1.119, giving

$$M_1 = \frac{RT(s_1)_V [1 + c_1 (\partial \ln y_1 / \partial c_1)_{T,P}]}{(D_{11})_V (1 - \bar{v}_1 \rho)} \tag{1.120}$$

In the limit of infinite dilution ($c_1 \to 0$), this reduces to

$$M_1 = \frac{RT(s_1)_V^0}{(D_{11})_V^0(1 - \bar{v}_1^0 \rho_0)} \tag{1.121}$$

where ρ_0 denotes the density of the pure solvent (component 0 in the pure state), and the superscript 0 indicates the values of respective quantities at infinite dilution. Equation 1.121 is called the *Svedberg relation*. It allows the evaluation of M_1 from measurements of ρ_0, $\bar{v}_1^0$, $(s_1)_V^0$, and $(D_{11})_V^0$. Values of $(D_{11})_V^0$ from separate free (or restricted) diffusion experiments at atmospheric pressure are usually substituted into the Svedberg relation to calculate M_1. For an accurate determination of M_1 it is required in this case that use be made of a value of $(s_1)_V^0$ which has been extrapolated or reduced to atmospheric pressure. Consideration of the procedures for this extrapolation or reduction leads to the study of pressure effects upon sedimentation processes in the ultracentrifuge cell. Approximate treatments of such effects are described in Chapter 2, Section V.

A similar procedure may be employed to find an equation for M_1 which applies to ternary solutions. Baldwin[25] has investigated this problem, and has derived the following equation:

$$(s_1)_V^0 = \frac{M_1(1 - \bar{v}_1^0 \rho_0')(D_{11})_V^0}{RT} \left\{ 1 + c_2 \left[\frac{M_2(1 - \bar{v}_2^0 \rho_0')}{M_1(1 - \bar{v}_1^0 \rho_0')} \right] \right.$$

$$\left. \times \left[\frac{(1/(D_{11})_V^0)(\partial(D_{12})_V/\partial c_1)_{T,P,c_2}^0 - (\partial \ln y_1/\partial c_2)_{T,P,c_1}^0}{1 + c_2(\partial \ln y_2/\partial c_2)_{T,P,c_1}^0} \right] \right\} \tag{1.122}$$

Here ρ_0' is the density of the *mixed solvent*, that is, component $0 +$ component 2, and the superscript 0 refers to the limit $c_1 \to 0$; thus ρ_0' and all the quantities with the superscript 0 are still functions of c_2. In the absence of component 2, the second term in the braces vanishes, and the equation reduces to the Svedberg relation. Creeth and Pain[26] state in their review article that no case is known to them for which this second term is significant. As long as this statement is valid, it follows that the Svedberg relation may be applied formally to evaluate the molecular weight of a macromolecule from transport experiments (sedimentation velocity and diffusion experiments) with a mixed solvent. Here the mixed solvent may be a mixture of two liquid solvents (as is often the case for synthetic polymers) or an aqueous solution of simple electrolytes (as is usually employed for proteins). In fact, many workers have adapted the Svedberg relation formally to ternary solutions or even more complicated solutions, always with some reservation about the significance of molecular weights so obtained. It is hoped that further investigations will elucidate how important the "correction" term in the Baldwin equation 1.122 is under given experimental conditions. Peller[27] and Schönert[28] have extended Baldwin's treatment to a general multicomponent system and have derived the limiting forms corresponding to equation 1.122. However, these forms are so complicated that their value appears to be no more than theoretical at present.

J. THE DIFFERENTIAL EQUATIONS FOR THE ULTRACENTRIFUGE

If we substitute equation 1.105 into the continuity equation 1.36 for $k = 1$, there results

$$\frac{\partial c_1}{\partial t} = \frac{1}{r}\frac{\partial}{\partial r}\left[rD\left(\frac{\partial c_1}{\partial r}\right) - s\omega^2 r^2 c_1 \right] \qquad (1.123)$$

where, for convenience of the mathematical development given in Chapter 2, $(s_1)_V$ and $(D_{11})_V$ have been replaced by the simpler notations s and D. In general, s and D are functions of T, P, and concentration c_1. Since the pressure dependence of these coefficients is usually quite small, the primary factor which has to be considered in dealing with equation 1.123 is their dependence on c_1. The necessary conditions for the validity of this equation are the same as those for equation 1.105; that is, the solution must be incompressible and at thermal and mechanical equilibrium.

Equation 1.123 is generally called the *Lamm differential equation* for the ultracentrifuge.[13] This is a partial differential equation of the parabolic type, being of first order with respect to time t and of second order with respect to space variable r. Its integration requires that s and D be assigned as functions of c_1 and P, and that one initial condition and two boundary conditions for c_1 be supplied. When s and D depend on P, equation 1.69 describing the pressure distribution in the cell has to be coupled with equation 1.123. Even when pressure effects on s and D are absent the mathematical treatment of equation 1.123 is not a simple matter, because the dependence of these coefficients on c_1 makes the equation nonlinear. One of the major tasks in the theory of sedimentation analysis is to investigate properties of this differential equation and thereby to derive procedures which may be employed to evaluate s and D from relevant experiments. Chapter 2 is devoted to its detailed discussions.

By substituting equation 1.102 into equation 1.36 we obtain a set of q differential equations for the q solute components, that is,

$$\frac{\partial c_k}{\partial t} = \frac{1}{r}\frac{\partial}{\partial r}\left[r \sum_{j=1}^{q} (D_{kj})_V \left(\frac{\partial c_j}{\partial r}\right) - (s_k)_V \omega^2 r^2 c_k \right] \tag{1.124}$$

$$(k = 1, 2, \ldots, q)$$

This set of equations is a generalization of the Lamm differential equation to a multicomponent solution, which holds when the solution is incompressible and is at thermal and mechanical equilibrium. It is convenient to refer to each equation of the set as a *generalized Lamm equation* for the ultracentrifuge.

K. ELECTROLYTE SOLUTIONS

So far we have explicitly or implicitly confined our consideration to systems of nonelectrolyte components. However, if we consider that the current application of ultracentrifugal methods is primarily directed to systems of biological interest, it is highly desirable to extend the theory to systems which contain electrolytes as components. Unfortunately, to the best of the author's knowledge, it appears that the general treatment of transport phenomena in such systems is not yet fully established. It is beyond the scope of this treatise to survey the existing literature and give relevant comments on the present status of research in this field. Here we shall be content with a very simple system to illustrate how the flow processes in electrolyte solutions can be formulated in terms of nonequilibrium thermodynamics.

Our simple system consists of a solvent (water) and two strong electrolytes having a common cationic or anionic species. It is assumed that the two electrolytes ionize completely in the solvent and that each of them is binary, that is, it gives on dissociation one cationic species and one anionic species. Our system is therefore described by $NaCl-KCl-H_2O$, sodium salt of polystyrene sulfonic acid–$NaCl-H_2O$, and so on. Treatment of a more general case in which two electrolytes dissociate to form ions different from one another becomes more complicated than that described below. Several authors, for example, Schönert[28] and Fitts,[5] have described treatments of transport processes in solutions which contain an arbitrary number of electrolytes and in which there is no restriction on the kind of dissociating ions. However, it is not yet clear to the present author whether their theories represent really general solutions of the problem.

1. Flow Equations for Ions

Besides the solvent, the actual kinetic units in an electrolyte solution are not neutral molecules, but the individual ions. Therefore, it is natural to start with flow equations for respective ionic species when we wish to formulate transport processes in electrolyte solutions.

Consider two binary electrolytes E_1 and E_2 which have a common anion A (designated as ion 3) with signed valence z_3. Their cations C and D are designated as ion 1 and ion 2, and the signed valences of these ions are denoted by z_1 and z_2. Finally, it is assumed that upon dissociation one molecule of electrolyte E_i $(i=1,2)$ gives n_i^+ cations and n_i^- anions.

Then from the condition that an electrolyte as a whole must be electrically neutral it follows that

$$n_1^+ z_1 + n_1^- z_3 = 0 \tag{1.125}$$

$$n_2^+ z_2 + n_2^- z_3 = 0 \tag{1.126}$$

The molecular weights M_1 and M_2 of electrolytes E_1 and E_2 are represented by

$$M_1 = n_1^+ M_1' + n_1^- M_3' \tag{1.127}$$

$$M_2 = n_2^+ M_2' + n_2^- M_3' \tag{1.128}$$

where M_k' stands for the molar weight of ion k $(k=1,2,3)$.

The electrochemical potential $\tilde{\mu}_k'$ per gram of ion k is defined by

$$\tilde{\mu}_k' = \mu_k' + \frac{z_k F}{M_k'} \phi \tag{1.129}$$

where μ_k' is the chemical potential per gram of ion k, F is the Faraday constant, and ϕ is the *electric potential*. One should not confuse ϕ with the *electrostatic potential* which appears in the Debye-Hückel theory of electrolyte solutions. In accordance with practice in electrochemistry, the chemical potentials μ_1 and μ_2 per gram of electrolytes E_1 and E_2 are represented by

$$\mu_1 = \frac{n_1^+ M_1' \tilde{\mu}_1' + n_1^- M_3' \tilde{\mu}_3'}{M_1} \tag{1.130}$$

$$\mu_2 = \frac{n_2^+ M_2' \tilde{\mu}_2' + n_2^- M_3' \tilde{\mu}_3'}{M_2} \tag{1.131}$$

Substituting equation 1.129 and using equations 1.125 through 1.128 properly, we obtain

$$\mu_1 = \frac{z_3 M_1' \mu_1' - z_1 M_3' \mu_3'}{z_3 M_1' - z_1 M_3'} \tag{1.132}$$

$$\mu_2 = \frac{z_3 M_2' \mu_2' - z_2 M_3' \mu_3'}{z_3 M_2' - z_2 M_3'} \tag{1.133}$$

For an electrolyte solution subject to a centrifugal force, the force X_k' acting (per gram) on ion k may be represented by equation 1.65 if μ_k in the equation is replaced by the electrochemical potential $\tilde{\mu}_k'$. Thus for the solution considered here

$$X_k' = r\omega^2 - \frac{\partial \mu_k'}{\partial r} - \frac{z_k F}{M_k'} \frac{\partial \phi}{\partial r} \qquad (k=1,2,3) \tag{1.134}$$

Now we assume that an equation of the same form as equation 1.85 holds between the solvent-fixed flow $(J_k')_0$ of ion k and a set of forces X_k' defined by equation 1.134. That is,

$$(J_k')_0 = \sum_{j=1}^{3} (L_{kj}')_0 \left(r\omega^2 - \frac{\partial \mu_j'}{\partial r} - \frac{z_j F}{M_j'} \frac{\partial \phi}{\partial r} \right) \qquad (k=1,2,3) \tag{1.135}$$

where coefficients $(L_{kj}')_0$ are assumed to obey Onsager's reciprocal relations.

It is a fundamental requirement in electrochemistry that in any volume element of an electrolyte solution the condition of electric neutrality be satisfied. This leads to

$$\sum_{k=1}^{3} \frac{z_k c_k'}{M_k'} = 0 \tag{1.136}$$

where c_k' is the c-scale concentration of ion k. In ordinary sedimentation and diffusion experiments, no electric field is applied externally to the cell, so that there must be no electric current flowing relative to the cell. As shown in Appendix C, this condition together with condition 1.136 leads to

$$\sum_{k=1}^{3} \frac{z_k}{M_k'} (J_k')_0 = 0 \tag{1.137}$$

Introducing equation 1.135 into equation 1.137 and solving for $\partial\phi/\partial r$, we obtain

$$F\frac{\partial\phi}{\partial r} = \frac{\sum_{j=1}^3\sum_{m=1}^3(z_j/M_j')(L_{jm}')_0(r\omega^2 - \partial\mu_m'/\partial r)}{\sum_{j=1}^3\sum_{m=1}^3(z_j z_m/M_j'M_m')(L_{jm}')_0} \qquad (1.138)$$

Substitution of this back into equation 1.135 and appropriate manipulation of the sums yield

$$(J_k')_0 = \sum_{i=1}^3\sum_{j=1}^3\sum_{m=1}^3 (L_{ki}')_0(L_{jm}')_0 \frac{z_j}{M_j'}$$

$$\times \frac{(z_m/M_m')Y_i' - (z_i/M_i')Y_m'}{\sum_{j=1}^3\sum_{m=1}^3(z_j z_m/M_j'M_m')(L_{jm}')_0} \qquad (1.139)$$

$$(k = 1,2,3)$$

where Y_m' is defined by

$$Y_m' = r\omega^2 - \frac{\partial\mu_m'}{\partial r} \qquad (1.140)$$

Although Y_m' involves the unmeasurable single-ion chemical potential, fortunately all the indicated combinations of Y_m' and Y_i' in equation 1.139 are reduced to the measurable ones by making use of the following relations:[29]

$$\left(\frac{z_3}{M_3'}\right)Y_1' - \left(\frac{z_1}{M_1'}\right)Y_3' = \frac{z_1 z_3}{M_1'M_3'}\left(\frac{M_1'}{z_1} - \frac{M_3'}{z_3}\right)X_1 \qquad (1.141)$$

$$\left(\frac{z_3}{M_3'}\right)Y_2' - \left(\frac{z_2}{M_2'}\right)Y_3' = \frac{z_2 z_3}{M_2'M_3'}\left(\frac{M_2'}{z_2} - \frac{M_3'}{z_3}\right)X_2 \qquad (1.142)$$

$$\left(\frac{z_2}{M_2'}\right)Y_1' - \left(\frac{z_1}{M_1'}\right)Y_2' = \frac{z_1 z_2}{M_1'M_2'}\left(\frac{M_1'}{z_1} - \frac{M_3'}{z_3}\right)X_1 - \frac{z_1 z_2}{M_1'M_2'}\left(\frac{M_2'}{z_2} - \frac{M_3'}{z_3}\right)X_2$$

$$(1.143)$$

where X_1 is defined by

$$X_i = r\omega^2 - \frac{\partial\mu_i}{\partial r} \qquad (i = 1,2) \qquad (1.144)$$

which represents the force acting (per gram) on electrolyte E_i as a whole. In this way, we obtain the following flow equations for ions 1, 2, and 3:

$$(J_k')_0 = \sum_{i=1}^{2} \sum_{j=1}^{3} \sum_{m=1}^{3} \frac{z_i z_j z_m}{M_i' M_j' M_m'} [(L_{ki}')_0 (L_{jm}')_0$$

$$- (L_{km}')_0 (L_{ji}')_0] Z_i / \sum_{j=1}^{3} \sum_{m=1}^{3} \frac{z_j z_m}{M_j' M_m'} (L_{jm}')_0 \qquad (1.145)$$

$$(k = 1, 2, 3)$$

where

$$Z_1 = \left(\frac{M_1'}{z_1} - \frac{M_3'}{z_3} \right) X_1 \qquad (1.146)$$

$$Z_2 = \left(\frac{M_2'}{z_2} - \frac{M_3'}{z_3} \right) X_2 \qquad (1.147)$$

It should be noted that since we have condition 1.137, only two of the three flow equations 1.145 are independent.

2. Flow Equations for Electrolyte Components

For the case considered here there is only one way of combining ions present in any volume element of the solution into neutral molecules by use of the condition of electric neutrality. The electrolytes so obtained are the given electrolytes E_1 and E_2. The number of moles of ion 1 which crosses per second 1 cm^2 of the solvent-fixed frame must, in this case, equal the number of moles of electrolyte E_1 which passes in the same time through the same area of the same reference frame. This fact yields the following relation:

$$\frac{(J_1)_0}{M_1} = \frac{(J_1')_0}{n_1^+ M_1'} \qquad (1.148)$$

where $(J_1)_0$ denotes the solvent-fixed flow of electrolyte E_1. From a similar consideration we obtain for the solvent-fixed flow of electrolyte E_2

$$\frac{(J_2)_0}{M_2} = \frac{(J_2')_0}{n_2^+ M_2'} \qquad (1.149)$$

Substituting for $(J_1')_0$ and $(J_2')_0$ from equations 1.145 and using equations 1.125 through 1.128, we obtain

$$(J_1)_0 = (L_{11})_0 X_1 + (L_{12})_0 X_2 \tag{1.150}$$

$$(J_2)_0 = (L_{21})_0 X_1 + (L_{22})_0 X_2 \tag{1.151}$$

where the $(L_{ij})_0$'s stand for

$$(L_{11})_0 = \frac{1}{A}\left(\frac{M_1'}{z_1} - \frac{M_3'}{z_3}\right)^2 \left(\frac{z_1}{M_1'}\right)^2$$

$$\times \sum_{j=1}^{3} \sum_{m=1}^{3} \left(\frac{z_j z_m}{M_j' M_m'}\right)[(L_{11}')_0(L_{jm}')_0 - (L_{1m}')_0(L_{j1}')_0] \tag{1.152}$$

$$(L_{12})_0 = \frac{1}{A}\left(\frac{M_1'}{z_1} - \frac{M_3'}{z_3}\right)\left(\frac{M_2'}{z_2} - \frac{M_3'}{z_3}\right)\frac{z_1 z_2}{M_1' M_2'}$$

$$\times \sum_{j=1}^{3} \sum_{m=1}^{3} \left(\frac{z_j z_m}{M_j' M_m'}\right)[(L_{12}')_0(L_{jm}')_0 - (L_{1m}')_0(L_{j2}')_0] \tag{1.153}$$

$$(L_{21})_0 = \frac{1}{A}\left(\frac{M_2'}{z_2} - \frac{M_3'}{z_3}\right)\left(\frac{M_1'}{z_1} - \frac{M_3'}{z_3}\right)\frac{z_1 z_2}{M_1' M_2'}$$

$$\times \sum_{j=1}^{3} \sum_{m=1}^{3} \left(\frac{z_j z_m}{M_j' M_m'}\right)[(L_{21}')_0(L_{jm}')_0 - (L_{2m}')_0(L_{j1}')_0] \tag{1.154}$$

$$(L_{22})_0 = \frac{1}{A}\left(\frac{M_2'}{z_2} - \frac{M_3'}{z_3}\right)^2 \left(\frac{z_2}{M_2'}\right)^2$$

$$\times \sum_{j=1}^{3} \sum_{m=1}^{3} \left(\frac{z_j z_m}{M_j' M_m'}\right)[(L_{22}')_0(L_{jm}')_0 - (L_{2m}')_0(L_{j2}')_0] \tag{1.155}$$

with A defined by

$$A = \sum_{j=1}^{3} \sum_{m=1}^{3} \frac{z_j z_m}{M_j' M_m'}(L_{jm}')_0 \tag{1.156}$$

It is readily shown that if $(L'_{12})_0 = (L'_{21})_0$ as has been assumed above, Onsager's reciprocal relation holds between $(L_{12})_0$ and $(L_{21})_0$, that is,

$$(L_{12})_0 = (L_{21})_0 \tag{1.157}$$

If we notice that X_k is given by equation 1.144 and that Onsager's reciprocal relation applies between the cross-term coefficients, the set of equations given above for the solvent-fixed flows of electrolytes E_1 and E_2 is precisely the special case of the set of equations for solvent-fixed flows of q nonelectrolyte solutes. In other words, for the electrolyte solution considered the formalism developed previously for nonelectrolyte solutions applies without modification if we treat the given electrolytes as individual flow components even though the actual flow units are individual ions. It is of interest to examine whether this rule can be extended to more complicated cases such as the one in which ionic species formed on dissociation from electrolytes E_1 and E_2 are all different. In this case, there are four ways of grouping the species to obtain neutral electrolytes. Two of the four electrolytes obtained are different from the ones which have been used to make up the given solution. For example, when a solution is prepared by dissolving NaCl and KBr in water, four ionic species—Na^+, K^+, Cl^-, and Br^-—appear in the solution, and we may combine them to give the following four kinds of electrolytes: NaCl, KCl, NaBr, and KBr. For given concentrations of the originally dissolved electrolytes, NaCl and KBr, the concentration of any one of the four electrolytes is automatically determined if those of the other three are assigned. Therefore, the number of independent electrolyte components in the solution is three. It is possible to obtain equations for solvent-fixed flows of these three independent electrolytes by proper combinations of the corresponding flow equations for Na^+, K^+, Cl^-, and Br^- which would be derived in a manner similar to that described above. Thus, taking NaCl, KCl, and NaBr as independent electrolytes E_1, E_2, and E_3, we will obtain the following set of flow equations:

$$(J_i)_c = (L_{i1})_0 X_1 + (L_{i2})_0 X_2 + (L_{i3})_0 X_3 \qquad (i = 1, 2, 3) \tag{1.158}$$

Here X_k $(k = 1, 2, 3)$ is defined by

$$X_k = r\omega^2 - \frac{\partial \mu_k}{\partial r} \tag{1.159}$$

and the coefficients $(L_{ki})_0$ $(k, i = 1, 2, 3)$ satisfy Onsager's reciprocal relations: $(L_{13})_0 = (L_{31})_0$, $(L_{23})_0 = (L_{32})_0$, and $(L_{12})_0 = (L_{21})_0$.

In general, if all ionic species formed on dissociation from given q neutral electrolytes are different from one another, the appropriate set of phenomenological flow equations for the system would be

$$(J_k)_0 = \sum_{i=1}^{2q-1} (L_{ki})_0 X_i \qquad (k=1,2,\ldots,2q-1) \qquad (1.160)$$

The coefficients $(L_{ki})_0$ should obey Onsager's reciprocal relations. Here $2q-1$ is the *minimum* number of neutral electrolytes which have to be considered independent flow components. By proper grouping we may have q^2 kinds of electrolytes (binary) from q cationic species and q anionic species which are dissociated from the given q electrolytes. We may choose any $2q-1$ of them as those required for the consideration of flow processes in terms of equation 1.160.

When the given q electrolytes have an anion in common, the number of electrolytes acting as independent flow components is q, and the appropriate set of flow equations becomes

$$(J_k)_0 = \sum_{i=1}^{q} (L_{ki})_0 X_i \qquad (k=1,2,\ldots,q) \qquad (1.161)$$

The system we have discussed above corresponds to a special case of this general one for $q=2$. Miller[29,30] investigated equation 1.161 for $q=2$ and for arbitrary q and gave general expressions for the coefficients in this equation. Our analysis for the case of $q=2$ is similar to that presented by him.

L.   TESTS OF ONSAGER'S RECIPROCAL RELATION

In the absence of a centrifugal force field, that is, $\omega=0$, equations 1.108 and 1.109 reduce to

$$(J_1)_c = -(D_{11})_V \frac{\partial c_1}{\partial r} - (D_{12})_V \frac{\partial c_2}{\partial r} \qquad (1.162)$$

$$(J_2)_c = -(D_{21})_V \frac{\partial c_1}{\partial r} - (D_{22})_V \frac{\partial c_2}{\partial r} \qquad (1.163)$$

Substituting equation 1.90 for $(L_{ki})_V$ into equation 1.111 and solving the resulting expressions for $(L_{12})_0$ and $(L_{21})_0$, we obtain

$$(L_{12})_0 = \frac{1}{S}[\mu_{11}(D_{12})_0 - \mu_{12}(D_{11})_0] \qquad (1.164)$$

$$(L_{21})_0 = \frac{1}{S}[\mu_{22}(D_{21})_0 - \mu_{21}(D_{22})_0] \qquad (1.165)$$

where

$$\mu_{ij} = \left(\frac{\partial \mu_i}{\partial c_j}\right)_{T,P,c_m} \qquad (i,j,m=1,2) \qquad (1.166)$$

$$S = \mu_{11}\mu_{22} - \mu_{12}\mu_{21} \qquad (1.167)$$

and

$$(D_{11})_0 = \left(1 + \frac{c_1\bar{v}_1}{c_0\bar{v}_0}\right)(D_{11})_V + \left(\frac{c_1\bar{v}_2}{c_0\bar{v}_0}\right)(D_{21})_V \qquad (1.168)$$

$$(D_{12})_0 = \left(1 + \frac{c_1\bar{v}_1}{c_0\bar{v}_0}\right)(D_{12})_V + \left(\frac{c_1\bar{v}_2}{c_0\bar{v}_0}\right)(D_{22})_V \qquad (1.169)$$

$$(D_{21})_0 = \left(1 + \frac{c_2\bar{v}_2}{c_0\bar{v}_0}\right)(D_{21})_V + \left(\frac{c_2\bar{v}_1}{c_0\bar{v}_0}\right)(D_{11})_V \qquad (1.170)$$

$$(D_{22})_0 = \left(1 + \frac{c_2\bar{v}_2}{c_0\bar{v}_0}\right)(D_{22})_V + \left(\frac{c_2\bar{v}_1}{c_0\bar{v}_0}\right)(D_{12})_V \qquad (1.171)$$

It can easily be shown that the $(D_{ij})_0$ defined here represent the set of solvent-fixed diffusion coefficients of the ternary solution considered. Since Onsager's reciprocal relation holds between $(L_{12})_0$ and $(L_{21})_0$, we find from equations 1.164 and 1.165 that the following relation must be satisfied:

$$\mu_{11}(D_{12})_0 - \mu_{12}(D_{11})_0 = \mu_{22}(D_{21})_0 - \mu_{21}(D_{22})_0 \qquad (1.172)$$

Obviously, this relation should be obeyed by any ternary solution consisting of nonelectrolyte components, because originally equations 1.162 and 1.163 were derived for such a system. It also should be valid for a two-electrolyte solution of the type considered in the preceding sections, that is, the system which consists of water and two strong electrolytes having an anion or a cation in common. Thus when applied to such an electrolyte solution, subscripts 1 and 2 in equation 1.172 should be referred to the given two electrolytes, not to the ions dissociated from them. Finally, we remark that equation 1.172 also applies to a ternary solution in which one of the two solute components is a nonelectrolyte and the other is a strong electrolyte.

In order to test the validity of equation 1.172, very precise data are required for the quantities $(D_{ij})_0$ and μ_{ij}. At present, the relevant data for the latter are available only for a limited number of systems, but they could be derived from appropriate thermodynamic measurements. Data for the former are calculable from the measurements of the partial specific volumes $\bar{v}_1$ and $\bar{v}_2$ and those of the four volume-fixed diffusion coefficients $(D_{ij})_V$.

Methods of evaluating $(D_{ij})_V$ from isothermal diffusion experiments have been developed by Fujita and Gosting.[21-23] Recently, Kim[24] has described the corresponding methods for quaternary solutions. These methods require that extremely precise diffusion measurements be performed with the aid of a very elaborate diffusiometer such as the one designed and constructed by Gosting and collaborators.[16] So far they have been applied to raffinose–KCl–H_2O,[31] raffinose–urea–H_2O,[32] LiCl–KCl–H_2O,[33] LiCl–NaCl–H_2O,[33] NaCl–KCl–H_2O,[34,35,22] Na_2SO_4–H_2SO_4–H_2O,[36] KCl–HCl–H_2O,[37] KBr–HBr–H_2O,[37] NaCl–HCl–H_2O,[37] H_2PO_4–$Ca(H_2PO_4)_2$–H_2O,[38] polystyrene-cyclohexane-toluene,[39] and sodium polyacrylate–NaCl–H_2O.[40] The results demonstrated that equation 1.172, hence Onsager's reciprocal relation, was satisfied with agreement better than a few percent when the experimental data used were considered sufficiently accurate.[22,35-37,39,41] The magnitudes of the cross-term diffusion coefficients were not always small compared with those of the main diffusion coefficients.[36-40] This latter fact implies that the inclusion of the terms involving cross-term diffusion coefficients is essential for complete description of isothermal diffusion in ternary solutions. Evidently, the same should be true for description of any transport processes in multicomponent solutions involving diffusion. However, the relative importance of the coupling diffusion may depend on the type of transport process considered. Thus, in the theoretical treatments of ultracentrifugation, it has been assumed that the coupling diffusion, if any, would play a negligible role in comparison with the sedimentation transport due to centrifugal forces, and the terms involving cross-term diffusion coefficients have been neglected from the flow equations. Throughout the present monograph we follow this assumption, but it must be kept in mind that because of this the results derived therefrom are approximate in a strict sense.

M. CHEMICALLY REACTING SYSTEMS

So far our discussion has been concerned with systems made up of components which do not react chemically with each other. It can be shown that all the flow equations derived above are applicable, as they stand, to chemically reacting solutes, but the continuity equations have to be corrected for the production or disappearance of matter when solutes undergo chemical reactions with themselves or with the others. The consideration of this problem is deferred to Chapter 4.

APPENDIX A

Let us define a set of new flows $\{J_k'\}$ and a set of new forces $\{X_k'\}$ by

$$J_k' = \sum_{i=0}^{q} B_{ki}(J_i)_M \tag{A.1}$$

$$X_k' = \sum_{i=0}^{q} C_{ki}X_i \tag{A.2}$$

$$(k = 0, 1, \ldots, q)$$

with the coefficients C_{ki} chosen so that

$$\mathbf{C} = \mathbf{B_T}^{-1} \tag{A.3}$$

Here $\mathbf{C}$ is the matrix of C_{ki} and $\mathbf{B_T}^{-1}$ is the inverse transpose of the matrix $\mathbf{B}$ of B_{ki}. If we are to be able to form its inverse, the matrix $\mathbf{B}$ must be nonsingular; that is, the determinant of $\mathbf{B}$ must not vanish. As far as this condition is satisfied, B_{ki} can be assigned arbitrarily.

From equations A.1 and A.2 we have

$$\sum_{k=0}^{q} J_k'X_k' = \sum_{k=0}^{q} \sum_{i=0}^{q} \sum_{j=0}^{q} B_{ki}C_{kj}(J_i)_M X_j \tag{A.4}$$

When the condition A.3 holds, it follows that

$$\sum_{k=0}^{q} B_{ki}C_{kj} = 1 \quad (i=j), \qquad \sum_{k=0}^{q} B_{ki}C_{kj} = 0 \quad (i \neq j)$$

Hence equation A.4 becomes

$$\sum_{k=0}^{q} J_k'X_k' = \sum_{i=0}^{q} (J_i)_M X_i = T\Phi_S \tag{A.5}$$

which indicates that J'_k and X'_k are conjugate.

Next, solving equation A.2 for X_i and substituting into equation 1.77, we obtain

$$(J_k)_M = \sum_{i=0}^{q} (L_{ki})_M \sum_{j=0}^{q} C_{ij}^{-1} X'_j \tag{A.6}$$

where C_{kj}^{-1} denotes the (k,j) element of the inverse matrix of $\mathbf{C}$. Substitution of equation A.6 into equation A.1 gives

$$J'_k = \sum_{n=0}^{q} L'_{kn} X'_n \tag{A.7}$$

where

$$L'_{kn} = \sum_{i=0}^{q} \sum_{j=0}^{q} B_{ki}(L_{ij})_M C_{jn}^{-1} \tag{A.8}$$

Thus we find that J'_k is written as a linear homogeneous function of X'_k. When the condition A.3 holds, we have the relation $C_{jn}^{-1} = B_{nj}$. Hence

$$L'_{kn} = \sum_{i=0}^{q} \sum_{j=0}^{q} B_{ki}(L_{ij})_M B_{nj} \tag{A.9}$$

So

$$L'_{nk} = \sum_{i=0}^{q} \sum_{j=0}^{q} B_{ni}(L_{ij})_M B_{kj}$$

which may be rewritten, by exchanging the dummy indices i and j,

$$L'_{nk} = \sum_{i=0}^{q} \sum_{j=0}^{q} B_{nj}(L_{ji})_M B_{ki} \tag{A.10}$$

Noticing $(L_{ij})_M = (L_{ji})_M$ $(i \neq j)$, it follows from equations A.9 and A.10 that

$$L'_{nk} = L'_{kn} \quad (k \neq n) \tag{A.11}$$

Thus the coefficients in equation A.7 obey Onsager's reciprocal relations.

Summarizing the results just obtained, we find that there is an infinite number of sets of conjugate flows and forces other than the set of $(J_k)_M$ and X_k for which the cross-term phenomenological coefficients satisfy Onsager's reciprocal relations.

APPENDIX B

We show that if the solution is isothermal and if the partial specific volumes of all components are independent of pressure and composition, the velocity of the local center of volume is zero, provided that at least one end of the cell is closed.

Differentiation of equation 1.14 with respect to time t gives

$$\sum_{k=0}^{q} \bar{v}_k \frac{\partial c_k}{\partial t} + \sum_{k=0}^{q} c_k \frac{\partial \bar{v}_k}{\partial t} = 0 \tag{B.1}$$

Substitution for $\partial c_k / \partial t$ from equation 1.35 yields

$$\sum_{k=0}^{q} \frac{\bar{v}_k}{r} \frac{\partial}{\partial r}[rc_k(u_k)_c] = \sum_{k=0}^{q} c_k \frac{\partial \bar{v}_k}{\partial t} \tag{B.2}$$

which, in turn, leads to

$$\frac{(u)_V}{r} + \frac{\partial (u)_V}{\partial r} = \sum_{k=0}^{q} c_k \left(\frac{\partial \bar{v}_k}{\partial t} + (u_k)_c \frac{\partial \bar{v}_k}{\partial r} \right) \tag{B.3}$$

where equation 1.15 has been considered. If the solution is isothermal and if the $\bar{v}_k$'s of all components are independent of pressure and composition, the right-hand side of equation B.3 vanishes. Thus we have

$$\frac{\partial [r(u)_V]}{\partial r} = 0 \tag{B.4}$$

which gives

$$(u)_V = \frac{f(t)}{r} \tag{B.5}$$

where $f(t)$ is an arbitrary function of time t. If, as in the ultracentrifuge, the cell is closed at one end (at least), the $(J_k)_c$ of all components become zero at this boundary for all values of t. Therefore, by equation 1.15, $(u)_V$ also vanishes at the same position for all t. This fact is combined with equation B.5 to yield the statement that $f(t)=0$ for all t and hence $(u)_V$ vanishes everywhere in the cell.

APPENDIX C

We apply equation 1.6 to ions 1, 2, and 3 with the primes attached to J and c. Multiplying both sides by z_k / M_k' ($k = 1, 2, 3$), summing over all ions, and using equation 1.136 and the condition that with no electric field applied externally there is no electric current flowing relative to the cell, that is,

$$\frac{z_1}{M_1'}(J_1')_c + \frac{z_2}{M_2'}(J_2')_c + \frac{z_3}{M_3'}(J_3')_c = 0 \qquad (C.1)$$

we obtain equation 1.137.

REFERENCES

1. G. J. Hooyman, "Thermodynamics of Irreversible Processes in Rotating Systems," Thesis, University of Leiden, Leiden, 1955.
2. S. R. de Groot, *Thermodynamics of Irreversible Processes*, North-Holland, Amsterdam, 1952.
3. S. R. de Groot and P. Mazur, *Non-equilibrium Thermodynamics*, North-Holland, Amsterdam, 1962.
4. R. Haase, *Thermodynamik der irreversiblen Prozesse*, Dietrich Steinkopff Verlag, Darmstadt, 1963.
5. D. D. Fitts, *Nonequilibrium Thermodynamics–A Phenomenological Theory of Irreversible Processes in Fluid Systems*, McGraw-Hill, New York, 1962.
6. H. J. V. Tyrrell, *Diffusion and Heat Flow in Liquids*, Butterworths, London, 1961.
7. J. Meixner and H. G. Reik, *Handbuch der Physik* (S. Flugge, Ed.), III/2, 413 (1959).
8. D. G. Miller, *Chem. Rev.* **60**, 15 (1960).
9. G. J. Hooyman, H. Holtan, Jr., P. Mazur, and S. R. de Groot, *Physica* **19**, 1095 (1953); see also Chap. 10 of Ref. 5.
10. As for the thermodynamic relations which appear in this treatise, see, for example, J. G. Kirkwood and I. Oppenheim, *Chemical Thermodynamics*, McGraw-Hill, New York, 1961.
11. B. D. Coleman and C. Truesdell, *J. Chem. Phys.* **33**, 28 (1960); see also Chap. 4 of Ref. 5.
12. J. G. Kirkwood, R. L. Baldwin, P. J. Dunlop, L. J. Gosting, and G. Kegeles, *J. Chem. Phys.* **33**, 1505 (1960).
13. O. Lamm, *Z. Phys. Chem.* (*Leipzig*) **A143**, 177 (1929); *Arkiv Mat. Astron. Fysik* **21**(B) (1929).
14. G. S. Hartley and J. Crank, *Trans. Faraday Soc.* **45**, 801 (1949).
15. L. J. Gosting, *Adv. Protein Chem.* **11**, 429 (1956).
16. L. J. Gosting, E. M. Hanson, G. Kegeles, and M. S. Morris, *Rev. Sci. Instrum.* **20**, 209 (1949).
17. D. F. Akeley and L. J. Gosting, *J. Am. Chem. Soc.* **75**, 5685 (1953).
18. P. J. Dunlop and L. J. Gosting, *J. Am. Chem. Soc.* **75**, 5073 (1953).
19. P. J. Dunlop, *J. Am. Chem. Soc.* **77**, 2994 (1955).

20. R. L. Baldwin, P. J. Dunlop, and L. J. Gosting, *J. Am. Chem. Soc.* **77**, 5235 (1955).
21. H. Fujita and L. J. Gosting, *J. Am. Chem. Soc.* **78**, 1099 (1956).
22. H. Fujita and L. J. Gosting, *J. Phys. Chem.* **64**, 1256 (1960).
23. H. Fujita, *J. Phys. Chem.* **63**, 242 (1959).
24. H. Kim, *J. Phys. Chem.* **70**, 562 (1966).
25. R. L. Baldwin, *J. Am. Chem. Soc.* **80**, 496 (1958).
26. J. M. Creeth and R. H. Pain, *Prog. Biophys. Mol. Biol.* **17**, 217 (1967).
27. L. Peller, *J. Chem. Phys.* **29**, 415 (1958).
28. H. Schönert, *J. Phys. Chem.* **64**, 733 (1960).
29. D. G. Miller, *J. Phys. Chem.* **71**, 616 (1967).
30. D. G. Miller, *J. Phys. Chem.* **71**, 3588 (1967).
31. P. J. Dunlop, *J. Phys. Chem.* **61**, 994 (1957).
32. P. J. Dunlop, *J. Phys. Chem.* **61**, 1619 (1957).
33. P. J. Dunlop and L. J. Gosting, *J. Am. Chem. Soc.* **77**, 5238 (1955).
34. I. J. O'Donnell and L. J. Gosting, in *The Structure of Electrolyte Solutions* (W. J. Hamer, Ed.), Wiley, New York, 1959, Chap. 11, pp. 160–182; in connection with this paper, see L.-O. Sundelöf, *Arkiv Kemi* **20**, 369 (1963).
35. P. J. Dunlop, *J. Phys. Chem.* **63**, 612 (1959).
36. R. P. Wendt, *J. Phys. Chem.* **66**, 1279 (1962).
37. H. Kim, private communication, 1972.
38. O. W. Edwards, R. L. Dunn, J. D. Hatfield, E. O. Huffman, and K. L. Elmore, *J. Phys. Chem.* **70**, 217 (1966).
39. E. L. Cussler, Jr., and E. N. Lightfoot, *J. Phys. Chem.* **69**, 1135 (1965).
40. V. Vitagliano, R. Laurentino, and L. Costantino, *J. Phys. Chem.* **73**, 2456 (1969).
41. P. J. Dunlop and L. J. Gosting, *J. Phys. Chem.* **63**, 86 (1959).

2

SEDIMENTATION TRANSPORT IN TWO-COMPONENT SYSTEMS

I. INTRODUCTION

A. THE SCOPE OF THIS CHAPTER

Comprehension of the theory of sedimentation processes in two component systems is not only essential for a rational interpretation and analysis of ultracentrifugal measurements on binary solutions but also a prerequisite for the study of similar problems on multicomponent solutions. Its basis is provided by the Lamm differential equation for the ultracentrifuge, equation 1.123, although for the reason mentioned in Chapter 1 this equation is approximate for compressible systems. Since the pioneering work by Faxén[1] in 1929 and by Archibald[2-4] around 1940 there have been many contributions to the mathematics of the Lamm equation. These have been concerned with solving the equation under conditions pertinent to actual sedimentation experiments with the analytical ultracentrifuges. Thus to date most of the theoretical features of sedimentation transport in two-component system have been elucidated, and there have been derived a variety of useful relations which allow such fundamental quantities as sedimentation and diffusion coefficients to be evaluated from optical measurements of sedimentation patterns.

This chapter aims at a systematic survey of the typical results evolved from these investigations of the Lamm equation. In doing this, the context and notation of the original authors are not always followed, but are sometimes modified so that the description of the whole chapter may become consistent both in logic and in appearance.

B. INITIAL AND BOUNDARY CONDITIONS

Two types of initial conditions can be set up for experiments with the current ultracentrifuge cells, which are the *conventional cell* and the *synthetic boundary cell*. In experiments with a conventional cell, initially a

47

uniform solution of a given concentration, say c_0, is admitted into the cell cavity. Thus* (Fig. 2.1)

$$c = c_0 \qquad (r_1 < r < r_2, \quad t = 0) \qquad (2.1)$$

where, as defined in Chapter 1, Section B, r_1 and r_2 denote, respectively, the radial positions of the air-liquid meniscus and the cell bottom.

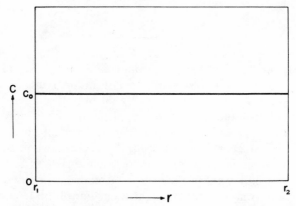

Fig. 2.1. Diagrammatic representation of the initial concentration distribution in the experiment with a conventional cell.

The synthetic boundary cell was introduced as a new tool for ultracentrifugal studies by Kegeles[5] and independently by Pickels et al.[6] in 1952. When it is used, initially the given solution of a concentration c_0 is separated from the solvent by a sharp boundary formed somewhere in the central part of the cell cavity. In practice, it is impossible to form a perfectly sharp boundary between solution and solvent. Even if it were possible to set up such a boundary, the boundary would diffuse to some extent before the solution was centrifuged. In the theoretical development given below, we neglect this spreading effect, because its inclusion makes mathematical analysis quite complicated. To maintain dynamical stability of the solution during centrifugation, a more dense liquid must be placed farther away from a less dense one in the direction of centrifugal acceleration. Therefore, usually, the initial condition for an experiment with a

*Throughout this chapter, the symbol c is used for the c-scale concentration of the solute component.

synthetic boundary cell is mathematically represented by (Fig. 2.2)

$$c = \begin{cases} 0 & (r_1 < r < r_0, \quad t=0) \\ c_0 & (r_0 < r < r_2, \quad t=0) \end{cases} \tag{2.2}$$

where r_0 denotes the radial position of the initial boundary between solution and solvent. With a synthetic boundary cell a sharp boundary may be formed between solutions of the same solute at different concentrations, say c_1 and c_2. In this case, the mathematical representation of the initial condition is

$$c = \begin{cases} c_1 & (r_1 < r < r_0, \quad t=0) \\ c_2 & (r_0 < r < r_2, \quad t=0) \end{cases} \tag{2.3}$$

This type of initial condition is often used in free-diffusion experiments on solutions, and sometimes it is referred to as the *differential* initial boundary.[7,8] Obviously, equation 2.2 is the special case of equation 2.3 in which $c_1 = 0$ and $c_2 = c_0$.

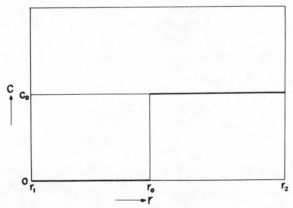

Fig. 2.2. Diagrammatic representation of the initial concentration distribution in the experiment with a synthetic boundary cell.

The boundary conditions are the same for both the conventional cell and the synthetic boundary cell. They derive from the physical requirement that no solute can pass through the air-liquid meniscus or the cell bottom. Of course, similar conditions must hold for the solvent component, but those are not explicitly needed for the subsequent development. The conditions for the solute are mathematically such that $(J_1)_c = 0$ at $r = r_1$ and $r = r_2$. Substitution of equation 1.105 for $(J_1)_c$ yields

$$D \frac{\partial c}{\partial r} = s\omega^2 r_1 c \qquad (r = r_1, \quad t > 0)$$

$$D \frac{\partial c}{\partial r} = s\omega^2 r_2 c \qquad (r = r_2, \quad t > 0)$$

(2.4)

if $r_1 \neq 0$ and $r_2 \neq \infty$. Here $(s_1)_V$ and $(D_{11})_V$ have been replaced by s and D, respectively, for the convenience of presentation. This convention is used throughout Chapter 2.

Either a combination of equations 2.1 and 2.4 or of equations 2.2 or 2.3 and 2.4 is sufficient for unique determination of a solution to the Lamm equation. In order to effect an actual solution, s and D must be given as functions of c and P, and the rotor speed ω as a function of t. When s and D depend on P, the Lamm equation is not sufficient, but it must be coupled with the pressure equation 1.69. In general, and especially for aqueous solutions, the dependence of s and D on pressure is considered quite small. Therefore, in all that follows, it is neglected unless otherwise stated. Experimentally, some nonzero time, ranging 5 to 15 min, is required before the rotor is brought to the desired operational speed. Introduction of this transient effect makes the mathematics of the Lamm equation extremely complex.[9] Hence, in what follows, we shall adopt the usual approximation that ω is independent of time during the entire course of a centrifugation.

C. CONCENTRATION DEPENDENCE OF s AND D

Many experimental results have demonstrated that the sedimentation coefficients of macromolecular solutes generally *decrease* with the increase in the solute concentration.[10] The degree of dependence varies with the kind of solute as well as the solvent conditions. At present, no satisfactory theory is available which explains observed results for $s(c)$. Empirically, however, most of the observed s versus c relations for dilute macromolecular solutions, with which sedimentation experiments are usually concerned, are described quite accurately either by

$$s = \frac{s_0}{1 + k_s c}$$

(2.5)

or by

$$s = s_0(1 - k_s c)$$

(2.6)

Here s_0 is the value of s at infinite dilution of the solution, and k_s is a positive parameter characteristic of a given solute-solvent pair. In this treatise, the quantity s_0 is referred to as the *limiting sedimentation coefficient*. The term "sedimentation constant" of early use soon became obsolete, following observations that s is ordinarily concentration dependent.

Published data indicate that equation 2.5 is generally adequate for synthetic polymers of high molecular weight and also for various biological macromolecules, except for globular proteins. Vinograd and Bruner[11] cite the following values for k_s of these macromolecules of biological interest: 1 to 3 for elongated proteins and viruses, such as myosin and tobacco mosaic virus; 10 to 20 for ribosomal DNA; and $(0.5-5)\times 10^3$ for high-molecular-weight DNA all in deciliters per gram. The enormously large k_s values for the DNA are worthy of note, suggesting that the s_0 of such solutes are made available only by sedimentation measurements at extremely low solute concentrations. For numerical values of s_0 and k_s for synthetic polymers the reader is referred to an extensive table prepared by Ende.[12]

Data on globular proteins and spherical viruses usually fit equation 2.6, although sometimes a quadratic term $k_s' c^2$ has to be added for a more accurate representation of s versus c. The values of k_s for these substances are generally as small as 0.1 to 0.2 dl/g, but it is hazardous to accept this fact as implying that the sedimentation coefficients of globular proteins are little affected by the solute concentration. As is quoted in Chapter 4, many of the native proteins in aqueous buffers tend to associate with each other to form dimers, trimers, and even higher oligomers, in a way depending on the environmental conditions such as pH, ionic strength, and temperature. It is shown (Chapter 4) that when this effect is present, there arises a perturbation which tends to suppress the normal decrease of s with c. Thus for certain protein systems, in which the association effect is appreciable, one may obtain sedimentation coefficients which are essentially independent of or even *increase* with c; in other words, $k_s \approx 0$ or $k_s < 0$, at least in the region of high dilutions (see Chapter 4, Section III.D).

Up to date the Lamm equation with a concentration-dependent sedimentation coefficient has been treated only for the two types of $s(c)$ represented by equations 2.5 and 2.6. However, inasmuch as these equations are empirically adequate for a good many cases of practical interest, the Lamm equations subject to them are not only theoretically but also practically useful enough for studying the effects which are exerted by the concentration dependence of s on c upon sedimentation transport processes in the ultracentrifuge. It is important to recognize that when s depends on c in any fashion, the Lamm equation becomes nonlinear in c and no longer feasible to analytic solution except for certain special cases of $s(c)$.

In general, the diffusion coefficient of a binary solution also depends on the solute concentration. For dilute solutions of macromolecules, either synthetic or biological, this dependence is represented well by an empirical relation:

$$D = D_0(1 + k_D c) \qquad (2.7)$$

Here D_0 is the value of D at infinite dilution of the solution, and k_D is a parameter characteristic of a given system. As in the case of s, the quantity D_0 will be referred to as the *limiting diffusion coefficient*. The parameter k_D is, in general, smaller in absolute value than k_s, and its sign for a given macromolecular solute depends on various factors such as the shape and dimensions of the solute, the kind of solvent, and temperature. Ende[12] prepared an extensive table for D_0 and k_D of a great variety of polymer-solvent systems.

1. Limiting Sedimentation and Diffusion Coefficients

Experimental determinations of the quantities s_0 and D_0 for binary solutions have two kinds of physical significance. First, they provide data for the evaluation of molecular weights by the Svedberg relation (equation 1.121). However, the s_0–D_0 method for molecular weight determination is no longer as popular as before in the current studies of macromolecular substances, but is rapidly being replaced by the sedimentation equilibrium method. Second, the translational friction coefficient of a solute molecule at infinite dilution, f_0, is related to s_0 and D_0 by[13]

$$s_0 = \frac{M(1 - \bar{v}^0 \rho_0)}{N_A f_0} \qquad (2.8)$$

$$D_0 = \frac{RT}{N_A f_0} \qquad (2.9)$$

Here M is the molecular weight of the solute, $\bar{v}^0$ is its partial specific volume at infinite dilution, ρ_0 is the density of the solvent, N_A is Avogadro's number, R is the gas constant, and T is the absolute temperature of the system. The quantity f_0 is defined as the force required to move a molecule at unit velocity in an unbounded solvent. Hydrodynamic theories of macromolecules in solution show that f_0 is related to the shape and dimensions of the solute molecule and the degree of solvation.[13] Thus, the measurements of s_0 and D_0, when converted to f_0 and compared with appropriate theories, provide information about such molecular quantities. In this monograph, we do not enter into a discussion of the molecular theoretical aspects of sedimentation and diffusion, but instead we are exclusively concerned with a phenomenological theoretical *description* of these transport processes.

Usually, when one studies proteins or other macromolecules of biological interest, the actually measured values of s_0 and D_0 in a buffer solution at temperature T are reduced to some standard state, often taken as "water at 20°C," by making use of the relations

$$(s_0)_{20,w} = \frac{(1 - \bar{v}^0 \rho_0)_{20,w} \eta_{T,b}}{(1 - \bar{v}^0 \rho_0)_{T,b} \eta_{20,w}} (s_0)_{T,b} \tag{2.10}$$

$$(D_0)_{20,w} = \frac{293.1}{T} \frac{\eta_{T,b}}{\eta_{20,w}} (D_0)_{T,b} \tag{2.11}$$

Here η denotes the solvent viscosity. These reductions depend on the assumption that f_0 is accurately proportional to η. It also should be noted that these equations correct only for variations of η, $\bar{v}^0$, ρ_0 with temperature and not for variations in molecular conformations which may accompany the change in temperature or in solvent.

The absolute values of s_0 and D_0 (more generally s and D) of macromolecular substances are of the order of 10^{-13} sec and 10^{-7} cm^2/sec, respectively. This general feature plays a fundamental role in developing phenomenological theories of sedimentation, because, as will be seen below in many places, its use makes various simplifications and approximations of the mathematical treatments feasible.

II. THE CASE OF NO DIFFUSION

A. THE GENERAL SOLUTION TO THE LAMM EQUATION WITH $D = 0$

Before studying the complete Lamm equation, let us consider its special case in which $D = 0$. Experimental data exhibit the general trend that s increases and D decreases as the molecular size of the solute becomes greater. Accordingly, the results from the Lamm equation with no diffusion term are of great significance in describing the limiting behavior of very high-molecular-weight substances such as DNA and viruses.

Now, with $D = 0$ the Lamm equation 1.123 reduces to

$$\frac{\partial c}{\partial t} = -\frac{\partial}{r \, \partial r}(r^2 \omega^2 s c) \tag{2.12}$$

This partial differential equation of the first order can be solved by the method of characteristics. For this purpose equation 2.12 is rewritten in the form

$$\frac{\partial c}{\partial t} + r\omega^2 \frac{d[s(c)c]}{dc}\frac{\partial c}{\partial r} + 2\omega^2 s(c)c = 0 \qquad (2.13)$$

Then the characteristic differential equations for this are found to be

$$\frac{dt}{1} = \frac{dr}{r\omega^2 d[s(c)c]/dc} = -\frac{dc}{2\omega^2 s(c)c} \qquad (2.14)$$

which have two independent solutions, given by

$$2s_0\omega^2 t + \int^c \frac{d\theta}{f(\theta)\theta} = a \qquad \text{and} \qquad r^2 f(c)c = b \qquad (2.15)$$

where a and b are arbitrary constants, and θ is an integration variable. The function $f(c)$ is a "reduced" sedimentation coefficient defined by

$$f(c) = \frac{s(c)}{s_0} \qquad (2.16)$$

where s_0 is the limiting sedimentation coefficient. Thus $f(0) = 1$.

From equations 2.15 the general solution to equation 2.12 may be written down in various ways. The most convenient for the development which follows is

$$\exp\left[\int^c \frac{d\theta}{f(\theta)\theta}\right] = G\left\{r^2 f(c)c \exp\left[-2s_0\omega^2 t - \int^c \frac{d\theta}{f(\theta)\theta}\right]\right\}\exp(-2s_0\omega^2 t)$$

$$(2.17)$$

where G stands for an arbitrary function of its argument. The form of G can be determined by the initial and boundary conditions imposed on c.

B. SOLUTIONS FOR TYPICAL CASES OF $s(c)$

CASE I. $s = \text{constant} = s_0$. In this simplest case, $f(c)$ is identically equal to unity, so that equation 2.17 gives

$$c = G[r^2 \exp(-2s_0\omega^2 t)]\exp(-2s_0\omega^2 t) \qquad (2.18)$$

Let us first consider the case in which the initial condition is given by equation 2.1. Then it follows from equation 2.18 that

$$c_0 = G(r^2) \qquad (r_1 < r < r_2) \tag{2.19}$$

Hence

$$G(y) = c_0 \qquad (r_1^2 < y < r_2^2) \tag{2.20}$$

Therefore, equation 2.18 gives

$$c = c_0 \exp(-2s_0\omega^2 t) \qquad [r_1 \exp(s_0\omega^2 t) < r < r_2 \exp(s_0\omega^2 t), \quad t > 0] \tag{2.21}$$

Because the physically meaningful region of r is $r_1 < r < r_2$, it is reasonable to write equation 2.21 as

$$c = c_0 \exp(-2s_0\omega^2 t) \qquad [r_1 \exp(s_0\omega^2 t) < r \leqslant r_2, \quad t > 0] \tag{2.22}$$

Since the Lamm equation with $D = 0$ is of first order with respect to r, its solutions cannot satisfy two boundary conditions. For $D = 0$ equations 2.4 give

$$c = 0 \qquad (r = r_1 \text{ and } r_2, \quad t > 0) \tag{2.23}$$

Because equation 2.22 does not satisfy the condition that $c = 0$ at $r = r_2$, we have to be content with fulfillment of the condition that $c = 0$ at $r = r_1$. Substitution of this boundary condition into equation 2.18 yields

$$G[r_1^2 \exp(-2s_0\omega^2 t)] = 0 \qquad (t > 0) \tag{2.24}$$

Hence

$$G(y) = 0 \qquad (0 < y < r_1^2) \tag{2.25}$$

Therefore we obtain from equation 2.18

$$c = 0 \qquad [r_1 \leqslant r < r_1 \exp(s_0\omega^2 t), \quad t > 0] \tag{2.26}$$

Combination of equations 2.22 and 2.26 completes the solution to the problem, that is,

$$c = \begin{cases} 0 & [r_1 \leqslant r < r_1 \exp(s_0\omega^2 t), \quad t > 0] \\ c_0 \exp(-2s_0\omega^2 t) & [r_1 \exp(s_0\omega^2 t) < r \leqslant r_2, \quad t > 0] \end{cases} \tag{2.27}$$

These expressions give a concentration distribution which is represented by a step function. The step point, denoted here by $r_*(t)$, moves toward the cell bottom, in accordance with the relation

$$r_*(t) = r_1 \exp(s_0 \omega^2 t) \tag{2.28}$$

Differentiation of equations 2.27 with respect to r yields a distribution of concentration gradients which is characterized by an infinitely sharp line (spike) located at the moving boundary r_*.

Equation 2.28 may be rewritten

$$\ln r_* = \ln r_1 + s_0 \omega^2 t \tag{2.29}$$

which shows that a plot of $\ln r_*$ against $\omega^2 t$ gives a straight line of slope s_0. Thus when $D = 0$, $s = \text{constant} = s_0$, and the initial condition of the experiment is given by equation 2.1, the value of s_0 can be determined from the measurement of r_* as a function of time.

The second equation of 2.27 indicates that the concentration in the solution phase *decreases* exponentially with the time of centrifugation. It might be supposed that this concentration would increase with time because of the accumulation of solute molecules due to sedimentation in the bottom region of the cell. The theoretical consequence that the solution undergoes dilution rather than concentration in accordance with centrifugation arises from the fact that both the cross section of the cell and the centrifugal acceleration increase with radial distance. It should be noted, however, that its derivation is based on $D = 0$. As will be shown later, the situation becomes complex when diffusion is not negligible.

Combination of equation 2.28 with the second equation of 2.27 gives the following interesting relation:

$$\frac{c}{c_0} = \frac{r_1^2}{r_*^2} \tag{2.30}$$

Thus the concentration in the solution phase, c, is diluted in inverse proportion to the square of the radial distance from the axis of the rotor to the solution-solvent separation boundary r_*^2. In other words, the product of c and r_*^2 remains constant during the course of a centrifugation. This feature is called the *square-dilution rule* of the sector-shaped cell.

When the initial condition is given by equation 2.2, application of the same procedure as above leads to the following solution:

$$c = \begin{cases} 0 & [r_1 \leqslant r < r_*(t), \quad t > 0] \\ c_0 \exp(-2s_0\omega^2 t) & [r_*(t) < r \leqslant r_2, \quad t > 0] \end{cases} \quad (2.31)$$

$$r_*(t) = r_0 \exp(s_0\omega^2 t) \quad (2.32)$$

$$c(r_*)^2 = c_0(r_0)^2 \quad (2.33)$$

The concentration distribution for this case is also a step function, and the position of the solution-solvent separation boundary r_* moves toward the cell bottom in exactly the same fashion as it does when the initial condition is given by equation 2.1. The square-dilution rule applies also to this case.

CASE II. $s = s_0(1 - k_s c)$, $k_s > 0$. In this case,

$$f(c) = 1 - k_s c \quad (2.34)$$

Introduction into equation 2.17 gives

$$\frac{c}{1 - k_s c} = \exp(-2s_0\omega^2 t) G\left[r^2(1 - k_s c)^2 \exp(-2s_0\omega^2 t)\right] \quad (2.35)$$

Applying the initial condition given by equation 2.1, we obtain

$$G(y) = \frac{c_0}{1 - k_s c_0} \quad \left[r_1^2(1 - k_s c_0)^2 < y < r_2^2(1 - k_s c_0)^2\right] \quad (2.36)$$

Hence equation 2.35 yields

$$c = \frac{c_0 \exp(-2s_0\omega^2 t)}{1 - k_s c_0[1 - \exp(-2s_0\omega^2 t)]}$$

$$\left[r_1\{1 - k_s c_0[1 - \exp(-2s_0\omega^2 t)]\} \exp(s_0\omega^2 t) < r \leqslant r_2, \quad t > 0\right] \quad (2.37)$$

where the solution for $r_2 < r$ has been discarded for the reason explained in Case I.

On the other hand, substitution of the boundary condition $c = 0$ at $r = r_1$ into equation 2.35 leads to

$$G(y) = 0 \quad (0 < y < r_1^2) \quad (2.38)$$

Therefore it follows from equation 2.35 that

$$c = 0 \qquad [r_1 \leqslant r < r_1 \exp(s_0\omega^2 t), \quad t > 0] \tag{2.39}$$

Here again the solution for $r < r_1$ has been discarded.

The solutions represented by equations 2.37 and 2.39 are shown graphically in Fig. 2.3. Because $k_s > 0$, as has been assumed above, the following inequality holds for $s_0\omega^2 t > 0$:

$$r_1 \exp(s_0\omega^2 t)\left\{1 - k_s c_0[1 - \exp(-2s_0\omega^2 t)]\right\} < r_1 \exp(s_0\omega^2 t) \tag{2.40}$$

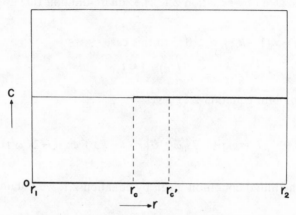

Fig. 2.3. Concentration distributions given by equations 2.37 and 2.39. The upper and lower thick lines represent equations 2.37 and 2.39, respectively. Note that the concentration becomes double-valued in the region between $r_c = r_1 \exp(s_0\omega^2 t)$ and $r'_c = r_1 \exp(s_0\omega^2 t) \times \{1 - k_s c_0[1 - \exp(-2s_0\omega^2 t)]\}$.

Thus the solution obtained becomes *doubled-valued* in the region of r such that

$$r_1 \exp(s_0\omega^2 t)\left\{1 - k_s c_0[1 - \exp(-2s_0\omega^2 t)]\right\} < r < r_1 \exp(s_0\omega^2 t) \tag{2.41}$$

Physically, the concentration must be a single-valued function of position r and time t. Therefore the desired solution must be made up from equations 2.37 and 2.39 so that this requirement may be satisfied. The relevant solution is

$$c = \begin{cases} 0 & [r_1 \leqslant r < r_*(t), \quad t > 0] \\ \dfrac{c_0 \exp(-2s_0\omega^2 t)}{1 - k_s c_0[1 - \exp(-2s_0\omega^2 t)]} & [r_*(t) < r \leqslant r_2, \quad t > 0] \end{cases} \tag{2.42}$$

where $r_*(t)$ is an arbitrary function of t which fulfills an inequality

$$r_1 \exp(s_0 \omega^2 t) \{ 1 - k_s c_0 [1 - \exp(-2s_0 \omega^2 t)] \} < r_*(t) < r_1 \exp(s_0 \omega^2 t) \quad (2.43)$$

It is readily confirmed that equations 2.42 define a single-valued function of r and t and satisfy not only the basic differential equation but also the boundary condition $c = 0$ at $r = r_1$. Furthermore, if we note from equation 2.43 that $r_*(t)$ has a limiting property

$$\lim_{t \to 0} r_*(t) = r_1 \quad (2.44)$$

equations 2.42 are found to satisfy the initial condition 2.1. Although in this way the function $c(r,t)$ defined by equations 2.42 satisfies all the required conditions of the problem, it is still left with an arbitrariness of $r_*(t)$. Because the physical solution of the problem should be unique, there must be an additional condition that allows this arbitrariness to be eliminated. It can be obtained from the requirement that the solution-solvent separation boundary must move with time in such a way that the amount of solute swept out by the boundary for an infinitesimally short interval of time, Δt, be equal to the amount of solute which will sediment through the same boundary for the same interval of time. Mathematically, this requirement is expressed as

$$c_*(t) \frac{dr_*(t)}{dt} \Delta t = s \omega^2 c_*(t) r_*(t) \Delta t \quad (2.45)$$

where $c_*(t)$ is the value of c at the moving boundary. The s in this equation refers to the value for $c = c_*(t)$. Equation 2.45 gives the following differential equation for $r_*(t)$:

$$\frac{dr_*(t)}{dt} = s \omega^2 r_*(t) \quad (2.46)$$

The s in this equation can be obtained by substituting the second equation of 2.42 into $s = s_0(1 - k_s c)$. With this s, integration of equation 2.46 and determination of the integration constant by equation 2.44 yield

$$r_*(t) = r_1 \exp(s_0 \omega^2 t) \{ 1 - k_s c_0 [1 - \exp(-2s_0 \omega^2 t)] \}^{1/2} \quad (2.47)$$

Note that this $r_*(t)$ satisfies the inequality 2.43. Combination of equations 2.42 and 2.47 determines the desired solution to the problem uniquely.

Equation 2.46 may be derived directly from the Lamm equation with $D = 0$ in the following manner. After both sides of equation 2.12 have been multiplied by r, the result is integrated over a range $r_*(t) - \delta < r < r_*(t) + \delta$ to give

$$\int_{r_*-\delta}^{r_*+\delta} \frac{\partial c}{\partial t} r\, dr = -s\omega^2 r^2 c \Big|_{r=r_*-\delta}^{r=r_*+\delta} \tag{2.48}$$

This may be rewritten

$$\frac{d}{dt} \int_{r_*-\delta}^{r_*+\delta} cr\, dr - cr\left(\frac{dr_*}{dt}\right)\Big|_{r=r_*+\delta} = -s\omega^2 r^2 c\big|_{r=r_*+\delta} \tag{2.49}$$

if $r_*(t)$ represents a solution-solvent separation boundary. Letting δ tend to zero, the first term on the left-hand side vanishes, giving a relation which agrees with equation 2.46.

From equations 2.42 and 2.47 it follows that

$$c(r_*)^2 = c_0(r_1)^2 \tag{2.50}$$

Thus the square-dilution rule applies even when s decreases linearly with c. Equation 2.47 yields

$$\ln r_* = \ln r_1 + s_0\omega^2 t + \tfrac{1}{2}\ln\left\{1 - k_s c_0[1 - \exp(-2s_0\omega^2 t)]\right\} \tag{2.51}$$

which may be compared to equation 2.29 for the case of constant s (Case I). Figure 2.4 illustrates curves of $\ln r_*$ versus $s_0\omega^2 t$ for three values of $k_s c_0$ calculated from equation 2.51. The curves for $k_s c_0 \neq 0$ are *concave upward*, but the curvature is small unless $k_s c_0$ is too large. Equation 2.51 yields

$$\lim_{t \to 0} \frac{d\ln r_*}{dt} = s_0(1 - k_s c_0)\omega^2 \tag{2.52}$$

which indicates that the initial slope of each curve in Fig. 2.4 equals $1 - k_s c_0$. Thus when $D = 0$, $s = s_0(1 - k_s c)$, and the initial condition is of the type given by equation 2.1, the value of $s_0(1 - k_s c_0)$ may be determined as the initial tangent of $\ln r_*$ plotted against $\omega^2 t$. Repetition of similar determinations at differing c_0, followed by graphing the resulting values of $s_0(1 - k_s c_0)$ versus c_0, allows s_0 and k_s to be evaluated from the ordinate intercept and slope of the plot. In practice, the initial tangent of $\ln r_*$ versus t may be determined most reliably by use of a procedure which is described in Section V.D.2.

For the initial condition given by equation 2.2 one can derive exactly the same results as above, with the only difference that r_1 in the equations given above is replaced by r_0.

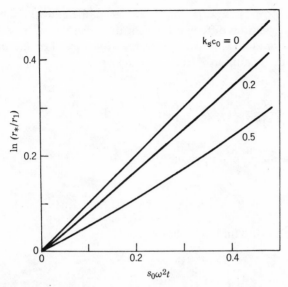

Fig. 2.4. Relations between $\ln r_*$ and $s_0\omega^2 t$ for three fixed values of $k_s c_0$ in the system where $s = s_0(1 - k_s c)$. Actual sedimentation velocity experiments are usually limited to times for which the $s_0\omega^2 t$ is 0.2 or less.

CASE III. $s = s_0/(1 + k_s c)$. The desired solution to the case with s of this form can be found in a way essentially similar to that described for Case II. For example, when the initial condition is given by equation 2.1, the concentration profile is represented by

$$c = 0 \qquad [r_1 \leqslant r < r_*(t), \quad t > 0]$$

$$\ln\left(\frac{c_0}{c}\right) + k_s(c_0 - c) = 2s_0\omega^2 t \qquad [r_*(t) < r \leqslant r_2, \quad t > 0]$$

$$(2.53)$$

This gives a step function similar in character to those in Cases I and II. Note that the second equation of 2.53 cannot be solved explicitly for c. It is shown that the square-dilution rule holds for this case too, that is,

$$c(r_*)^2 = c_0(r_1)^2 \qquad (2.54)$$

The solution-solvent boundary position $r_*(t)$ is given by the second equation of 2.53 and equation 2.54 with c as an intermediate parameter. Numerical calculations give a family of $\ln r_*$ versus $\omega^2 t$ curves which are quite similar to those shown in Fig. 2.4. The initial tangent of each curve equals $s_0/(1 + k_s c_0)$.

CASE IV.   It is possible to carry out a similar mathematical analysis of the Lamm equation with no diffusion term for the general case in which $s(c)$ is an arbitrary, monotonically decreasing function of c. For the experiment initiated with the condition 2.1 equation 2.17 gives

$$\int_c^{c_0} \frac{d\theta}{f(\theta)\theta} = 2s_0\omega^2 t \quad \left(r_1 \left[\frac{f(c_0)c_0}{f(c)c} \right]^{1/2} < r \leqslant r_2', \quad t > 0 \right) \quad (2.55)$$

Application of the boundary condition $c = 0$ at $r = r_1$ ($t > 0$) to equation 2.17 yields

$$c = 0 \quad [r_1 \leqslant r < r_1 \exp(s_0\omega^2 t), \quad t > 0] \quad (2.56)$$

because the following limiting relation holds:

$$\int_0^c \frac{d\theta}{f(\theta)\theta} \to \ln c \quad \text{as} \quad c \to 0$$

In these derivations, as in the preceding cases, the solutions for the regions $r < r_1$ and $r_2 < r$ have been discarded. We define a quantity Δ by

$$\Delta = 2s_0\omega^2 t - \ln\left[\frac{f(c_0)c_0}{f(c)c} \right] \quad (2.57)$$

where c is given by equation 2.55. Substitution for $2s_0\omega^2 t$ from equation 2.55 and expression of the second term in an integral form yield

$$\Delta = \int_c^{c_0} \frac{1 - f(\theta) - \theta(df/d\theta)}{f(\theta)\theta} \, d\theta \quad (2.58)$$

If s is a decreasing function of c, $f(\theta) < 1$ and $df/d\theta < 0$ for $\theta > 0$, and hence the integrand of equation 2.58 is positive for $\theta > 0$. On the other hand, it follows from equation 2.55 that $c < c_0$ for $t > 0$, in the range of c in which $cf(c)$ is positive. From these facts it is seen that for $t > 0$ the quantity Δ is positive, so that

$$r_1 \exp(s_0\omega^2 t) > r_1 \left[\frac{f(c_0)c_0}{f(c)c} \right]^{1/2} \quad (t > 0) \quad (2.59)$$

if s is a decreasing function of c. This indicates that for this type of $s(c)$, the solution of the problem gives double-valued c in the range of r such that

$$r_1 \left[\frac{f(c_0)c_0}{f(c)c} \right]^{1/2} < r < r_1 \exp(s_0\omega^2 t) \qquad (2.60)$$

where c is represented by equation 2.55. The situation is entirely similar to the one encountered in Case II. The same procedure as employed there allows the desired single-valued solution to be constructed from equations 2.55 and 2.56, yielding

$$c = 0 \qquad [r_1 \leqslant r < r_*(t), \quad t > 0]$$

$$(2.61)$$

$$\int_c^{c_0} \frac{d\theta}{f(\theta)\theta} = 2s_0\omega^2 t \qquad [r_*(t) < r \leqslant r_2, \quad t > 0]$$

where $r_*(t)$ satisfies the relation

$$\frac{c}{c_0} = \frac{r_1^2}{r_*^2} \qquad (2.62)$$

with c given by the second equations of 2.61. From these it is seen that if s is a decreasing function of c, the Lamm equation with no diffusion term gives a concentration profile of a step-function type and the square-dilution rule applies. It also can be verified that the curve of $\ln r_*$ versus $\omega^2 t$ is concave upward and that its initial tangent equals $s(c_0)$. As is shown in Chapter 4, however, these conclusions are no longer applicable if $s(c)$ is an increasing function of c. In this case, the concentration profile has a certain finite width even in the absence of diffusion.

III. SOLUTIONS OF THE FAXÉN TYPE

A. INTRODUCTION

For a systematic survey of the reported solutions to the Lamm equation it is convenient to classify them into two categories, which, in this treatise, are referred to as the Faxén type and the Archibald type. Solutions of the

Faxén type are those which take advantage of the approximations to the boundary conditions introduced by Faxén in his 1929 paper.[1] Because of these approximations they are restrictive to the sedimentation behavior at early stages of a centrifugation. On the other hand, solutions of the Archibald type are free of such a restriction. However, since the infinite series constituting them converge more rapidly for larger values of time, they are actually relevant for a description of the sedimentation behavior at later stages of a centrifugation. Thus they are often very useful for theoretical investigations of the rate at which sedimentation equilibrium is approached.

This section is concerned with a detailed discussion of typical solutions of the Faxén type, together with the ways in which they may be applied to ultracentrifugal analysis of binary solutions. In view of its historic and fundamental importance the classic problem treated by Faxén is given a particularly detailed consideration. This problem deals with the sedimentation of a solute in a hypothetical cell, which we refer to as the infinite sectorial cell. Despite this cell condition it is shown that the solution obtained can be adapted to a description of concentration or concentration gradient distributions which are observed at early stages of actual sedimentation velocity experiments if the solution is used under proper conditions of parameters. The reader is advised to study with care the logical process in which this adaptation is made.

B.   FAXÉN'S EXACT SOLUTION

In the case when s and D are independent of c, the Lamm equation becomes linear in c and may be amenable to analytic solution. Faxén's first solution was obtained for this simplest case. Another point to note is that Faxén treated the problem under an initial condition which corresponded to a special case of equation 2.2. It is interesting that in his day no synthetic boundary cell was available.

As is well known, the treatment of any equation of mathematical physics is facilitated by transforming it to a dimensionless form with the appropriate "reduced" variables and parameters. For the Lamm equation with constant s and D and subject to the initial condition 2.2, such reduced quantities are

$$\theta = \frac{c}{c_0}, \qquad x = \left(\frac{r}{r_0}\right)^2, \qquad \tau = 2s\omega^2 t, \qquad \epsilon = \frac{2D}{s\omega^2 r_0^2} \qquad (2.63)$$

In terms of these, the Lamm equation 1.123 becomes

$$\frac{\partial \theta}{\partial \tau} = \frac{\partial}{\partial x}\left[x\left(\epsilon\frac{\partial \theta}{\partial x} - \theta\right)\right] \tag{2.64}$$

and the initial condition 2.2 and the boundary conditions 2.4 are written

$$\theta = \begin{cases} 0 & (x_1 < x < 1, \quad \tau = 0) \\ 1 & (1 < x < x_2, \quad \tau = 0) \end{cases} \tag{2.65}$$

$$\epsilon\frac{\partial \theta}{\partial x} = \theta \qquad (x = x_1, \quad \tau > 0) \tag{2.66}$$

$$\epsilon\frac{\partial \theta}{\partial x} = \theta \qquad (x = x_2, \quad \tau > 0) \tag{2.67}$$

where $x_1 = (r_1/r_0)^2$ and $x_2 = (r_2/r_0)^2$.

With the substitutions

$$\theta = \Theta e^{-\tau} \tag{2.68}$$

$$\xi = 2(xe^{-\tau})^{1/2} \tag{2.69}$$

$$\eta = \epsilon(1 - e^{-\tau}) \tag{2.70}$$

the set of dimensionless equations given above is transformed to

$$\frac{\partial \Theta}{\partial \eta} = \frac{\partial^2 \Theta}{\partial \xi^2} + \frac{1}{\xi}\frac{\partial \Theta}{\partial \xi} \tag{2.71}$$

$$\Theta(\xi, 0) = \begin{cases} 0 & \left(2\sqrt{x_1} < \xi < 2\right) \\ 1 & \left(2 < \xi < 2\sqrt{x_2}\right) \end{cases} \tag{2.72}$$

$$\epsilon\frac{\partial \Theta}{\partial \xi} = \Theta\left[\frac{x_1}{1 - (\eta/\epsilon)}\right]^{1/2} \qquad \left(\xi = 2\left[x_1\left(1 - \frac{\eta}{\epsilon}\right)\right]^{1/2}, \quad \eta > 0\right) \tag{2.73}$$

$$\epsilon\frac{\partial \Theta}{\partial \xi} = \Theta\left[\frac{x_2}{1 - (\eta/\epsilon)}\right]^{1/2} \qquad \left(\xi = 2\left[x_2\left(1 - \frac{\eta}{\epsilon}\right)\right]^{1/2}, \quad \eta > 0\right) \tag{2.74}$$

It is seen that the boundaries $x = x_1$ and $x = x_2$ on the x-axis (which correspond to r_1 and r_2 in the original physical space) are now transformed to

$$\xi = 2\left[x_1\left(1 - \frac{\eta}{\epsilon}\right)\right]^{1/2} \quad \text{and} \quad \xi = 2\left[x_2\left(1 - \frac{\eta}{\epsilon}\right)\right]^{1/2}$$

on the ξ-axis. These new boundaries are not fixed but move with η and hence with τ. This implies that the transformations above make the problem more difficult, rather than easier, to solve.

The case investigated by Faxén is not this general case of the problem but its special case in which $x_1 = 0$ and $x_2 = \infty$. Since $x_1 = 0$ and $x_2 = \infty$ correspond, respectively, to $r_1 = 0$ and $r_2 = \infty$, Faxén's problem is concerned with an infinitely extended cell of sector shape as illustrated in Fig. 2.5. Actual centrifuge cells are placed about 6 cm away from the axis of

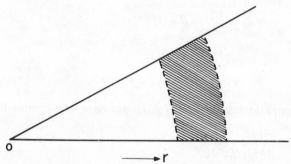

Fig. 2.5. An infinite sectorial cell. The shaded portion is the domain of an actual centrifuge cell.

the rotor, and their lengths are limited to about 1 cm or less. Therefore one may suspect that any solutions to the Lamm equation for such a hypothetical, infinite sectorial cell will have no practical meaning. It is shown in Section III.B.3, however, that such solutions are indeed very useful for an approximate description of sedimentation processes in the actual cell if they are applied under proper conditions.

Now with $x_1 = 0$ and $x_2 = \infty$, the initial condition 2.72 becomes

$$\Theta(\xi, 0) = \begin{cases} 0 & (0 < \xi < 2) \\ 1 & (2 < \xi < \infty) \end{cases} \tag{2.75}$$

The boundary conditions for this case cannot be deduced from equations

2.73 and 2.74, because equations 2.4, from which these equations were derived, had been obtained under the restriction that $r_1 \neq 0$ and $r_2 \neq \infty$. Actually, for the infinite cell shown in Fig. 2.5 no boundary condition of a specified form can be assigned, but instead we may simply require that $\Theta(\xi,\eta)$ and its derivative with respect to ξ be finite and continuous over the entire range of ξ, that is, from zero to infinity, for all positive values of η.

The solution to equation 2.71 subject to these auxiliary conditions may be obtained by various techniques known in the theory of linear partial differential equations of the second order. Here we take advantage of the method of Laplace transforms. The so-called subsidiary equation to 2.71 is given by

$$\frac{d^2\overline{\Theta}(\xi,p)}{d\xi^2} + \frac{1}{\xi}\frac{d\overline{\Theta}(\xi,p)}{d\xi} - p\overline{\Theta}(\xi,p) = -\Theta(\xi,0) \tag{2.76}$$

where $\overline{\Theta}(\xi,p)$ is the Laplace transform of $\Theta(\xi,\eta)$, that is,

$$\overline{\Theta}(\xi,p) = \int_0^\infty \Theta(\xi,\eta)e^{-p\eta}d\eta \tag{2.77}$$

Substitution of equation 2.75 into equation 2.76 gives

$$\frac{d^2\overline{\Theta}}{d\xi^2} + \frac{1}{\xi}\frac{d\overline{\Theta}}{d\xi} - p\overline{\Theta} = \begin{cases} 0 & (0 < \xi < 2) \\ -1 & (2 < \xi < \infty) \end{cases} \tag{2.78}$$

For convenience, we write

$$\overline{\Theta}_0(\xi,p) = \overline{\Theta}(\xi,p) \qquad (0 < \xi < 2) \tag{2.79}$$

$$\overline{\Theta}_\infty(\xi,p) = \overline{\Theta}(\xi,p) \qquad (2 < \xi < \infty) \tag{2.80}$$

The above-mentioned requirement for $\Theta(\xi,\eta)$ leads to the conditions that $\overline{\Theta}(\xi,p)$ and $d\overline{\Theta}(\xi,p)/d\xi$ must be finite and continuous for all ξ and $p > 0$. The solution of equation 2.78 subject to these conditions is

$$\overline{\Theta}_0(\xi,p) = AI_0(\xi\sqrt{p}) \tag{2.81}$$

$$\overline{\Theta}_\infty(\xi,p) = \frac{1}{p} + BK_0(\xi\sqrt{p}) \tag{2.82}$$

where A and B are arbitrary constants, and I_0 and K_0 are, respectively, the modified Bessel functions of the first and second kinds of zeroth order. Constants A and B are determined from the conditions that $\overline{\Theta}(\xi,p)$ and $d\overline{\Theta}(\xi,p)/d\xi$ must be continuous at $\xi=2$ and $p>0$, yielding

$$A = \frac{2}{\sqrt{p}} K_1(2\sqrt{p})\tag{2.83}$$

$$B = -\frac{2}{\sqrt{p}} I_1(2\sqrt{p})\tag{2.84}$$

Here I_1 and K_1 denote, respectively, the modified Bessel functions of the first and second kind of first order.

The desired solution for $\theta(\xi,\eta)$ is obtained by inverting equations 2.81 and 2.82 with the help of Bromwich's theorem. It can be shown that the inverted forms are expressed by a single equation valid for all $\xi\,(>0)$ and η (>0), that is,

$$\theta(\xi,\eta) = 1 - \int_0^{1/\eta} I_0\left(\xi\sqrt{\frac{q}{\eta}}\right)\exp\left(-\frac{\xi^2}{4\eta}-q\right)dq\tag{2.85}$$

Introduction of this into equation 2.68 yields the desired expression for θ and hence for c. Thus

$$c = c_0 e^{-\tau}J(\alpha,\beta)\tag{2.86}$$

where

$$J(\alpha,\beta) = 1 - \int_0^\alpha \exp(-\beta-q)I_0(2\sqrt{\beta q})\,dq\tag{2.87}$$

with α and β defined by

$$\alpha = \frac{1}{\eta}\tag{2.88}$$

$$\beta = \frac{\xi^2}{4\eta}\tag{2.89}$$

Substitution for ξ and η from equations 2.69 and 2.70 gives

$$\alpha = \frac{1}{\epsilon(1-e^{-\tau})}\tag{2.90}$$

$$\beta = \frac{xe^{-\tau}}{\epsilon(1-e^{-\tau})}\tag{2.91}$$

It can be shown that equation 2.86 stands in complete agreement with the solution obtained by Faxén,[1] though he did not rearrange his final result in this compact form. The function $J(\alpha, \beta)$ often appears in mathematical treatments of physical and engineering problems such as ion-exchange columns.[14] Its analytical and numerical properties have been fully explored by several authors.[15-17]

1. Asymptotic Forms of Faxén's Solutions

As mentioned in Section I.C, values of s and D for macromolecular solutes are, respectively, of the order of 10^{-13} and 10^{-7} when expressed in cgs units. To fix the idea we summarize typical values of $s_{20,w}$ and $D_{20,w}$ for some native proteins in Table 2.1.[18] If we take $\omega = 1000$ rev/sec and $r_0 = 6.5$ cm as typical experimental values, then for bovine serum albumin in water at 20°C, the magnitude of ϵ is about 1.5×10^{-3}, and for ribonuclease it is about $\epsilon = 7.1 \times 10^{-3}$. The values of ϵ for high-molecular-weight synthetic polymers are generally smaller than those for native proteins. Hence for most macromolecular solutes which may be studied ultracentrifugally with the machines and experimental conditions currently available, the values of ϵ are of the order of 10^{-2} to 10^{-3} except under special circumstances.

TABLE 2.1. Values of $s_{20,w}$ and $D_{20,w}$ for Some Native Proteins[18]

Protein	$s_{20,w} \times 10^{13}$ (sec)	$D_{20,w} \times 10^{7}$ (cm²/sec)
Bovine serum albumin	4.59	5.9
Aldolase	7.9	4.3
Ribonuclease	1.96	11.9
Myoglobin	1.96	10.3
Hemoglobin	4.60	6.0

Now with ϵ of this order, the magnitudes of α and β defined by equations 2.90 and 2.91 become quite large in comparison to unity unless x is too close to zero and also τ is too large. This feature suggests that the behavior of Faxén's solution for x not too close to zero and for τ not too large may be represented accurately by an asymptotic expansion of $J(\alpha, \beta)$ for large α and β. From the various asymptotic forms[14-16,19] for $J(\alpha, \beta)$ we select the one due to Hiester and Vermeulen,[16] who showed that if $(\alpha\beta)^{1/2}$ is larger than 6, $J(\alpha,\beta)$ may be expressed, with sufficient accuracy, as

$$J(\alpha,\beta) = \frac{1}{2}\left\{1 - \Phi\left(\sqrt{\alpha} - \sqrt{\beta}\right) + \frac{\exp\left[-\left(\sqrt{\alpha} - \sqrt{\beta}\right)^2\right]}{\sqrt{\pi}\left[\beta^{1/2} + (\alpha\beta)^{1/4}\right]}\right\} \quad (2.92)$$

where $\Phi(y)$ denotes the error function defined by*

$$\Phi(y) = \frac{2}{\sqrt{\pi}}\int_0^y e^{-y^2}\,dy \quad (2.93)$$

Note that there are several other definitions for the error function. In this monograph, we adhere to the definition just given. If equation 2.92 is inserted, equation 2.86 becomes

$$c = \frac{c_0 e^{-\tau}}{2}\left\{1 - \Phi\left(\frac{1 - (xe^{-\tau})^{1/2}}{[\epsilon(1 - e^{-\tau})]^{1/2}}\right)\right.$$

$$\left. + \frac{[2\epsilon\sinh(\tau/2)]^{1/2}}{\sqrt{\pi}\,(x)^{1/4}\left[1 + (xe^{-\tau})^{1/4}\right]}\exp\left(-\frac{\left[1 - (xe^{-\tau})^{1/2}\right]^2}{\epsilon(1 - e^{-\tau})}\right)\right\}$$

$$(2.94)$$

The condition $(\alpha\beta)^{1/2} > 6$, under which this expression for c holds, may be expressed in terms of x, τ, and ϵ as

$$\frac{1}{12} > \epsilon(x)^{-1/2}\sinh\left(\frac{\tau}{2}\right) \quad (2.95)$$

In actual centrifuge cells, the range of x is limited to a narrow region about $x = 1$, the initial position of the boundary between solvent and solution. For example, if $r_1 = 6.0$ cm, $r_0 = 6.5$ cm, and $r_2 = 7.0$ cm, then x varies from 0.852 to 1.16 between the air-liquid meniscus and the cell bottom. Thus, as far as the behavior of Faxén's solution in this region of x is concerned, condition 2.95 may be replaced by

$$\frac{1}{12} > \epsilon\sinh\left(\frac{\tau}{2}\right) \quad (2.96)$$

---

*The widely used notation for the error function is erf$y$.

Because ϵ is, in general, as small as 10^{-2} to 10^{-3}, this inequality is fully satisfied unless τ is too large in comparison to unity.

For all nonnegative values of x, τ, and ϵ we have

$$\left[1+\left(xe^{-\tau}\right)^{1/4}\right]^{-1}\exp\left\{-\frac{\left[1-\left(xe^{-\tau}\right)^{1/2}\right]^2}{\epsilon(1-e^{-\tau})}\right\}\leqslant 1 \qquad (2.97)$$

Therefore, if, in addition to the condition $\epsilon\ll 1$, we impose the restriction $\tau\ll 1$, equation 2.94 may be simplified to

$$c=\frac{c_0 e^{-\tau}}{2}\left[1-\Phi\left(\frac{1-\left(xe^{-\tau}\right)^{1/2}}{\left[\epsilon(1-e^{-\tau})\right]^{1/2}}\right)\right] \qquad (2.98)$$

in the approximation that a term of the order of $(\epsilon\tau)^{1/2}$ is ignored in comparison to unity.

2. Features of Concentration Distribution Curves

Since

$$0.99997<\Phi(z)<1 \qquad \text{for} \quad z>3$$

$$-0.99997>\Phi(z)>-1 \qquad \text{for} \quad z<-3$$

it follows from equation 2.98 that the following relations hold very accurately:

$$c=c_0 e^{-\tau} \qquad \text{for} \quad x>x_*\left\{1+3\left[\epsilon(1-e^{-\tau})\right]^{1/2}\right\}^2 \qquad (2.99)$$

$$c=0 \qquad \text{for} \quad x<x_*\left\{1-3\left[\epsilon(1-e^{-\tau})\right]^{1/2}\right\}^2 \qquad (2.100)$$

where x_* is a function of τ defined by

$$x_*=\exp(\tau) \qquad (2.101)$$

Thus equation 2.98 predicts a curve of c versus x which is almost completely flat outside the region of x such that

$$e^{\tau}\left\{1-3\left[\epsilon(1-e^{-\tau})\right]^{1/2}\right\}^2<x<e^{\tau}\left\{1+3\left[\epsilon(1-e^{-\tau})\right]^{1/2}\right\}^2 \qquad (2.102)$$

and which rises monotonically from zero to $c_0 \exp(-\tau)$ as the region defined by equation 2.102 is crossed from left to right. This region may be defined as the sedimentation boundary between solvent and solution. Its nonzero spread is due to diffusion. In fact, when $D=0$, the spread reduces to zero, giving rise to a discontinuous sedimentation boundary as shown in Section II. It readily follows from equation 2.102 that the boundary spreading, Δx, on the x-axis is given by

$$\Delta x = 12(\epsilon)^{1/2} e^{\tau} (1 - e^{-\tau})^{1/2} \qquad (2.103)$$

Hence the sedimentation boundary broadens as time elapses. The position x_* defined by equation 2.101 coincides approximately with the midpoint of the boundary region. Thus, while it is spreading, the sedimentation boundary moves as a whole in the direction of increasing radius, since x_* is an increasing function of time. Equation 2.99 indicates that the uniform concentration in the region ahead of the moving boundary decreases exponentially with time. It is interesting to note that this behavior agrees with the prediction deduced in Case I of Section II.B from the Lamm equation with no diffusion term. These features of the concentration distribution predicted by equation 2.98 are illustrated in Fig. 2.6.

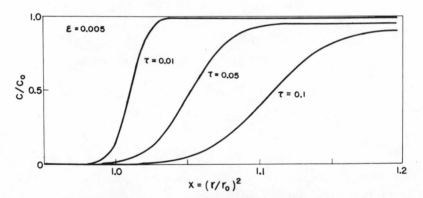

Fig. 2.6. Concentration distributions predicted by equation 2.98 for three different values of the reduced time $\tau = 2s\omega^2 t$ when the parameter $\epsilon = 2D/(s\omega^2 r_0^2)$ is 0.005.

The geometrical conditions of the cell used for the derivation of Faxén's solution differ so greatly from those of actual cells that one might consider such a solution to be of mere mathematical interest. The following argument shows, however, that if used under proper care, Faxén's solution or even its simplified form 2.98 serves the purpose of analyzing actual sedimentation data:

1. In actual cells, there occurs continuous accumulation of sedimenting solutes on the wall of the cell bottom. The accumulated solutes, however, tend to diffuse back to a less concentrated portion of the solution column. Thus, as sketched in Fig. 2.7, there is set up a field of increasing concentration in the vicinity of the cell bottom. This field spreads over the solution column until sedimentation-diffusion equilibrium is established at all points in the cell. The rate of spreading depends on the ratio of D to s, or more precisely on the magnitude of the parameter ϵ. If $\epsilon \ll 1$, it is so slow that the bottom effect may be confined to a quite narrow region adjacent to the bottom. In such a case, therefore, the solutes in almost an entire region of the solution column will sediment and diffuse as if they were in a cell unbounded in the direction of increasing radius. In other words, if $\epsilon \ll 1$, the solute distributions in such a hypothetical cell will not differ very much from those in actual cells with a rigid bottom, except for a region near the cell bottom.

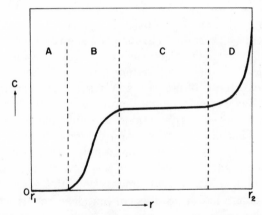

Fig. 2.7. Schematic representation of a concentration distribution in the actual centrifuge cell. A: solvent region; B: solvent-solution boundary; C: plateau region; D: bottom region.

2. It follows from equations 2.100 and 2.101 that if the initial boundary between solvent and solution is formed away from the meniscus so that the relation

$$\frac{r_1}{(1-9\epsilon)^{1/2}} < r_0 \qquad (\epsilon > 0) \tag{2.104}$$

is satisfied, the meniscus remains devoid of solute at all times during centrifugation. This suggests that if $\epsilon \ll 1$, the solute distribution is essentially the same as that which would be obtained if the centripetal side of the initial boundary were not truncated by an air-liquid meniscus but continued down to the center of rotation, unless r_0 is not too close to r_1.

3. According to equation 2.101, the distance on the x-axis traveled by the midpoint of the sedimentation boundary for a time t is $e^\tau - 1$. Taking $r_0 = 6.5$ cm and $r_2 = 7.0$ cm as a typical experimental case, we find the distance on the x-axis between the initial boundary and the cell bottom to be about 0.16. Hence, if the boundary region could move without being affected by the bottom, it would reach the cell end at a time corresponding to $\tau = 0.15$. This implies that the duration of time for which the cell may be regarded as unbounded in the centrifugal direction may be limited to values of τ much smaller than unity.

These three considerations ensure that under the conditions that $\epsilon \ll 1$, $\tau \ll 1$, and the initial boundary is formed suitably away from the meniscus, Faxén's exact solution for the hypothetical infinite sectorial cell can be adapted to a description of solute distributions in actual cells. For $\epsilon \ll 1$ and $\tau \ll 1$ it may be replaced, with a trivial error, by its approximate form 2.98. This conclusion explains why equation 2.98 has for years been the primary basis for sedimentation analyses of binary solutions.

In this monograph, the name "Faxén type" is given to solutions of the ultracentrifuge equations (the Lamm equation and its generalized forms derived in Chapter 1) if they are determined to fit the conditions for cells unbounded in the direction of increasing radius. It must be recognized that this definition of Faxén-type solution imposes no restriction on the magnitude of the parameter ϵ nor mentions anything about the boundary condition at the other end of the solution column, that is, the air-liquid meniscus. For experiments with the conventional cell we have $r_0 = r_1$ so that equation 2.104 is not obeyed. Thus, in this case, we must be concerned with having the solution satisfy the boundary condition at the meniscus. A Faxén-type solution relevant to the conventional cell is presented in Section III.D.

The utmost theoretical merit of Faxén-type solutions is that their determination avoids the mathematical complications which arise in attempts to fulfill the boundary condition at the cell bottom. In obtaining a solution of this type we simply require that the concentration at the limit of infinite radial distance remains finite for all values of time.

It is almost intuitively evident that any Faxén-type solution predicts a uniform distribution of concentration in the region beyond the sedimentation boundary between solvent and solution. Customarily, such a region of constant concentration is called the *plateau region* of a sedimentation boundary curve. In actual experiments, one always observes it for a certain duration of time after initiation of a centrifuge run. However, as sedimentation proceeds, its span will progressively diminish because, on one hand, the sedimentation boundary moves as a whole toward the cell bottom and, on the other hand, the bottom effect spreads toward the other end of the solution column. Eventually, it will disappear at a certain time, and after that time the infinite medium approximation to actual cells ceases to be valid; in other words, Faxén-type solutions become invalid at least in the region ahead of the sedimentation boundary. It is convenient to introduce the phrase "early stages" of a sedimentation experiment to mean the times at which a well-defined plateau region can be observed in the cell. Then Faxén-type solutions may also be called early-stage solutions to the ultracentrifuge equations.

This type of solution is more adequate for a system with a smaller ϵ, because, in such a system, the bottom effect will be confined to a narrower region, and hence the early stages will last longer. In fact, as is illustrated in Section III.G, if ϵ is as small as 10^{-3}, the plateau region is clearly visible until the midpoint of the sedimentation boundary travels more than two-thirds the entire solution column.

3. Features of Concentration Gradient Curves

The quantity usually measured in sedimentation velocity experiments is the distribution of refractive index gradients in a solution column.* As is shown in Chapter 3, Section I.B, it can be converted to the distribution of concentration gradients, $\partial c / \partial r$, if the solution is binary. One of the major purposes in the sedimentation analysis of binary solutions is to evaluate s and D from experimental data of this kind, usually taken at a number of times with suitable intervals. In what follows, we discuss how equation 2.98 can be utilized for this purpose.

From the various transformations in the preceding sections it follows that

$$\frac{\partial c}{\partial r} = 2r_0^{-1}(x)^{1/2} \frac{\partial c}{\partial x} \tag{2.105}$$

Substitution of equation 2.98 yields

$$\frac{\partial c}{\partial r} = \left(\frac{c_0}{r_0}\right) \frac{e^{-\tau}}{\left[\pi\epsilon(e^{\tau}-1)\right]^{1/2}} \exp\left\{ -\frac{\left[(r/r_0) - e^{\tau/2}\right]^2}{\epsilon(e^{\tau}-1)} \right\} \tag{2.106}$$

---

*The measurement is made in most cases by use of the schlieren optical system. Understanding of this and other optical methods of wide use in sedimentation studies is essential to critical analysis of published results.

which gives a Gaussian concentration gradient curve with its centroid at a point r_* given by

$$r_* = r_0 \exp\left(\frac{\tau}{2}\right) \tag{2.107}$$

On the x-axis this point corresponds to the x_* defined by equation 2.101. The maximum concentration gradient $(\partial c/\partial r)_{max}$ has the value

$$\left(\frac{\partial c}{\partial r}\right)_{max} = \left(\frac{c_0}{r_0}\right)\frac{e^{-\tau}}{\left[\pi\epsilon(e^{\tau}-1)\right]^{1/2}} \tag{2.108}$$

The spread of the gradient curve about its maximum, Δr, is proportional to $r_0(\epsilon)^{1/2}(e^{\tau}-1)^{1/2}$. Thus both r_* and Δr increase with time, while the maximum concentration gradient decreases with time. These time-dependent features of the concentration gradient curve predicted by equation 2.98 are schematically shown in Fig. 2.8. The region of zero gradient beyond each maximum corresponds to the plateau region defined previously, and the region of nonzero concentration gradients spans the sedimentation boundary between solvent and solution.

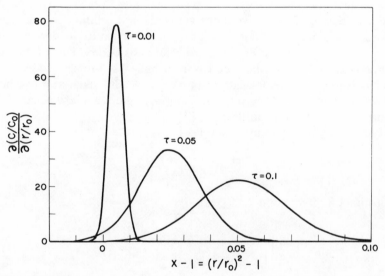

Fig. 2.8. Concentration gradient distributions corresponding to the concentration distribution curves of Fig. 2.6.

Concentration gradient curves (or more generally, refractive index gradient curves) observed during ultracentrifugation are called *sedimentation boundary curves*, *boundary curves*, *gradient curves*, and so on. These terms are not restricted to a region where the concentration undergoes appreciable change, but they apply to the entire solution column.

Equation 2.107 may be rewritten in the form

$$\ln r_* = \ln r_0 + s\omega^2 t \tag{2.109}$$

which provides a means of evaluating s from experimental observations of r_* as a function of t. This relation is identical to equation 2.29. Thus it might be considered that the maximum position would not be affected by diffusion. However, this conclusion is not correct, because a term of the order of $(\epsilon\tau)^{1/2}$ was dropped in deriving equation 2.98 and hence equation 2.109. If a more rigorous expression 2.94 is used to calculate the maximum position, it will be found that diffusion shifts the boundary peak by a distance proportional to $(\epsilon\tau)^{1/2}$ from the position predicted by equation 2.109.

The line on which the concentration gradient is zero is called the *base line* or the reference line of a sedimentation boundary curve. Equation 2.106 indicates that the base line is approached as r goes to positive or negative infinity. Thus the area A enclosed by a sedimentation boundary curve above its base line is represented by

$$A = \int_{-\infty}^{\infty} \frac{\partial c}{\partial r}\, dr \tag{2.110}$$

or

$$A = (c)_{r=\infty} - (c)_{r=-\infty} \tag{2.111}$$

Substitution for c from equation 2.98 gives

$$A = c_0 \exp(-\tau) \tag{2.112}$$

Because equation 2.98 yields zero concentration for $r = -\infty$, it follows from equation 2.111 that the area A is actually equal to the concentration at a place far ahead of the sedimentation boundary, that is, in the plateau region. In fact, equation 2.112 is seen to agree with the expression for c in such a region, equation 2.99.

Elimination of τ from equations 2.107 and 2.112 yields

$$A(r_*)^2 = c_0(r_0)^2 \tag{2.113}$$

This relationship is equivalent to the square-dilution rule introduced in Section II, and may be used to estimate the initial concentration c_0 from experimental measurements of A and r_*. However, as will be seen in Section III.H.1, equation 2.113 is not exact.

The *height-area ratio* of a sedimentation boundary curve is defined as the maximum height of the curve, H, divided by the area A. Use of equations 2.108 and 2.112 for H and A, respectively gives

$$\frac{H}{A} = r_0^{-1} [\pi \epsilon (e^\tau - 1)]^{-1/2} \tag{2.114}$$

from which we obtain

$$\left(\frac{A}{H}\right)^2 = D \left(\frac{2\pi}{s\omega^2}\right) [\exp(2s\omega^2 t) - 1] \tag{2.115}$$

This relation, due originally to Lamm,[20] shows that plots of $(A/H)^2$ versus $(2\pi/s\omega^2)[\exp(2s\omega^2 t) - 1]$ follow a straight line of slope D. Since all data necessary to construct this type of plot are obtainable from a series of sedimentation boundary curves measured as a function of time, here we have a method of determining the desired diffusion coefficient from ultracentrifugal experiment alone. In this monograph, this is called Lamm's method.

The practical validity of equation 2.98, from which all the relations presented above have been derived, is limited to values of τ much smaller than unity, as has been discussed in the previous section. Therefore, equation 2.115 is essentially equivalent to

$$\left(\frac{A}{H}\right)^2 = 4\pi Dt(1 + s\omega^2 t) \tag{2.116}$$

or even to a simpler form

$$\left(\frac{A}{H}\right)^2 = 4\pi Dt \tag{2.117}$$

Use of the latter equation has the advantage that no separate determination of s is required, although the resulting D value may be less accurate.

It was shown[21] that these traditional expressions for the height-area ratio cease to be valid for systems in which s depends on c. Thus it appeared that there would be little chance for the Lamm method to be of actual use, because sedimentation coefficients of macromolecular solutes are generally concentration dependent. In a recent paper, however, Kawahara[18] has shown that equation 2.116 or 2.117 still can be of effective use even for such solutes if applied to height-area ratios obtained at relatively low speeds of rotation. His theory is discussed in Section III.F.2.

C. A FAXÉN-TYPE SOLUTION RELEVANT TO THE SYNTHETIC BOUNDARY CELL

In the development just given we spent a considerable number of pages to verify that Faxén's exact solution for the hypothetical infinite sectorial cell can be adapted to an approximate description of the sedimentation behavior at early stages of experiments with a synthetic boundary cell under the condition that the parameter ϵ is sufficiently small in comparison with unity. In this section, it is shown that there is a way in which a Faxén-type solution of similar approximation relevant for experiments with a synthetic boundary cell is available more directly.

The substitutions

$$\theta = e^{-\tau}\Theta \qquad (2.118)$$

$$x = e^{z} \qquad (2.119)$$

transform the reduced Lamm equation 2.64 to

$$\frac{\partial \Theta}{\partial \tau} = \epsilon e^{-z} \frac{\partial^{2}\Theta}{\partial z^{2}} - \frac{\partial \Theta}{\partial z} \qquad (2.120)$$

The values of z corresponding to those of x which appear in actual cells are restricted to a quite narrow range about zero. For example, if $r_1 = 6.0$ cm, $r_0 = 6.5$ cm, and $r_2 = 7.0$ cm, z varies from -0.15 to 0.15 between the air-liquid meniscus and the cell bottom. This geometrical feature allows the factor $\exp(-z)$ in equation 2.120 to be replaced by unity. Thus

$$\frac{\partial \Theta}{\partial \tau} = \epsilon \frac{\partial^{2}\Theta}{\partial z^{2}} - \frac{\partial \Theta}{\partial z} \qquad (2.121)$$

This approximated Lamm equation should be more useful at earlier stages of centrifugation, because the sedimentation boundary is less removed from its initial position $z = 0$ at such stages. Its mathematical merit is that it can be handled much more easily by available techniques than the original Lamm equation. In fact, it can be reduced to an equation of the diffusion (or heat conduction) type:

$$\frac{\partial v}{\partial \tau} = \epsilon \frac{\partial^{2}v}{\partial z^{2}} \qquad (2.122)$$

by the substitution

$$\Theta = v \exp\left(\frac{2z - \tau}{4\epsilon}\right) \quad (2.123)$$

For the reason mentioned in Section III.B.2, if $\epsilon \ll 1$ and the initial boundary between solvent and solution is formed not too close to the meniscus, the presence of the meniscus may be ignored. Therefore, when these conditions are obeyed, the appropriate initial condition for the derivation of the Faxén-type solution relevant to a sedimentation experiment using a synthetic boundary cell may be

$$c = \begin{cases} 0 & (0 < r < r_0, \quad t = 0) \\ c_0 & (r_0 < r < \infty, \quad t = 0) \end{cases} \quad (2.124)$$

In terms of v, z, and τ this condition is written

$$v = \begin{cases} 0 & (-\infty < z < 0, \quad \tau = 0) \\ \exp\left(\frac{-z}{2\epsilon}\right) & (0 < z < \infty, \quad \tau = 0) \end{cases} \quad (2.125)$$

At this place it is worthy of noting that by means of equation 2.119 the range of r from zero to plus infinity, the infinite sectorial medium in Fig. 2.5, is transformed to the range of z from minus infinity to plus infinity. In mathematics, such a range of z is often called one-dimensional infinite domain.

Equation 2.122 can be integrated exactly under condition 2.125 to give[22]

$$v = \frac{1}{2}\left[1 - \Phi\left(\frac{\tau - z}{2\sqrt{\epsilon\tau}}\right)\right] \exp\left(\frac{\tau - 2z}{4\epsilon}\right) \quad (2.126)$$

where Φ denotes the error function defined by equation 2.93. If this solution for v is expressed in terms of c and z by use of the transformations indicated above, we obtain as a Faxén-type solution to the approximate Lamm equation 2.121

$$c = \frac{c_0 e^{-\tau}}{2}\left[1 - \Phi\left(\frac{z - \tau}{2\sqrt{\epsilon\tau}}\right)\right] \quad (2.127)$$

As noted in Section III.B.2, the values of τ for which Faxén-type solutions may be a useful approximation to experiments using a synthetic boundary cell would be restricted to a range quite close to zero. Also, the range of an actual cell usually corresponds to values of z much smaller than unity. Thus equation 2.127 may in effect be replaced, with a trivial error, by

$$c = \frac{c_0 e^{-\tau}}{2}\left[1 - \Phi\left(\frac{1 - e^{-(\tau-z)/2}}{[\epsilon(1 - e^{-\tau})]^{1/2}}\right)\right] \tag{2.128}$$

This agrees with equation 2.98 derived from Faxén's exact solution under the conditions that $\epsilon \ll 1$ and $\tau \ll 1$.

D.   A FAXÉN-TYPE SOLUTION RELEVANT TO THE CONVENTIONAL CELL

Faxén-type solutions appropriate for experiments with a conventional cell are of more practical interest than those for experiments with a synthetic boundary cell, because the cells employed in usual sedimentation studies are of the conventional type. They need fulfillment of the boundary condition at the air-liquid meniscus. The first approach to them was reported by Faxén[23] in 1936, but the result obtained was too complex to be of use for numerical examinations. Later, Fujita and MacCosham[24] attempted to derive a simpler solution by resort to an approximation similar to that described in Section III.C. Independently, Yphantis[25] investigated the same problem, using the "rectangular cell" approximation to the sectorial cell. This approximation is discussed in Section IV.B.1 in relation to solutions of the Archibald type to the Lamm equation.

The initial condition for the calculation of Faxén-type solutions relevant to the conventional cell is obtained by letting r_2 in equation 2.1 go to plus infinity. Thus

$$c = c_0 \quad (r_1 < r < \infty, \quad t = 0) \tag{2.129}$$

With a variable x redefined as

$$x = \left(\frac{r}{r_1}\right)^2 \tag{2.130}$$

equation 2.129 may be written

$$\theta = 1 \quad (1 < x < \infty, \quad \tau = 0) \tag{2.131}$$

where θ and τ have the same meaning as in equation 2.63. The boundary condition at the meniscus, the first equation of 2.4, may be expressed in terms of θ, x, and τ as

$$\epsilon' \frac{\partial \theta}{\partial x} = \theta \qquad (x = 1, \quad \tau > 0) \qquad (2.132)$$

where ϵ' is a new dimensionless parameter defined by

$$\epsilon' = \frac{2D}{s\omega^2 r_1^2} \qquad (2.133)$$

The quantity ϵ' differs by a factor $(r_0/r_1)^2$ from the previously defined similar parameter ϵ. Hence its magnitudes under usual conditions of sedimentation velocity experiments on macromolecular solutes are of the order of 10^{-2} to 10^{-3}.

If the same substitutions as equations 2.118 and 2.119 are made (with the recognition that x is here defined by equation 2.130) and then, on the consideration given in Section III.C, the factor $\exp(-z)$ is replaced by unity, equation 2.64 is approximated to the form

$$\frac{\partial \Theta}{\partial \tau} = \epsilon' \frac{\partial^2 \Theta}{\partial z^2} - \frac{\partial \Theta}{\partial z} \qquad (2.134)$$

In terms of Θ and z conditions 2.131 and 2.132 are written

$$\Theta = 1 \qquad (0 < z < \infty, \quad \tau = 0) \qquad (2.135)$$

and

$$\epsilon' \frac{\partial \Theta}{\partial z} = \Theta \qquad (z = 0, \quad \tau > 0) \qquad (2.136)$$

It is to be noted that, by equation 2.119 with the x-redefined as above, a sectorial medium truncated by an air-liquid meniscus at a certain radial position but unbounded in the direction of increasing radius is transformed to the range of z extending from zero to infinity, the one often called one-dimensional semiinfinite domain. Thus, the approximation $\exp(-z) \cong 1$ (and hence the accuracy of equation 2.134) becomes worse as one proceeds further away from the meniscus.

Equation 2.134 can be integrated in closed form under auxiliary conditions 2.135 and 2.136, yielding for the distribution of solute concentrations

$$c = \frac{c_0 e^{-\tau}}{2} \left\{ 1 - \Phi\left(\frac{\tau - z}{2\sqrt{\epsilon' \tau}}\right) - \frac{2}{\sqrt{\pi}} \left(\frac{\tau}{\epsilon'}\right)^{1/2} \exp\left[-\frac{(\tau - z)^2}{4\epsilon' \tau} \right] \right.$$

$$\left. + \left(1 + \frac{\tau + z}{\epsilon'}\right) \left[1 - \Phi\left(\frac{\tau + z}{2\sqrt{\epsilon' \tau}}\right) \right] \exp\left(\frac{z}{\epsilon'}\right) \right\} \qquad (2.137)$$

Differentiation with respect to r gives

$$\frac{\partial c}{\partial r} = \left(\frac{c_0}{r_1}\right)\frac{e^{-\tau}}{\epsilon'}\left\{\left[1 - \Phi\left(\frac{\tau + z}{2\sqrt{\epsilon'\tau}}\right)\right]\left(2 + \frac{\tau + z}{\epsilon'}\right)\exp\left(\frac{z}{\epsilon'}\right)\right.$$

$$\left. - \frac{2}{\sqrt{\pi}}\left(\frac{\tau}{\epsilon'}\right)^{1/2}\exp\left[-\frac{(\tau - z)^2}{4\epsilon'\tau}\right]\right\} \tag{2.138}$$

where a multiplicative factor $\exp(-z/2)$ has been dropped to be consistent with the approximation used in the derivation of equation 2.134. The position of the maximum gradient, r_*, may be determined from the condition $\partial^2 c / \partial r^2 = 0$, which gives

$$\sqrt{T}(3 + T + Z)\left[1 - \Phi\left(\frac{Z + T}{2\sqrt{T}}\right)\right] = \frac{2}{\sqrt{\pi}}(1 + T)\exp\left[-\frac{(Z + T)^2}{4T}\right]$$

$$\tag{2.139}$$

where Z and T are defined by

$$Z = \frac{z_*}{\epsilon'} \tag{2.140}$$

$$T = \frac{\tau}{\epsilon'} \tag{2.141}$$

with

$$z_* = 2\ln\frac{r_*}{r_1} \tag{2.142}$$

The derivation just given has imposed no restriction on the magnitude of ϵ'. However, this quantity must be sufficiently small in comparison to unity in order that the solution obtained may be one of much practical significance, because otherwise the effects of restricted diffusion near the cell bottom may spread so quickly to the central region of the solution column that the infinite-cell approximation will cease to be valid at quite early stages of the experiment.

1. Effects of the Air-Liquid Meniscus

The Fujita-MacCosham solution allows examination of the effects which will be exerted by the air-liquid meniscus on the solute distributions at early stages of a sedimentation experiment. Figures 2.9 and 2.10 show the gradient curves calculated from equation 2.138 for two largely different values of ϵ': $\epsilon' = 0.001$ for Fig. 2.9 and $\epsilon' = 0.02$ for Fig. 2.10. The set of curves in Fig. 2.9 may be taken as typical of high-molecular-weight solutes and that in Fig. 2.10 as typical for low-molecular-weight solutes, both compared at similar rotor speeds. Or the former may be considered to represent results of a high-speed experiment and the latter those of a low-speed experiment, both for comparable high-molecular-weight solutes.

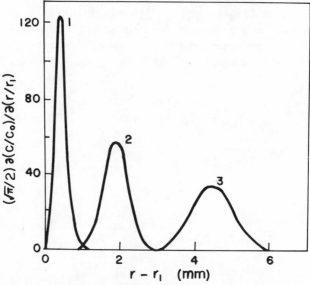

Fig. 2.9. Concentration gradient curves for a system characterized by a very small value of $\epsilon'(= 0.001)$. Curves 1, 2, and 3 correspond to times of centrifugation of 400, 1600, and 3600 sec, respectively. The boundary condition at the meniscus is taken rigorously into calculation.

Figure 2.9 shows that when ϵ' is very small, the maximum of the boundary curve separates from the meniscus shortly after the rotor is set in motion. The gradient at the meniscus soon drops to zero, and the entire curve moves toward the cell bottom while maintaining, at all times, its distinct, bell-shaped form. Thus the general character of the gradient curves for this case is essentially similar to that which will be observed in experiments with synthetic boundary cells. On the other hand, Fig. 2.10 shows that when ϵ' is relatively large, the boundary curve is markedly affected by the presence of the meniscus. Even after prolonged centrifugation, the maximum of the curve remains close to the meniscus, and the gradient at the meniscus decreases slowly with time.

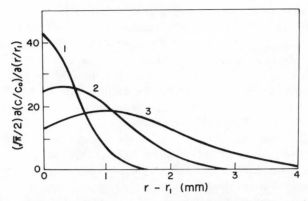

Fig. 2.10. Concentration gradient curves for a system characterized by a relatively large value of $\epsilon'(=0.02)$. Curves 1, 2, and 3 correspond to times of centrifugation of 900, 2500, and 4900 sec, respectively. The boundary condition at the meniscus is taken rigorously into calculation.

The relationship between Z and T given by equation 2.139 is shown graphically in Fig. 2.11. For T values less than 0.436 the sedimentation boundary curves exhibit no maxima in the solution phase. This time interval, Δt_M, which may be termed the time lag for a sedimentation experiment with a conventional cell, is given by

$$\Delta t_M = \frac{0.436 D}{s^2 \omega^4 r_1^2} \tag{2.143}$$

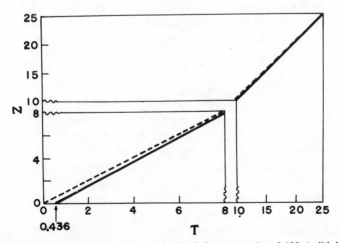

Fig. 2.11. Relations between Z and T calculated from equation 2.139 (solid line) and equation 2.144 (dashed line). The maximum concentration gradient first appears in the solution column at a time corresponding to $T=0.436$.

Let us take $r_1 = 6.0$ cm, $s = 5 \times 10^{-13}$ sec, $D = 5 \times 10^{-7}$ cm^2/sec, and ω = 1000 rev/sec as a typical operational case for a macromolecular solute. Then we have $\Delta t_M = 0.3$ min. If we take $s = 0.5 \times 10^{-13}$ sec and $D = 10 \times 10^{-7}$ cm^2/sec as typical values of a relatively small solute, Δt_M becomes 52 min. This implies that for relatively small solutes the maximum gradient may adhere to the meniscus during the entire period of a sedimentation velocity run (which, in many cases, is limited to 1 to 2 hr) even if the rotor is spun at the maximum operational speed.

Archibald[26] showed that the molecular weight of a solute can be evaluated from measurements of the concentration and the concentration gradient at either end of the solution column. To obtain the necessary data with accuracy it is required that the experimental conditions be chosen so that the maximum gradient remains at or close to the air-liquid meniscus during the entire period of a sedimentation run. Equation 2.143 may be used as a guideline in selecting the required conditions. The Archibald method for molecular weight determination is discussed in Chapter 3.

Figure 2.11 shows that once a maximum of the gradient curve has made its appearance in the solution phase at $t = \Delta t_M$, Z increases almost linearly with time. The dashed line, which represents the asymptote to the calculated Z versus T curve, is described by the relation

$$\ln r_* = \ln r_1 + s\omega^2 t \qquad (2.144)$$

This agrees with equation 2.109, except for r_0 in the latter being replaced here by r_1. Thus the dashed line depicts the relation which would be obtained if a sharp boundary between solution and solvent were formed initially at $r = r_1$ in an infinite sectorial cell. The negligibly small difference from the solid line implies that the movement of the maximum gradient for high-molecular-weight solutes is scarcely disturbed by the meniscus unless rotor speeds are too low.

E. BAND CENTRIFUGATION

Sedimentation experiments of the types discussed so far in this chapter are usually called *boundary* (or *moving boundary*) *centrifugation*, because they are characterized by the appearance of a moving boundary, be it sharp or spreading, which separates a solution from its solvent. Sometimes, these experiments are referred to as the conventional type. There is available another type of sedimentation experiment called *band* or *density-gradient centrifugation*. The original idea of this new technique was introduced by Brakke[27] when he attempted to resolve a macromolecular mixture into components in the preparative centrifuge. Later it was elaborated by Vinograd et al.[28] to the scheme convenient for use in the analytical ultracentrifuge.

In this type of experiment, a thin lamella of a solution of macromolecules is layered onto a denser solvent poured into the cell while the ultracentrifuge is in operation,* and the migration and spreading of the band of macromolecules then produced in the solution column is optically recorded as a function of time. Thus the band centrifugation differs in its initial condition from the boundary centrifugation, as illustrated in Fig. 2.12. The negative density gradient at the leading edge of the band must be compensated by a positive density gradient in the solvent medium in order to prevent convection. If the positive gradient is inadequate, limited convective transport causes forward spreading of the band. Positive gradients may be set up by allowing a low-molecular-weight substance (such as sucrose, D_2O, and simple electrolytes) added to the solvent to redistribute. In one case, as in the original work of Brakke, one may establish the necessary gradient prior to the layering of a sample solution. Rosenbloom and Schumaker[30] described a procedure for this operation. In the other, that is, when one starts with a uniform solvent medium, it can be self-generated during the subsequent centrifugation. Two mechanisms are responsible for this. First, the difference in chemical potentials of the added small molecules between the lamella and the bulk solvent medium causes the small molecules to diffuse in such a way that positive density gradients are generated. These "diffusion gradients," as termed by Vinograd and Bruner,[31] advance from the lamella toward the cell bottom. Second, the centrifugal force causes redistribution of the small molecules, which is the sedimentation effect. The generated gradients, called "field gradients" by Vinograd and Bruner,[31] spread from both ends of the cell toward its central region as the centrifugation continues, and they eventually assume an equilibrium distribution. Thus field gradients become more appreciable at relatively late stages of a band centrifugation. On the other hand, diffusion gradients gradually tend to vanish as time goes on. Hence they play a stabilizing role at relatively early stages of the centrifugation. Vinograd and Bruner[31] examined these two kinds of gradient theoretically, in the hope of establishing guidelines which would enable the investigator to select experimental conditions relevant for convection-free band sedimentation.

The adaptation of Brakke's idea to the analytical ultracentrifuge was motivated by the desire to find a way which facilitates sedimentation analysis of such biological macromolecules as nucleic acids and viruses,

*Band-forming centerpieces for the analytical ultracentrifuge were designed by Vinograd and associates.[29]

which are generally available only in a very limited quantity. These substances have the distinct advantage that they can be studied at very low concentrations by use of the light-absorption method which has acquired a greater sensitivity in recent years.[32] For example, Vinograd et al.[28] reported successful band sedimentation experiments with DNA at polymer concentrations as low as a few micrograms per milliliter. With so low a lamellar concentration, one may adjust the solvent density to a value only slightly higher than the sample solution. For example, a density difference of 0.02 to 0.04 g/ml is sufficient to ensure stability when one deals with solutions of DNA as dilute as 0.05 to 0.1 mg/ml. Under such conditions the inhomogeneity of solvent density produced by redistribution of the added small molecules is so slight that one may treat the solvent medium as if it were a liquid of uniform density. Thus, as far as sedimentation and diffusion of the banded macromolecules are concerned, the system may be treated as a binary solution, and the concentration distribution of the macromolecular solute may be calculated on the basis of the Lamm equation. This approximation ceases to be valid when the assigned density difference is large. In such cases, there arise various complications which make the theoretical analysis of band centrifugation almost prohibitive.

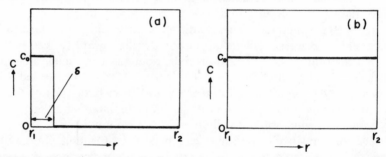

Fig. 2.12. Initial concentration distributions for band sedimentation (*a*) and boundary sedimentation (*b*). The quantity δ represents the thickness of a macromolecular solution initially layered on a solvent.

1. The Rubin-Katchalsky Theory

Gehatia and Katchalski[33] were the first to describe a mathematical theory of band centrifugation. They treated an idealized case in which the initial lamella of a macromolecular solute is infinitely thin. Rubin and Katchalsky[34] extended the Gehatia-Katchalski theory to a more realistic case in which the initial lamella has a finite width and, moreover, the presence of the air-liquid meniscus may not be ignored. On the other hand, Vinograd et al.[28] developed a theoretical treatment of band centrifugation by use of the method of moments (see Section III.H), and their theory was further elaborated by Vinograd and Bruner[11,31] and also by Schumaker and Rosenbloom.[35]

In this section, an outline of the Rubin-Katchalsky theory is described. This theory needs a Faxén-type solution to the Lamm equation subject to an initial condition of the form

$$c = \begin{cases} c_0 & (r_1 < r < r_1 + \delta, \quad t=0) \\ 0 & (r_1 + \delta < r < \infty, \quad t=0) \end{cases} \qquad (2.145)$$

and the boundary condition

$$D \frac{\partial c}{\partial r} = s\omega^2 r_1 c \qquad (r=r_1, \quad t>0) \qquad (2.146)$$

The quantity δ denotes the thickness of the initial band. The neglect of the cell bottom explicit in these auxiliary conditions is entirely permissible for theoretical treatments of band centrifugation. The reason is that since in this case the entire region of the solution column is initially devoid of macromolecular solutes except for a very thin layer at its air-liquid meniscus, no concentration gradient is set up in the region near the cell bottom during the subsequent centrifugation. In this respect, band centrifugation is contrasted to boundary centrifugation; in the latter the restricted diffusion produced by the presence of the cell bottom gradually spreads over the remaining part of the solution column and affects the movement and shape of the solution-solvent moving boundary.

In solving the Lamm equation with conditions 2.145 and 2.146, Rubin and Katchalsky took advantage of the approximation used by Fujita and MacCosham for boundary centrifugation (Section III.D), and obtained for $c(r,t)$

$$
\begin{aligned}
c(z,\tau) = \frac{c_0 e^{-\tau}}{2} \Bigg\{ & \Phi\left(\frac{\beta+\tau-z}{2\sqrt{\epsilon'\tau}}\right) - \Phi\left(\frac{\tau-z}{2\sqrt{\epsilon'\tau}}\right) \\
& + e^{z/\epsilon'}\left(1 + \frac{\tau+z}{\epsilon'}\right)\left[\Phi\left(\frac{\beta+\tau+z}{2\sqrt{\epsilon'\tau}}\right) - \Phi\left(\frac{\tau+z}{2\sqrt{\epsilon'\tau}}\right)\right] \\
& - \frac{\beta}{\epsilon'}\left[1 - \Phi\left(\frac{\beta+\tau+z}{2\sqrt{\epsilon'\tau}}\right)\right] \\
& + 2\sqrt{\frac{\tau}{\pi\epsilon'}}\left[\exp\left(-\frac{(\beta+\tau-z)^2}{4\epsilon'\tau}\right) - \exp\left(-\frac{(\tau-z)^2}{4\epsilon'\tau}\right)\right]\Bigg\} \quad (2.147)
\end{aligned}
$$

where

$$z = 2\ln\left(\frac{r}{r_1}\right), \qquad \tau = 2s\omega^2 t, \qquad \epsilon' = \frac{2D}{s\omega^2 r_1^2}, \qquad \beta = 2\ln\left(\frac{r_1 + \delta}{r_1}\right) \qquad (2.148)$$

It was assumed that both s and D are independent of c, but no assumption was made for the magnitude of ϵ'. In the limit of indefinitely large β (which corresponds to $\delta = \infty$), equation 2.147 reduces to Fujita-MacCosham's solution for boundary centrifugation, equation 2.137, as should be expected.

For very small ϵ' equation 2.147 predicts that except for a few minutes after initiation of the experiment, the concentration at the meniscus is effectively zero and the concentration distribution in the solution column is described accurately by a simpler equation:

$$c(z,\tau) = \frac{c_0 e^{-\tau}}{2}\left[\Phi\left(\frac{\beta + \tau - z}{2\sqrt{\epsilon'\tau}}\right) - \Phi\left(\frac{\tau - z}{2\sqrt{\epsilon'\tau}}\right)\right] \qquad (2.149)$$

It is expected that this expression is good enough for DNA, RNA, and viruses, because these macromolecules are characterized by very small values of ϵ' unless rotor speeds are unusually low.

It follows from equation 2.149 that

$$\frac{\partial c}{\partial r} = \frac{c_0 e^{-\tau}}{r\sqrt{\pi\epsilon'\tau}}\left\{\exp\left[-\frac{(\tau - z)^2}{4\epsilon'\tau}\right] - \exp\left[-\frac{(\beta + \tau - z)^2}{4\epsilon'\tau}\right]\right\} \qquad (2.150)$$

from which it can be shown that the concentration distribution has a maximum at the position $r_*(t)$ given by

$$r_*(t) = [r_1(r_1 + \delta)]^{1/2}\exp(s\omega^2 t) \qquad (2.151)$$

The corresponding maximum concentration $c_{max}(t)$ is found to be

$$c_{max}(t) = c_0 e^{-\tau}\Phi\left(\frac{\beta}{4\sqrt{\epsilon'\tau}}\right) \qquad (2.152)$$

Equations 2.151 and 2.152 may be used to determine s and D from measurements of $r_*(t)$ and $c_{max}(t)$ as function of time. For this purpose it is advantageous to rewrite equation 2.151 as

$$\ln r_* = \tfrac{1}{2}\ln\left[r_1(r_1+\delta)\right]+s\omega^2 t \tag{2.153}$$

which indicates that the desired s can be evaluated as a slope of plots of $\ln r_*$ versus $\omega^2 t$. A convenient equation for the estimate of D is obtained by combination of equations 2.151 and 2.152, yielding

$$D=\frac{1}{t}\left\{\frac{r_1\ln\left[(r_1+\delta)/r_1\right]}{4\Phi^{-1}\left[c_{max}r_*^2/c_0 r_1(r_1+\delta)\right]}\right\}^2 \tag{2.154}$$

Here Φ^{-1} denotes the inverse function of Φ. For this equation to be put in actual use the thickness of the initial band must be known. Rubin and Katchalsky[34] proposed a method useful for the case in which the exact amount of banded material is unknown.

If the error functions in equation 2.149 are expanded in Taylor's series about the argument $[\tau+(\beta/2)-z]/2\sqrt{\epsilon'\tau}$ and only the first term of the series is retained, we obtain for $c(r,t)$

$$c=\frac{c_0 e^{-\tau}}{2\sqrt{\pi\epsilon'\tau}}\exp\left\{-\frac{[\tau+(\beta/2)-z]^2}{4\epsilon'\tau}\right\} \tag{2.155}$$

which is a Gaussian distribution on the z-axis with its peak located at $z=\tau+(\beta/2)$. The maximum concentration at the peak is higher than that given by the more exact equation 2.149; the relative difference becomes larger as the initial band gets wider.[34]

Figure 2.13 illustrates the development of the sedimenting band with time calculated for the case in which $s=12\times10^{-13}$ sec, $D=5.896\times10^{-7}$ cm^2/sec, $r_1=6.00$ cm, $\delta=0.02$ cm, and $\omega=47,750$ rev/min. It is seen that the trailing edge of the concentration distribution curve completely clears the meniscus after 5 min of centrifugation and that the maximum concentration of the banded solutes decreases first quite sharply and then gradually as the band travels toward the cell bottom.

The advantage of the band centrifugation method is not only in its capability of determining s and D with a very small quantity of sample but also in that when applied to multicomponent solutions it allows the solute components to be separated very effectively in accordance with their sedimentation coefficients. Probably, the latter feature is the major attraction to biochemists and molecular biologists.

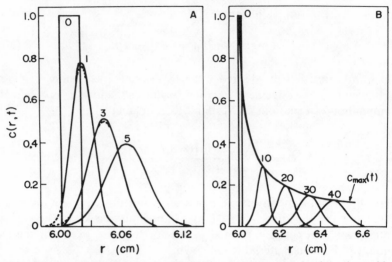

Fig. 2.13. Band spreading, migration, and dilution for a macromolecular solute with s $= 12 \times 10^{-13}$ sec and $D = 5.896 \times 10^{-7}$ cm^2/sec, under the conditions that $r_1 = 6.00$ cm, $\delta = 0.02$ cm, and $\omega = 47,750$ rev/min. A, concentration distributions on expanded scale after 0, 1, 3, and 5 min of centrifugation; solid line, by equation 2.147, and dashed line, by equation 2.149. B, band patterns after 0, 10, 20, 30, and 40 min of centrifugation, calculated by equation 2.149.

F.   CONCENTRATION-DEPENDENT SYSTEMS

Although it greatly simplifies mathematical treatments of the Lamm equation, the condition that s and D are independent of solute concentration c is seldom obeyed by actual systems, especially by solutions of macromolecular substances. As mentioned in Section I.C, in general s decreases and D increases with c, at least in the region of low concentrations. Analytical solution of the Lamm equation for such concentration-dependent systems becomes an exceedingly difficult task because the equation then becomes nonlinear in c.

Empirically, it has long been known that observed sedimentation boundary curves of a system for which there is reason to believe that the s decreases with c are much sharper than the curves expected theoretically with constant s and D. Furthermore, there is much experimental evidence which shows that even when the "sharpening" is apparent, the boundary curves are essentially symmetric about their maxima during the early stage of a sedimentation velocity run. Until 1956 when Fujita[21] presented an approximate Faxén-type solution of the Lamm equation with an s decreasing linearly with c (see equation 2.6), no attempt had been reported for a quantitative interpretation of these boundary features of concentration-dependent systems.

In this paper, apart from the various approximations required for the derivation of a Faxén-type solution, Fujita assumed that D was independent of c. This additional assumption is not only less realistic but also apparently inconsistent with allowing for a variation in s with c, because the concentration dependences of s and D have their molecular origins in common, that is, the variation of the molecular friction factor with c. Fujita pointed out, however, that the dependence of D on c is generally less than that of s on c and that when $\epsilon \ll 1$, the term multiplied by D in the Lamm equation becomes a correction to the term multiplied by s so that the effect of a change in D with c on the shape of a boundary curve should be less than the effect of a change in s. Thus for the purpose of obtaining Faxén-type solutions one may treat D as a constant parameter even when s is allowed to vary with c.

In what follows, Fujita's approximate solution with $s = s_0(1 - k_s c)$ and $D = $ constant will be outlined, although mathematically it is more relevant to present the work by Billick and Weiss,[38] who verified that Fujita's problem could be solved exactly. In this connection, the reader also should consult a paper by Weiss and Yphantis,[39] in which an Archibald-type solution to the same problem was derived by use of the "rectangular cell" approximation.

Since equation 2.5 finds wider applicability than equation 2.6, it is certainly more desirable to work out the Lamm equation with a concentration-dependent s of this form. However, at present, there is little hope that this case can be treated without resort to numerical methods of integration. Recently, very extensive numerical solutions to the Lamm equation have been made available by Dishon et al.[40-42] for systems obeying equations 2.5 and 2.6. Some important results of their investigations are presented in Section III.G.

1. An Approximate Faxén-Type Solution with $s = s_0(1 - k_s c)$

The case treated by Fujita[21] is the same as that discussed in Section III.B, except that s is here assumed to vary with c in accordance with equation 2.6. Thus the problem is here concerned with the sedimentation transport in an infinite sectorial cell in which a sharp boundary between solvent and solution (of solute concentration c_0) is initially formed at a certain radial position r_0.

With the reduced variables defined by

$$\theta = \frac{c}{c_0}, \qquad x = \left(\frac{r}{r_0}\right)^2, \qquad \tau = 2s_0\omega^2 t, \qquad \epsilon = \frac{2D}{s_0\omega^2 r_0^2} \qquad \alpha = k_s c_0 \quad (2.156)$$

where α is assumed to be less than unity, the Lamm equation for the present case can be written

$$\frac{\partial \theta}{\partial \tau} = \frac{\partial}{\partial x}\left\{ x\left[\epsilon \frac{\partial \theta}{\partial x} - (1 - \alpha\theta)\theta \right] \right\} \tag{2.157}$$

The initial condition of the problem is expressed as

$$\theta = \begin{cases} 1 & (1 < x < \infty, \quad \tau = 0) \\ 0 & (0 < x < 1, \quad \tau = 0) \end{cases} \tag{2.158}$$

The task is to solve equation 2.157 with condition 2.158 and the additional requirement that θ and its derivatives with respect to x be finite and continuous for all positive values of x and τ.

To do this we first note that equation 2.157, which is nonlinear in θ, can be linearized to give

$$\frac{\partial u}{\partial \zeta} = \epsilon \frac{\partial^2 u}{\partial \eta^2} + \frac{1}{2 - \eta} \frac{\partial u}{\partial \eta} \tag{2.159}$$

if the following substitutions are made:

$$\theta = \frac{\epsilon}{\alpha} \frac{\partial \ln u}{\partial x} \tag{2.160}$$

$$\eta = 2\left[1 - (xe^{-\tau})^{1/2} \right] \tag{2.161}$$

$$\zeta = 1 - e^{-\tau} \tag{2.162}$$

The initial boundary between solution and solvent is transformed to $\eta = 0$ on the η-axis. This fact suggests that to obtain a Faxén-type solution relevant to the present problem, the factor $2 - \eta$ in equation 2.159 may be replaced by 2. Then

$$\frac{\partial u}{\partial \zeta} = \epsilon\left(\frac{\partial^2 u}{\partial \eta^2} + \frac{1}{2} \frac{\partial u}{\partial \eta} \right) \tag{2.163}$$

This approximation is essentially similar to the one used in Sections III.C and III.D, where the factor $\exp(-z)$ was replaced by unity. The initial condition 2.158 may be expressed in terms of u, η, and ζ as

$$u(\eta, 0) = \begin{cases} \exp\left[\frac{\alpha}{\epsilon}\left(1 - \frac{\eta}{2} \right)^2 \right] & (\eta < 0) \\ \exp\left(\frac{\alpha}{\epsilon} \right) & (\eta > 0) \end{cases} \tag{2.164}$$

Equation 2.163 subject to condition 2.164 can be solved analytically by transformation to an equation of the diffusion type:

$$\frac{\partial u}{\partial \zeta} = \epsilon \frac{\partial^2 u}{\partial w^2} \tag{2.165}$$

with the substitution

$$w = \eta + \frac{\epsilon \zeta}{2} \tag{2.166}$$

The solution is substituted back into the series of transformations just given to yield the desired expression for $c(r,t)$.

Actually, the result has much practical significance for ϵ and τ sufficiently small in comparison to unity, because, as mentioned in previous sections, sedimentation behavior in the infinite sectorial cell can be a good approximation to that in the actual cell under these conditions. For this reason we ignore terms of the order of $(\epsilon\tau)^{1/2}$ from the derived expression for $c(r,t)$. Then the result gives for the concentration gradient $\partial c / \partial r$

$$\frac{\partial c}{\partial r} = \left(\frac{c_0}{r_0}\right) \frac{2 \exp(-\tau/2)}{(1-\lambda)[\pi\epsilon(e^\tau-1)]^{1/2}} \left(\frac{Q_1}{Q_2}\right) \tag{2.167}$$

where

$$Q_1 = [1 + \Phi(\xi)] \exp(\xi^2) + (1-\lambda)^{1/2} \left[1 - \Phi\left(\frac{\xi - \gamma(\lambda)^{1/2}}{(1-\lambda)^{1/2}}\right)\right]$$

$$\times \exp\left[\frac{(\xi - \gamma(\lambda)^{1/2})^2}{1-\lambda}\right] + \left(\frac{\pi\lambda}{1-\lambda}\right)^{1/2} [1 + \Phi(\xi)][\gamma - \xi(\lambda)^{1/2}]$$

$$\times \left[1 - \Phi\left(\frac{\xi - \gamma(\lambda)^{1/2}}{(1-\lambda)^{1/2}}\right)\right] \exp\left[\xi^2 + \frac{(\xi - \gamma(\lambda)^{1/2})^2}{1-\lambda}\right] \tag{2.168}$$

$$Q_2 = \left\{(1-\lambda)^{1/2}[1 + \Phi(\xi)] \exp(\xi^2)\right.$$

$$\left. + \left[1 - \Phi\left(\frac{\xi - \gamma(\lambda)^{1/2}}{(1-\lambda)^{1/2}}\right)\right] \exp\left[\frac{(\xi - \gamma(\lambda)^{1/2})^2}{1-\lambda}\right]\right\}^2 \tag{2.169}$$

The quantities ξ, λ, and γ have the following significance:

$$\xi = \frac{1 - (xe^{-\tau})^{1/2}}{[\epsilon(1 - e^{-\tau})]^{1/2}} \tag{2.170}$$

$$\lambda = \alpha(1 - e^{-\tau}) \tag{2.171}$$

$$\gamma = \left(\frac{\alpha}{\epsilon}\right)^{1/2} \tag{2.172}$$

In the limit of vanishingly small α (i.e., $k_s \approx 0$), Q_1 and Q_2 reduce, respectively, to $2\exp(\xi^2)$ and $4\exp(2\xi^2)$. Hence

$$\lim_{\alpha \to 0} \frac{\partial c}{\partial r} = \left(\frac{c_0}{r_0}\right) \frac{\exp(-\tau)}{[\pi\epsilon(e^{\tau} - 1)]^{1/2}} \exp(-\xi^2) \tag{2.173}$$

which agrees with equation 2.106, as should be expected.

Figure 2.14 shows concentration gradient curves for several values of α ($= k_s c_0$) computed from equation 2.167 with fixed values of τ and ϵ ($e^{-\tau} = 0.8$ and $\epsilon = 0.002$). The following features may be noted from this graph: With increasing concentration dependence of s, (a) the shape of the gradient curve is markedly sharpened; (b) the position of the maximum gradient noticeably shifts to the positive side of ξ; and (c) symmetry of the curve is practically maintained except at edges. Negative and positive ξ correspond, respectively, to the solution and solvent edges of the moving boundary region. Hence, feature (b) can be taken to mean that the rate of movement of the sedimentation boundary is slowed down as there is a linear decrease of s with c.

It was shown in Section III.B.3 that if s and D are constant, the maximum gradient occurs at the same position as that of the discontinuous sedimentation boundary which would be observed if D were zero. For the set of parameters chosen for the calculation of Fig. 2.14 it was confirmed[21] that this relationship holds quite accurately also for the present case. Thus we assume that equation 2.51 may be used to describe the position of the maximum gradient r_* for the gradient curve represented by equation 2.167. Then

$$\xi_* = \frac{\gamma(\lambda)^{1/2}}{1 + (1 - \lambda)^{1/2}} \tag{2.174}$$

where ξ_* denotes the value of ξ corresponding to $r = r_*$. The maximum height of the boundary curve, H, is obtained by evaluating equation 2.167

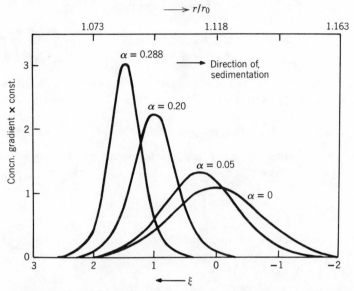

Fig. 2.14. Effects of concentration dependence on the boundary gradient curve when $s = s_0(1 - k_s c)$; $\alpha = k_s c_0$ (c_0: initial concentration of the solution). ξ is the reduced cell coordinate defined by equation 2.170. Negative and positive values of ξ correspond to the leading and trailing edges of the sedimentation boundary region.

at $\xi = \xi_*$, yielding

$$H = \left(\frac{c_0}{r_0}\right) \frac{\exp(-\tau/2)}{\alpha(e^\tau - 1)[1 - \alpha(1 - e^{-\tau})]} G(\xi_*) \qquad (2.175)$$

where

$$G(z) = z\left[2z + \frac{\Phi'(z)}{1 + \Phi(z)}\right] \qquad (2.176)$$

with

$$\Phi'(z) = \frac{2}{\sqrt{\pi}} \exp(-z^2) \qquad (2.177)$$

Integration of equation 2.167 between the two positions far removed from the sedimentation boundary in both centripetal and centrifugal directions gives the area A under the gradient curve. The result is

$$A = \frac{c_0 e^{-\tau}}{1 - \alpha(1 - e^{-\tau})} \qquad (2.178)$$

Combination of equations 2.175 and 2.178 yields

$$\frac{H}{A} = \frac{G(\xi_*)}{2r_0\alpha \sinh(\tau/2)} \tag{2.179}$$

This expression for the height-area ratio reduces to equation 2.114 when $k_s = 0$, as should be expected.

2. Evaluation of D from Height-Area Ratios

Solution of equation 2.179 for ξ_*, followed by the approximation $\sinh(\tau/2) \approx \tau/2$, gives

$$\xi_* = G^{-1}\left(2r_0\omega^2 k_s c_0 s_0 \left(\frac{H}{A}\right)t\right) \tag{2.180}$$

where G^{-1} denotes the inverse function of G. On the other hand, expanding equation 2.174 in powers of τ and neglecting terms higher than the order of τ, we obtain

$$\xi_* = \frac{r_0\omega^2 k_s c_0 s_0}{2\sqrt{D}}\left[1 - \frac{s_0\omega^2 t}{2}(1 - k_s c_0)\right](t)^{1/2} \tag{2.181}$$

Combination of these two equations leads to

$$G^{-1}\left(2r_0\omega^2 k_s c_0 s_0 \left(\frac{H}{A}\right)t\right) = \frac{r_0\omega^2 k_s c_0 s_0}{2\sqrt{D}}\left[1 - \frac{s_0\omega^2 t}{2}(1 - k_s c_0)\right](t)^{1/2} \tag{2.182}$$

which shows that plots of $G^{-1}(2r_0\omega^2 k_s c_0 s_0 t(H/A))$ against $[1-(s_0\omega^2 t/2)$ $\cdot(1 - k_s c_0)](t)^{1/2}$ yield a straight line, of slope $r_0\omega^2 k_s s_0 c_0/2(D)^{1/2}$, which passes through the origin. Thus, with s_0, k_s, c_0, r_0, and ω known separately, the value of D may be determined from the measurement of H/A as a function of time. To put this procedure in practice, the form of $G^{-1}(z)$ must be known. Figure 2.15 shows graphically part of it for $0 < z < 2$.

Tests of the method just given for D were first attempted by Baldwin,[43] using his own data for a preparation of bovine plasma albumin in an aqueous buffer. Figure 2.16 shows the values of D computed by equation 2.182 and those by Lamm's method as functions of time for two initial concentrations of the protein, $c_0 = 0.67$ g/dl and $c_0 = 1.35$ g/dl. The success of the new equation is apparent. In this connection, it is important to note that the k_s value of the protein system treated here was as small as 0.054

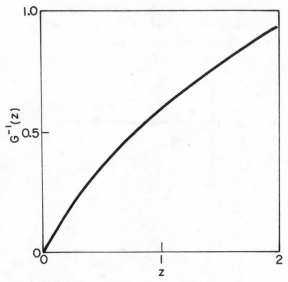

Fig. 2.15. Graphical representation of $G^{-1}(z)$ as a function of z.

dl/g, a magnitude presumably less than those of many other macromolecular solutes by an order of 10. Thus it will be learned how important it is to use an equation corrected for concentration dependence of s when one wishes to evaluate D for a binary solution from height-area ratios of sedimentation boundary curves. Inasmuch as the sedimentation coefficients of macromolecular solutions are generally concentration dependent, one may thus have the impression that there are few actual systems for which Lamm's method can be of effective use. However, as has been noted by Kawahara,[18] this is not necessarily correct.

The argument of the function G^{-1} in equation 2.182 tends to zero as ω is lowered. Hence, regardless of whether or not k_s is small, the limiting form of equation 2.182 for $k_s = 0$, that is,

$$\left(\frac{A}{H}\right)^2 = 4\pi Dt\left(1 + s_0\omega^2 t\right) \tag{2.183}$$

should apply for H/A data obtained at sufficiently low speeds of rotation. This means that Lamm's method (note that equation 2.183 agrees with equation 2.116) is useful even for concentration-dependent systems if applied to low-speed experiments. Kawahara[18] demonstrated it with data for various native and denatured proteins in aqueous media taken at as low a speed of rotation as 12,590 rev/min. However, Kawahara's idea should be accepted with some reservation, because as the rotor speed is reduced, the condition $\epsilon \ll 1$, on which equation 2.182 or 2.183 is founded, may no longer be obeyed with accuracy.

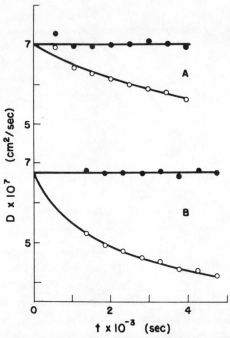

Fig. 2.16. Values of D computed by equation 2.182 (filled circles) and by Lamm's method (open circles) as a function of the time of centrifugation, from Baldwin's data[43] for a preparation of bovine plasma albumin in aqueous buffer. A, for $c_0 = 0.67$ g/dl. B, for $c_0 = 1.35$ g/dl.

G. NUMERICAL SOLUTIONS OF DISHON ET AL.

Dishon et al.[40–42] attempted numerical solution of the Lamm equation subject to the initial and boundary conditions for actual experiments. The computations were made for systems in which $s(c)$ is given by equation 2.5 or 2.6 but D is independent of c. In one paper,[42] the systems with very small values of ϵ (of the order of 10^{-3} or less) were treated, so the results obtained actually represented the Faxén-type solutions to the problems. In this section, some of them are reproduced in illustrations to show typical features of the exact boundary curves in these two concentration-dependent systems. In another paper,[41] Dishon et al. chose relatively large values for ϵ so that the calculations might fit conditions corresponding to sedimentation equilibrium runs, and examined how the rate of approach to equilibrium is affected by the dependence of s on c. Important results from this study are presented in Chapter 7.

The following is a summary of the numerical values that were assigned to various parameters for the study of small-ϵ systems.

a. $k_s c_0 = \frac{1}{11}$, 0.1, $\frac{1}{6}$, 0.2, and 0.5 for $s = s_0(1 - k_s c)$; $k_s c_0 = 0.1$, 0.2, 0.5, 1.0, 10, and 50 for $s = s_0/(1 + k_s c)$.

b. $r_1 = 5.9$ cm and $r_2 = 7.2$ cm for the conventional cell; $r_1 = 5.78$ and 5.6 cm, $r_0 = 5.9$ cm, and $r_2 = \infty$ for the synthetic boundary cell.

c. $\sigma = 5.782$, 23.49, and 117.44 cm^{-2}, where

$$\sigma = \frac{s_0 \omega^2}{D} \tag{2.184}$$

For $r_0 = 5.9$ cm these values of σ yield

$$\epsilon = 0.00994, \quad 0.00245, \quad 0.000489$$

Dishon et al. state that the values of σ selected correspond roughly to protein molecules of molecular weights 13,000, 50,000, and 260,000 in an aqueous medium, all at an ω of 60,000 rev/min.

Figure 2.17 depicts the gradient curves obtained for $\sigma = 23.5$ cm^{-2}; the upper graph, corresponding to $s = s_0$, shows typical symmetric curves of Gaussian shape, while the lower graph, corresponding to $s = s_0/[1 + 0.5(c/c_0)]$, indicates a slowing down and a resulting boundary sharpening (compare with Fig. 2.14 showing similar features). It is to be noted that a small asymmetry exists in the boundary curves shown on the lower panel. The asymmetry becomes pronounced as $k_s c_0$ is increased. In Fig. 2.18, the boundary curves for $\sigma = 23.5$ cm^{-2} and $s = s_0/[1 + (c/c_0)]$ are compared with the corresponding curves for $s = s_0[1 - 0.5(c/c_0)]$. Note that these two expressions for s have the same value of $0.5 s_0$ at $c = c_0$. It is seen that the boundary curves for linear c dependence of s maintain symmetry extremely well, while those for nonlinear c dependence of s exhibit a more marked asymmetry than in Fig. 2.17. Further examples of the slowing down and the consequent boundary sharpening for the case of nonlinear c dependence of s are presented in Fig. 2.19.

Dishon et al.[42] observed for a number of cases in which $k_s c_0$ was comparatively large that the calculated boundary curves manifested in common such behavior as illustrated in Fig. 2.20. Shortly after the trailing edge clears the meniscus, the boundary curve assumes an almost constant shape and travels down the cell roughly unchanged in both size and shape.

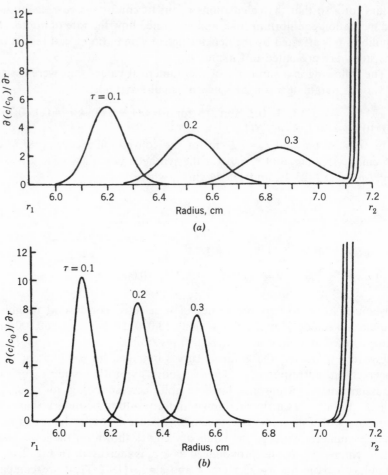

Fig. 2.17. Caluclated gradient curves for $\sigma = 23.5$ cm^{-2}: (a) $s = s_0$; (b) $s = s_0/[1 + 0.5(c/c_0)]$.

In many of these cases, the height-area ratio of the curve approaches a constant value. The occurrence of such an approximate "steady state" is the consequence of an almost exact compensation of boundary sharpening due to concentration dependence of s by boundary spreading due to diffusion. A true steady state, however, cannot occur because of the radial dilution caused by the sector shape of the cell and the radially increasing field of centrifugal force. Creeth[44] investigated the behavior of approximate steady-state boundary curves both theoretically and experimentally, and proposed a method for estimating D from measurements of such curves. Dishon et al.[42] confirmed the validity of Creeth's relation[44] numerically.

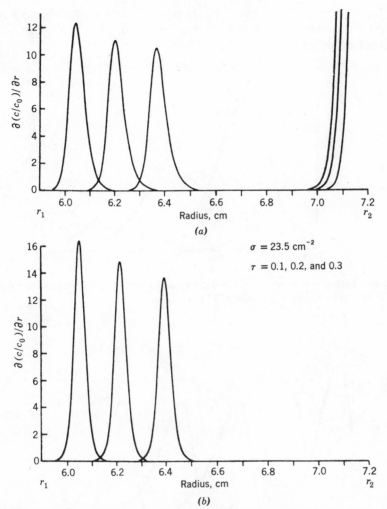

Fig. 2.18. Calculated gradient curves for $\sigma = 23.5$ cm^{-2}: (a) $s = s_0/[1+(c/c_0)]$; (b) $s = s_0[1 - 0.5(c/c_0)]$.

In Section III.B.3 it was shown that when s and D are constant, there holds a linear relation between $(A/H)^2$ and $\exp(\tau) - 1$ with a proportionality coefficient equal to $D(2\pi/s\omega^2)$ or $2\pi/\sigma$. Dishon et al. examined the effect of c-dependence of s on this relationship, and obtained for $\sigma = 23.5$ cm^{-2} the curves shown in Fig. 2.21. It is seen that even a slight c-dependence of s introduces a pronounced curvature into the initial portion of the relation. This suggests that when s varies with c, one would be led to a gross error if D were formally computed from the ratio of observed

$(A/H)^2$ to $\exp(\tau) - 1$. We have already pointed out a similar shortcoming of Lamm's method when Baldwin's data on bovine plasma albumin were analyzed in Section III.F.2.

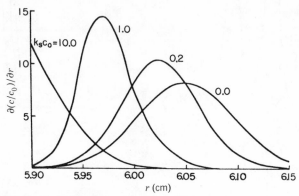

Fig. 2.19. Examples of slowing down and consequent boundary sharpening calculated for the case in which $s = s_0/(1 + k_s c)$ and $\sigma = 23.5$ cm^{-2}.

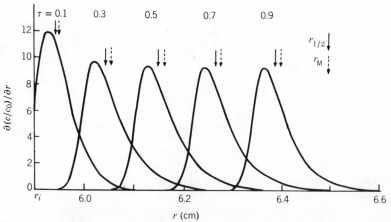

Fig. 2.20. Approximate "steady state" gradient curves, obtained with $\sigma = 23.5$ cm^{-2} and $s = s_0/[1 + 0.5(c/c_0)]$. r_M, the equivalent boundary position (see Section III.4.1). $r_{1/2}$, the radial position at which the concentration becomes half that in the plateau region.

Equation 2.182, valid as an approximation for the case of $s = s_0(1 - k_s c)$, may be written

$$G^{-1}(z) = \left(\frac{\sigma}{8}\right)^{1/2} r_0 k_s c_0 \left[1 - \frac{\tau}{4}(1 - k_s c_0)\right]\tau^{1/2} \qquad (2.185)$$

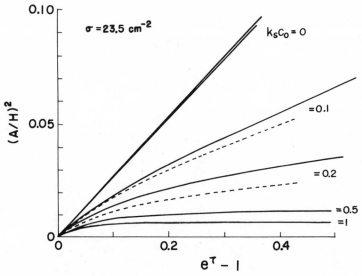

Fig. 2.21. Curves of $(A/H)^2$ as a function of $\exp(\tau)-1$ for $\sigma=23.5$ cm^{-2}. The upper curve for $k_s=0$ includes the meniscus effect. The solid lines are for $s=s_0/(1+k_s c)$, and the dashed lines are for $s=s_0(1-k_s c)$.

where

$$z = r_0 k_s c_0 \left(\frac{H}{A}\right)\tau \tag{2.186}$$

Dishon et al.[42] examined the validity of equation 2.185 in terms of their numerical data and found that in all cases treated with the initial condition for the synthetic boundary cell, plots of $G^{-1}(z)$ versus $r_0 k_s c_0[1-(\tau/4) \times(1-k_s c_0)]\tau^{1/2}$ gave straight lines through the origin and appropriate values of σ from their slopes. The data obtained with the initial condition for the conventional cell gave rise to a very small upward displacement of the plots, without changing their linearity, as illustrated in Fig. 2.22. The slopes of the resulting lines still led to accurate values of σ. The displacement of the plots, which becomes large as σ decreases, is a reflection of the restricted diffusion produced by the presence of the meniscus.

No analytical method for the estimate of D from height-area ratio data is available for systems having s of the form $s=s_0/(1+k_s c)$. From their numerical data Dishon et al.[42] found that if the k_s in the definition of the variable z (equation 2.186) is replaced by the k^* defined as

$$k^* = \frac{k_s}{1+k_s c_0} \tag{2.187}$$

values of $G^{-1}(z)$ for this type of $s(c)$ follow a straight line almost exactly when plotted against $r_0 k^* c_0 [1 - (\tau/4)(1 - k^* c_0)] \tau^{1/2}$, provided that $k_s c_0$ is smaller than 0.2. However, the values of D estimated from the slopes of the straight lines are a few percent larger than the correct ones.

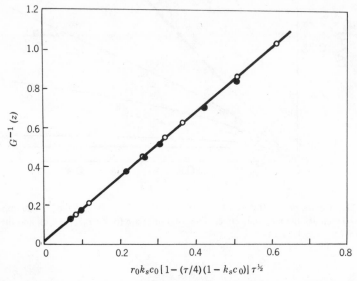

Fig. 2.22. Plots of $G^{-1}(z)$ versus $r_0 k_s c_0 [1 - (\tau/4)(1 - k_s c_0)] \tau^{1/2}$ for $\sigma = 23.5$ cm^{-2} Open circles, for $s = s_0 [1 - 0.2(c/c_0)]$; filled circles, for $s = s_0 / [1 + 0.2(c/c_0)]$.

H. THE METHOD OF MOMENTS

Although the method of moments can be developed for solutions containing any number of solute components, the present discussion is restricted to binary solutions. Baldwin[45] was the first to use this method for sedimentation analysis, but it should not be overlooked that in an earlier paper Goldberg[46] had introduced the concept of "equivalent boundary position" in terms of the second moment of the boundary curve about the center of rotation. The practical usefulness of the method of moments is limited to boundary curves which are not restricted by the meniscus and, at the same time, leave the plateau regions ahead of their maxima. For convenience in presentation, such a boundary curve is referred to as the *freely sedimenting boundary*.

1. Application to Boundary Centrifugation

The nth reduced moment μ_n of a freely sedimenting boundary curve is defined by

$$\mu_n = \int_{r_1}^{r_p} r^n \left(\frac{\partial c}{\partial r} \right) dr \Big/ \int_{r_1}^{r_p} \left(\frac{\partial c}{\partial r} \right) dr \qquad (n = 0, 1, \dots) \qquad (2.188)$$

where r_1 is the radial position of the air-liquid meniscus, and r_p is an arbitrary point in the plateau region ahead of the sedimentation boundary. Thus $\partial c/\partial r$ and its higher derivatives with respect to r are all zero at r_p. If the concentration at r_p is designated by c_p, equation 2.188 may be written

$$\mu_n = \int_{r_1}^{r_p} r^n \left(\frac{\partial c}{\partial r} \right) dr / c_p \qquad (2.189)$$

Note that for the boundary curve under consideration the concentration and its gradient at the meniscus vanish. This condition is used below without specific mention.

Both sides of the Lamm equation are multiplied by r^{n-1} and then integrated from r_1 to r_p to give

$$\frac{d}{dt} \int_{r_1}^{r_p} c r^{n-1} dr = \int_{r_1}^{r_p} r^{n-2} \frac{\partial}{\partial r} \left(rD \frac{\partial c}{\partial r} - r^2 s\omega^2 c \right) dr \qquad (2.190)$$

On the other hand, if it is applied to the plateau region, the Lamm equation reduces to the form

$$\frac{dc_p}{dt} = -2s(c_p)\omega^2 c_p \qquad (2.191)$$

where $s(c_p)$ denotes the value of s for the concentration c_p. Using equation 2.191, it can be shown that manipulations involving integration by parts transform equation 2.190 to

$$\frac{d\mu_n}{dt} = [2\mu_n + (n-2)r_p^n] s(c_p)\omega^2 + \frac{n(n-2)}{c_p} \int_{r_1}^{r_p} r^{n-2} \left(D \frac{\partial c}{\partial r} - s\omega^2 cr \right) dr \qquad (2.192)$$

This relation holds regardless of whether or not both s and D depend on c. If s and D are constant, use of equation 2.189 gives

$$\int_{r_1}^{r_p} r^{n-2} D \frac{\partial c}{\partial r} dr = Dc_p \mu_{n-2} \qquad (2.193)$$

$$\int_{r_1}^{r_p} r^{n-1} s\omega^2 c \, dr = \left(\frac{r_p^n}{n} - \frac{\mu_n}{n} \right) s\omega^2 c_p \qquad (2.194)$$

Therefore, equation 2.192 is reduced to

$$\frac{d\mu_n}{dt} = ns\omega^2\mu_n + n(n-2)D\mu_{n-2} \tag{2.195}$$

This linear differential difference equation for μ_n is due originally to Baldwin.[45] The definition of μ_n gives

$$\mu_0 = 1 \tag{2.196}$$

For $n=2$ equation 2.195 reduces to

$$\frac{d\mu_2}{dt} = 2s\omega^2\mu_2 \tag{2.197}$$

which, upon integration, yields

$$\mu_2 = K\exp(2s\omega^2 t) \tag{2.198}$$

where K is a constant of integration. Since freely sedimenting boundary curves are under consideration, the relevant initial condition is equation 2.2, which gives in terms of a normalized delta function δ

$$\frac{\partial c}{\partial r} = c_0\delta(r-r_0) \qquad (t=0) \tag{2.199}$$

Introduction of this into equation 2.188 yields

$$\mu_2 = r_0^2 \qquad (t=0) \tag{2.200}$$

Thus K in equation 2.198 is found to be r_0^2, so that

$$\mu_2 = r_0^2 \exp(2s\omega^2 t) \tag{2.201}$$

Baldwin[45] showed that the general solution to equation 2.195 subject to the conditions 2.196 and 2.200 is given by

$$\mu_{2n} = r_0^{2n}e^{2ns\omega^2 t}\sum_{m=1}^{n}\frac{n!(n-1)!}{m!(n-m)!(m-1)!}(2a)^{n-m} \tag{2.202}$$

where

$$a = \frac{D}{s\omega^2 r_0^2}(1 - e^{-2s\omega^2 t}) = \frac{\epsilon}{2}(1 - e^{-\tau}) \tag{2.203}$$

For example, μ_4 and μ_6 are

$$\mu_4 = r_0^4 e^{4s\omega^2 t}(1 + 4a)$$

$$\mu_6 = r_0^6 e^{6s\omega^2 t}(1 + 12a + 24a^2)$$

$$(2.204)$$

Since equation 2.202 is concerned with freely sedimenting boundary curves, it could be derived from the Faxén exact solution to the Lamm equation for the infinite sectorial cell, equation 2.86. In fact, this prediction was verified by Baldwin.[45]

If $n = 2$, equation 2.192 reduces to

$$\frac{d \ln r_M}{\omega^2 dt} = s(c_p) \tag{2.205}$$

where r_M is defined by

$$r_M = \left[\int_{r_1}^{r_p} r^2 \left(\frac{\partial c}{\partial r} \right) dr \bigg/ \int_{r_1}^{r_p} \left(\frac{\partial c}{\partial r} \right) dr \right]^{1/2} \tag{2.206}$$

and may be evaluated for each freely sedimenting boundary by numerical integration. Combination of equations 2.191 and 2.205 yields an important relation:

$$\frac{d \ln c_p(r_M)^2}{dt} = 0 \tag{2.207}$$

It is to be noted that this differential equation holds for a series of freely sedimenting boundary curves observed in a sedimentation run, regardless of the dependence of s on c and of the initial condition of the experiment. Thus if the initial condition for the conventional cell is chosen, integration of equation 2.207 gives

$$(r_M)^2 c_p = (r_1)^2 c_0 \tag{2.208}$$

This relation, first derived by Goldberg[46] in his commemorable paper of 1953, shows that the square-dilution rule is obeyed by any freely sedimenting boundary in a binary solution if r_M is taken as the position of an infinitely sharp solvent-solution boundary in the case of no diffusion (see Section II). In other words, r_M defines a hypothetical sharp boundary which would be obtained if the material of solute were rearranged so that the square-dilution rule holds for the concentration difference across the boundary. For this reason, r_M is usually called the *equivalent boundary position* of a freely sedimenting boundary curve.

Equation 2.205, another important relation, indicates that the slopes of a plot of $\ln r_M$ versus $\omega^2 t$ yield values of s which correspond to the concentrations in the plateau region.* Since c_p decreases with increasing r_M (or time t), this plot should have an upward curvature if, as is generally the case, s is a decreasing function of c. The initial slope of the plot gives the value of s corresponding to the initial concentration c_0, but, in actual cases, it must be determined by extrapolation. A useful method for this purpose is presented in Section V.D.2.

In Section III.B.3, it was shown that the maximum position r_* of a sedimentation boundary curve represented by equation 2.106 moves with time according to equation 2.109. Since equation 2.106 has been derived for constant s, equation 2.109 is equivalent to the relation

$$\frac{d\ln r_*}{\omega^2 dt} = s \tag{2.209}$$

Comparison with equation 2.205 shows that r_* is proportional to r_M. Since at $t=0$ both r_* and r_M tend to r_0, the proportionality constant must be equal to unity. Hence it might be concluded that for the boundary curve represented by equation 2.106 the position of the maximum gradient would coincide with the equivalent boundary position. This conclusion, however, is not correct. To verify it, we define a quantity σ_2 by

$$\sigma_2 = (r_M)^2 - (r_m)^2 \tag{2.210}$$

where r_m is a radial position given by

$$r_m = \int_{r_1}^{r_p} r\left(\frac{\partial c}{\partial r}\right) dr \bigg/ \int_{r_1}^{r_p} \left(\frac{\partial c}{\partial r}\right) dr \tag{2.211}$$

that is, it is the first reduced moment μ_1 of a freely sedimenting boundary curve.† With substitution of equation 2.206, we obtain

$$\sigma_2 = \int_{r_1}^{r_p} (r - r_m)^2 \left(\frac{\partial c}{\partial r}\right) dr \bigg/ \int_{r_1}^{r_p} \left(\frac{\partial c}{\partial r}\right) dr \tag{2.212}$$

Therefore, σ_2 is nonzero unless the boundary curve has the form of a delta function whose peak is at $r = r_m$. For any symmetric, freely sedimenting,

---

*This means that the equivalent boundary moves at rates corresponding to s for c_p.
† r_m may be termed the center of gravity of a sedimentation boundary curve.

and single-peaked boundary curve the position r_m coincides with the position of the maximum gradient r_*. Since the boundary curve represented by equation 2.106 has a Gaussian shape with its maximum at $r = r_0 \exp(\tau/2)$, it is an example of boundary curves of this type. Thus its σ_2 is nonzero, that is, $r_* \neq r_M$, which verifies the incorrectness of the above-mentioned conclusion. The difference between r_* and r_M can be estimated approximately by substituting equation 2.106 into equation 2.212, with r_m taken to be $r_0 \exp(\tau/2)$. The result gives[45]

$$r_M - r_* \approx \frac{r_0 \epsilon \tau}{4} = \frac{Dt}{r_0} \qquad (2.213)$$

Thus r_* is smaller than r_M by a distance approximately equal to $r_0(\epsilon\tau)/4$. A difference of this magnitude can be treated as negligible within the regime of Faxén-type solutions, because these solutions have actual validity only under the condition that $\epsilon \ll 1$ and $\tau \ll 1$.

The difference between r_* and r_M for concentration-dependent systems was examined numerically by Dishon et al.[42] Figure 2.23 shows their numerical results for the system with $s = s_0(1 - k_s c)$ and $\sigma = 23.5$ cm^{-2}. Actually, the curves shown were derived from the numerical solutions of the Lamm equation subject to the initial condition for the conventional cell. The rising portions of the curves at very small values of τ are thus attributed to the meniscus effects. It is seen that r_M approaches r_* very quickly and remains close to it after a certain time of centrifugation. Interestingly, the proximity between r_M and r_* is even closer for a larger $k_s c_0$ value, that is, for a more markedly concentration-dependent system. However, as was shown by Dishon et. al., this is not generally true. Their results for the system with $s = s_0/(1 + k_s c)$ and $\sigma = 23.5$ cm^{-2} indicate the reverse effect; the difference between r_M and r_* increases with increasing $k_s c_0$. This behavior is due to an increased asymmetry of the boundary curve with increasing $k_s c_0$ in the system characterized by s of this form. At any rate, it may be concluded from these results that r_M may be equated to r_* with a trivial error when the boundary curve is free from the meniscus effects and is not too asymmetric. This conclusion helps the actual determination of s, because the integration involved in the evaluation of r_M can be obviated.

2. Application to Band Centrifugation

Theoretical formulations of band centrifugation by the method of moments were made by Vinograd et al.,[28] Schumaker and Rosenbloom,[35] and Vinograd and Bruner,[11] in essentially the same manner as described above for boundary centrifugation. Some of the important relations given by these authors are presented below.

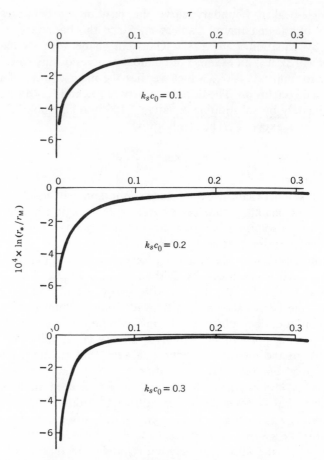

Fig. 2.23. Deviation plot for r_*/r_M when $\sigma = 23.5$ cm^{-2} and $s = s_0(1 - k_s c)$. r_*, the position of maximum gradient. r_M, the equivalent boundary position.

As before, we treat freely sedimenting bands. The nth reduced moment of such a band, ν_n, is defined by

$$\nu_n = \frac{\int_{r_1}^{r_2} r^n (cr)\,dr}{\int_{r_1}^{r_2} (cr)\,dr} \tag{2.214}$$

where r_1 and r_2 denote the radial positions of the meniscus and the cell bottom, respectively.

If s and D are constant, the following differential difference equation can be derived for v_n:

$$\frac{dv_n}{dt} = ns\omega^2 v_n + n^2 D v_{n-2} \qquad (2.215)$$

which may be compared to Baldwin's equation 2.195 for the boundary centrifugation. The first reduced moment v_1 defines the center of gravity of a band, r_m. Vinograd and Bruner showed that equation 2.215 with $n = 1$ becomes approximately

$$\frac{d \ln r_m}{dt} = s\omega^2 + \frac{D}{r_m^2} \qquad (2.216)$$

If an apparent sedimentation coefficient s_m is defined by

$$s_m \equiv \frac{d \ln r_m}{\omega^2 dt} \qquad (2.217)$$

then equation 2.216 is written

$$s_m = s\left(1 + \frac{D}{s\omega^2 r_m^2}\right) \qquad (2.218)$$

This shows how much s_m differs from the true s. The second term in the parentheses, $D/s\omega^2 r_m^2$, which remains approximately constant during a band sedimentation experiment, actually equals $\epsilon'/2$, where ϵ' is the parameter defined by equation 2.133. Thus for high-molecular-weight solutes the differences between s_m and s are generally quite small unless rotor speeds are too low. Vinograd and Bruner[11] quote 0.4, 0.1, and 0.005% for the differences with ribonuclease, bovine serum albumin, and ϕX virus, respectively. Because it is assumed here that s is constant, s_m also may be treated as constant. Hence equation 2.217 is equivalent to the integral form

$$\ln r_m = s_m \omega^2 t + \text{constant} \qquad (2.219)$$

Evaluation of r_m requires numerical integration of primary experimental data which, if obtained with a light-absorption method, must be linearly related to concentration. However, as Vinograd and Bruner state, this requirement is often not fulfilled in practice. Hence, the use of equation 2.219 for the determination of s_m is not necessarily convenient. For any single-peaked band we may define another apparent sedimentation coefficient s_* by the relation

$$\ln r_* = s_* \omega^2 t + \text{constant} \qquad (2.220)$$

where r_* stands for the radial position of the maximum concentration, which can be read directly from densitometer readings when the data are obtained with the aid of a light-absorption method. Unlike the case of boundary centrifugation, the position r_* does not coincide with the position r_m even for a symmetrical band. Vinograd and Bruner[11] showed for such a band that s_* is approximately related to the true s by

$$s_* = s\left(1 - \frac{D}{s\omega^2 r_*^2}\right) \tag{2.221}$$

The error in s_* is of the same magnitude as the error in s_m but is of the opposite sign, that is, $s_* < s$ and $s_m > s$.

Asymmetric bands may result from various transient effects which include nonideal layering of a given solution on the solvent medium, convection, restricted diffusion of the solute at the air-liquid meniscus, and concentration dependence of s. These effects, however, tend to vanish at later stages of the experiment. It is important to note that since the concentrations in the band are diluted due to increasing spreading as the band travels down the cell, effects of a concentration dependence of s, if any, become successively less important. Thus if the concentration of the layered solution is sufficiently low, the sedimenting band will sooner or later behave "ideally." However, as the theory of Rubin and Katchalsky[34] (see Section III.F.1) indicates, even the "ideal" band is not truly symmetric. Vinograd and Bruner[11] developed an approximate treatment of asymmetric bands, but it appears still too premature to be of use for actual sedimentation analysis.

Schumaker and Rosenbloom[35] showed that if s is constant, the following relation holds exactly for freely sedimenting bands:

$$\frac{d\langle \ln r \rangle}{\omega^2 dt} = s \tag{2.222}$$

where $\langle \ln r \rangle$ is defined by

$$\langle \ln r \rangle = \frac{\displaystyle\int_{r_1}^{r_2} (\ln r)(cr)\,dr}{\displaystyle\int_{r_1}^{r_2} (cr)\,dr} \tag{2.223}$$

They refer to the position corresponding to $\langle \ln r \rangle$ as the "center of sedimentation," which may be found by numerical integration of the experimentally determined concentration distribution in the band. In Fig. 2.24 are plotted the values of $\langle \ln r \rangle$ for a preparation of catalase as a function of time.[35] As required by equation 2.222, the plotted points fit a

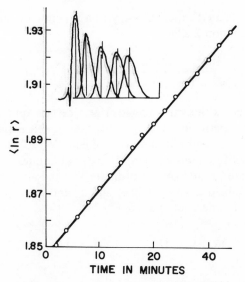

Fig. 2.24. Variation of $\langle \ln r \rangle$ (see equation 2.223) with time of centrifugation for a preparation of catalase.[35] The insert shows representative photodensitometer tracings of the concentration distributions in the migrating bands.

straight line very closely; the slope gives a value of 11.3×10^{-13} sec for $s_{20,w}$ of this enzyme. The treatment of Schumaker and Rosenbloom[35] yields for concentration-dependent systems

$$\frac{d\langle \ln r \rangle}{\omega^2 dt} = \langle s \rangle \qquad (2.224)$$

where $\langle s \rangle$ is defined by

$$\langle s \rangle = \frac{\int_{r_1}^{r_2} s(c)(cr)dr}{\int_{r_1}^{r_2} (cr)dr} \qquad (2.225)$$

For example, when $s(c) = s_0(1 - k_s c)$, equation 2.225 yields

$$\langle s \rangle = s_0(1 - k_s \langle c \rangle) \qquad (2.226)$$

where $\langle c \rangle$ is given by

$$\langle c \rangle = \frac{\int_{r_1}^{r_2} c(cr)dr}{\int_{r_1}^{r_2} (cr)dr} \qquad (2.227)$$

which may be evaluated experimentally as a function of time. In this case, integration of equation 2.224 gives

$$\frac{\langle \ln r \rangle - \langle \ln r \rangle_0}{\omega^2 (t - t_0)} = s_0 - \frac{k_s s_0}{t - t_0} \int_{t_0}^{t} \langle c \rangle dt \qquad (2.228)$$

where $\langle \ln r \rangle_0$ denotes the value of $\langle \ln r \rangle$ at a certain time t_0 elapsed after initiation of the experiment. The quantity on the left-hand side and that multiplied by $k_s s_0$ on the right-hand side of equation 2.228 are experimentally obtainable as functions of time. If these quantities corresponding to the same times are plotted on the ordinate and the abscissa, respectively, there will be obtained a straight line, and s_0 and k_s may be determined from their ordinate intercept and slope.

I. METHODS FOR EVALUATION OF SMALL SEDIMENTATION COEFFICIENTS

The peak methods for the determination of s from boundary sedimentation experiments fail under conditions in which freely sedimenting boundary curves are not observable. As the Fujita-MacCosham theory indicates (see Section III.D), the boundary curves characterized by relatively large values of the parameter ϵ' may have to spend a considerable interval of time before they are freed from the meniscus effects and become freely sedimenting. In some cases, the plateau region will disappear before the gradient at the meniscus decreases to zero. These situations are often encountered in experiments with relatively low-molecular-weight solutes which, in general, have smaller s and larger D so that their values of ϵ' are fairly large even at the maximum operational speed (see Section III.D.1).

Methods useful for the determination of s in such situations were worked out by Gutfreund and Ogston[47] and by Baldwin.[48] These are derivable on the sole assumption that there exists a well-defined plateau region in the solution column during an appropriate interval of time after initiation of the experiment.

The Lamm equation is multiplied by r and integrated from $r = r_1$ to $r = r_p$, where r_1 and r_p have the same significance as before. Then

$$\frac{d}{dt} \int_{r_1}^{r_p} cr \, dr = -s(c_p) c_p \omega^2 r_p^2 \qquad (2.229)$$

where the boundary condition at the air-liquid meniscus has been taken into account. Equation 2.229 is combined with equation 2.191 to give

$$2\frac{d}{dt}\int_{r_1}^{r_p}cr\,dr=r_p^2\frac{dc_p}{dt}\tag{2.230}$$

If this is integrated with respect to t from $t=0$ and the initial condition for the conventional cell, that is, equation 2.1, is applied, then

$$\int_{r_1}^{r_p}(c-c_0)r\,dr=\frac{r_p^2}{2}(c_p-c_0)\tag{2.231}$$

which, upon integration by parts, may be rearranged to yield

$$c_p=c_0-\frac{1}{r_1^2}\int_{r_1}^{r_p}(r^2-r_1^2)\frac{\partial c}{\partial r}dr\tag{2.232}$$

This formula may be used to calculate c_p for a given boundary curve. Substitution of equation 2.232 into equation 2.191 gives

$$s(c_p)=-\frac{1}{2\omega^2}\frac{d}{dt}\ln\left[1-\frac{1}{r_1^2c_0}\int_{r_1}^{r_p}(r^2-r_1^2)\frac{\partial c}{\partial r}dr\right]\tag{2.233}$$

If s is constant, this yields, after integration from 0 to t,

$$2\omega^2st=-\ln\left[1-\frac{1}{r_1^2c_0}\int_{r_1}^{r_p}(r^2-r_1^2)\frac{\partial c}{\partial r}dr\right]\tag{2.234}$$

which is an expression first derived by Baldwin.[48] For systems with concentration-dependent s, equation 2.233 must be used, but unless the c dependence of s is too pronounced, a plot of the quantity on the right-hand side of equation 2.234 against $2\omega^2t$ would be almost exactly linear. One may then equate the slope of the resulting line to the value of s for the initial concentration c_0. Inasmuch as the derivation just described only assumes the existence of the plateau region, equation 2.233 is applicable regardless of whether or not the boundary curve exhibits a maximum. According to experimental evidence, this method is advantageous for solutes whose s are smaller than 2×10^{-13} sec.

Equation 2.234 may be transformed, by applying integration by parts and using the relation $c_p=c_0\exp(-2s\omega^2t)$ which follows from equation

2.191 with constant s, to the relation

$$2\omega^2 st = -\ln\left[\left(\frac{r_1}{r_p}\right)^2 + 2\int_{r_1}^{r_p}\left(\frac{r}{r_p}\right)\left(\frac{c}{c_0}\right)\frac{dr}{r_p}\right] \tag{2.235}$$

This equation was derived by Gutfreund and Ogston[47] through a somewhat different route. It can be of use for the determination of s when primary experimental data are given in the form of concentration (more generally refractive index) as a function of r.

Neither Baldwin's method nor that of Gutfreund and Ogston can be of effective use for systems in which ϵ' is so large that the plateau region soon disappears after start of the experiment. Even for such systems there must be an interval of time in which the plateau region may be observed. However, if the length of this period is too short, no reliable determination of s may be made by these methods, and another approach must be sought. This problem was investigated by Yphantis and Waugh,[49] but the method they proposed appears to have received scant attention from experimentalists.

IV. SOLUTIONS OF THE ARCHIBALD TYPE

A. ARCHIBALD'S EXACT SOLUTION

For the Lamm equation with constant s and D, Archibald[2] obtained for the first time a solution which rigorously satisfies the boundary conditions for actual cells. Unfortunately, his solution, written in the form of an infinite series, was too complex to be of immediate use for actual sedimentation analysis. Thus, differing from Faxén's exact solution for the infinite sectorial medium, its significance seems to be little more than historical. We therefore present here only his final expression for the concentration distribution $c(r,t)$. It reads

$$\frac{c(r,t)}{c_0} = \frac{(z_2 - z_1)\exp(z)}{\exp(z_2) - \exp(z_1)} + \sum_{n=1}^{\infty} M(\alpha_n, 1, z)\exp[(\alpha_n - 1)\tau]$$

$$\times \frac{\int_{z_1}^{z_2} M(\alpha_n, 1, z)e^{-z}\,dz}{\int_{z_1}^{z_2}[M(\alpha_n, 1, z)]^2 e^{-z}\,dz} \tag{2.236}$$

where

$$\tau = 2s\omega^2 t, \qquad z = \left(\frac{1}{\epsilon'}\right)\left(\frac{r}{r_1}\right)^2, \qquad z_1 = \frac{1}{\epsilon'}, \qquad z_2 = \left(\frac{1}{\epsilon'}\right)\left(\frac{r_2}{r_1}\right)^2 \qquad (2.237)$$

with

$$\epsilon' = \frac{2D}{s\omega^2 r_1^2} \qquad (2.238)$$

The function $M(\alpha_n, 1, z)$ is an eigenfunction associated with a confluent hypergeometric differential equation of the form

$$\frac{d^2 M}{dz^2} + \left(\frac{1}{z} - 1\right)\frac{dM}{dz} - \left(\frac{\alpha_n}{z}\right)M = 0 \qquad (2.239)$$

and the α_n is the associated eigenvalue of nth order determined from the relations

$$\frac{d}{dz} M(\alpha_n, 1, z_1) = M(\alpha_n, 1, z_1)$$

$$(2.240)$$

$$\frac{d}{dz} M(\alpha_n, 1, z_2) = M(\alpha_n, 1, z_2)$$

The evaluation of α_n involves very laborious computations even if we confine ourselves only to the first few of them. Thus Archibald[4] obtained the values only for $n = 1$. Later, Waugh and Yphantis[50] computed α_1, α_2, and α_3 for certain sets of z_1 and z_2. Since their work referred to some specific values of parameters, these numerical data may be of limited use in general.

B. APPROXIMATE SOLUTIONS

In view of the computational difficulty involved in Archibald's series solution, various attempts have been made toward approximate solutions which permit simpler and quicker calculation of solute distributions with accuracy. The first of them was reported by Archibald himself,[51] who introduced the condition $z_2 - z_1 \approx 1$. Later it was shown by Waugh and Yphantis[50] that this approximate solution of Archibald can be applied even for larger differences of z_1 and z_2 which range from 3 to 9. However, one still needs a considerable amount of computational work to derive the desired results with it. We therefore do not discuss these earlier studies in the present treatise.

1. The Rectangular-Cell Approximation

For the ultracentrifuge cells in current use the sector angle is very small (2–4°), and the length of solution column relative to the distance between the air-liquid meniscus and the center of rotation is quite small in comparison to unity, that is, $(r_2 - r_1)/r_1 \ll 1$. From these geometric conditions it may be assumed that the physical processes which occur in an actual sector-shaped cell are approximately identical to those in a rectangular cell with a constant field of force. This assumption is usually called the *rectangular-cell* (or simply *rectangular*) *approximation*. It was first introduced by Yphantis and Waugh[50] in the mathematics of ultracentrifuge equations. As will be seen below at many places, it plays a very important role in theoretical treatments of sedimentation behavior.

Yphantis and Waugh[50] showed that if the Lamm differential equation and the associated boundary conditions are properly simplified by the rectangular-cell approximation, the sedimentation problem for a binary solution with constant s and D can be solved by a simple transcription of the solution which had been obtained by Mason and Weaver[52] in 1924 for the settling of spherical particles under gravity. In this manner, they deduced for the concentration distribution $c(r,t)$

$$\frac{c}{c_0} = \frac{\exp(y/\alpha)}{\alpha[\exp(1/\alpha) - 1]} + \exp\left(\frac{y - \frac{1}{2}\gamma\tau}{2\alpha}\right)$$

$$\times \sum_{m=1}^{\infty} E_m(\tau)(\sin m\pi y + 2\pi m\alpha \cos m\pi y) \qquad (2.241)$$

where

$$E_m(\tau) = \frac{16\pi m\alpha^2[1 - (-1)^m \exp(-1/2\alpha)]}{(1 + 4\pi^2 m^2 \alpha^2)^2} \times \exp(-\pi^2 m^2 \alpha\gamma\tau) \qquad (2.242)$$

$$y = \frac{r - r_1}{r_2 - r_1} \qquad (2.243)$$

$$\alpha = \frac{D}{s\omega^2 \bar{r}(r_2 - r_1)} \qquad (2.244)$$

and

$$\gamma = \frac{\bar{r}}{2(r_2 - r_1)} \qquad (2.245)$$

The quantity $\bar{r}$ is left undetermined in their treatment, but, in practice, it may be chosen as the arithmetic mean of r_1 and r_2, that is,

$$\bar{r} = \tfrac{1}{2}(r_1 + r_2) \tag{2.246}$$

The simplification of the basic equations for the ultracentrifuge by the rectangular-cell approximation can be carried out in the following way. If s and D are independent of c, the Lamm equation may be written

$$\frac{\partial c}{\partial \tau} = \frac{1}{2r} \frac{\partial}{\partial r} \left[r^2 \left(\frac{D}{s\omega^2 r} \frac{\partial c}{\partial r} - c \right) \right] \tag{2.247}$$

Under the geometric condition $(r_2 - r_1)/r_1 \ll 1$ the variation of r in the cell relative to r_1 is so small that r in the equation above may be replaced by some mean value $\bar{r}$ between r_1 and r_2. Then

$$\frac{\partial c}{\partial \tau} = \frac{\bar{r}}{2} \frac{\partial}{\partial r} \left(\frac{D}{\bar{r}s\omega^2} \frac{\partial c}{\partial r} - c \right) \tag{2.248}$$

which, in terms of y, α, and γ, may be written

$$\frac{\partial c}{\partial (\gamma\tau)} = \frac{\partial}{\partial y} \left(\alpha \frac{\partial c}{\partial y} - c \right) \tag{2.249}$$

This is the Lamm equation subject to the rectangular-cell approximation, and is equivalent to the sedimentation equation of Mason and Weaver. The initial condition for the conventional cell is written in terms of y and τ as

$$c = c_0 \qquad (0 < y < 1, \quad \tau = 0) \tag{2.250}$$

The boundary conditions, equations 2.4, may be written

$$\left(\frac{\bar{r}}{r_i} \right) \alpha \frac{\partial c}{\partial y} = c \qquad (y = 0, \quad i = 1; \quad y = 1, \quad i = 2, \quad \tau > 0) \tag{2.251}$$

The rectangular-cell approximation replaces r_i by $\bar{r}$ to give

$$\alpha \frac{\partial c}{\partial y} = c \qquad (y = 0, 1, \quad \tau > 0) \tag{2.252}$$

which are also equivalent to the boundary conditions in the Mason-Weaver problem. In this way, it is found that an appropriate transcription of the solution of Mason and Weaver for sedimentation under gravity should yield the solution to the rectangular-cell approximation of the Lamm equation with constant s and D subject to the initial condition for the conventional cell. It is to be noted that not only the basic differential equation but also the boundary conditions must be approximated to obtain the result of Yphantis and Waugh.

The rectangular-cell approximation has been used by many investigators. To mention a few, Pasternak et al.[53] applied it to obtain an Archibald-type solution which satisfies the initial condition for the synthetic boundary cell. Weiss and Yphantis[39] showed that the case treated by Pasternak et al. is analytically tractable even if s varies with c according to equation 2.6. Furthermore, Hexner et al.[54] and Klenin et al.[55] investigated by this approximation the case in which the rotor speed is varied in a stepwise fashion during the course of a sedimentation experiment. Some of these studies are discussed in Chapter 7 in relation to the approach to sedimentation equilibrium.

The approximated Lamm equation 2.248 may be written

$$\frac{\partial c}{\partial \tau} = \frac{D}{2s\omega^2}\left(\frac{\partial^2 c}{\partial r^2}\right) - \frac{\bar{r}}{2}\left(\frac{\partial c}{\partial r}\right) \tag{2.253}$$

The corresponding form of the exact Lamm equation is

$$\frac{\partial c}{\partial \tau} = \frac{D}{2s\omega^2}\left(\frac{\partial^2 c}{\partial r^2}\right) - \left(\frac{r}{2} - \frac{D}{2rs\omega^2}\right)\left(\frac{\partial c}{\partial r}\right) - c \tag{2.254}$$

Comparison shows that equation 2.248 is incorrect roughly by an amount

$$\frac{D}{2rs\omega^2}\left(\frac{\partial c}{\partial r}\right) - c$$

If r in the term $2rs\omega^2$ is replaced by $\bar{r}$, this becomes

$$\left(\frac{\alpha}{2}\right)\frac{\partial c}{\partial y} - c$$

which, according to equation 2.252, gives $-c/2$ at $y=0$ and $y=1$. As sedimentation proceeds, the concentration near the meniscus ($y=0$) decreases, while that in the region near the cell bottom ($y=1$) increases appreciably due to continuous accumulation of solutes. Thus the Lamm equation as approximated above should be less accurate in the latter region, especially at later stages of a centrifugation. It also should be noted that inasmuch as this equation is equivalent to the Mason-Weaver equation for the sedimentation in a uniform gravitational field, any solution to it does not exhibit effects associated with the sectorial cell and the radially varying field of centrifugal force.

2. Higher Approximations

Despite the shortcomings stated above, the rectangular-cell approximation is presumably accurate enough for mathematical investigations of the sedimentation behavior in actual centrifuge cells. It is nonetheless of theoretical interest to seek ways in which a solution of higher approximation may be derived. Two such ways, one due to Fujita[56] and the other due to Nazarian,[57] are described briefly in the following.

Fujita's method starts from the Lamm equation written in the form of equation 2.64, with ϵ replaced by ϵ' and with x redefined as $x = (r/r_1)^2$. With the substitution

$$c = c_0 u \exp\left(\frac{x}{\epsilon'}\right) \tag{2.255}$$

this equation becomes

$$\frac{\partial u}{\partial \tau} = \epsilon' x \frac{\partial^2 u}{\partial x^2} + (\epsilon' + x) \frac{\partial u}{\partial x} \tag{2.256}$$

and the initial condition for the conventional cell is written

$$u = \exp\left(\frac{-x}{\epsilon'}\right) \qquad (1 < x < X, \quad \tau = 0) \tag{2.257}$$

where X is defined by

$$X = \left(\frac{r_2}{r_1}\right)^2 \tag{2.258}$$

The boundary conditions 2.4 reduce to

$$\frac{\partial u}{\partial x} = 0 \qquad (x = 1 \text{ and } X, \quad \tau > 0) \tag{2.259}$$

Since, in actual cells, the quantity X is very close to unity, it is reasonable to replace the term x in equation 2.256 by its arithmetic mean (or other suitable means) between the air-liquid meniscus and the cell bottom, that is, $(1 + X)/2$. Then

$$2 \frac{\partial u}{\partial \tau} = \epsilon'(1 + X) \frac{\partial^2 u}{\partial x^2} + (1 + X + 2\epsilon') \frac{\partial u}{\partial x} \tag{2.260}$$

This approximated Lamm equation takes roughly into account the terms which have been lost in the rectangular-cell approximation, and can be integrated, under the conditions 2.257 and 2.259, in terms of elementary functions. It should be noted that no approximation need be introduced into the equations for the boundary conditions, unlike the case of rectangular-cell approximation. The resulting expression for $c(r,t)$ reads

$$\frac{c}{c_0} = \frac{1+a}{a}\left(\frac{e^\mu - e^\nu}{e^\mu - e^{-\nu}}\right)\exp[-\lambda(\lambda - x)]$$

$$+ 2\lambda(X-1)\exp\left[\left(\frac{\lambda}{2}\right)(1-a)(x-1)-(1+a)^2\frac{\tau}{4a}\right]$$

$$\times \sum_{m=1}^{\infty} \frac{[1-(-1)^m e^{-\nu}](m\pi)^2}{[(m\pi)^2+\nu^2][(m\pi)^2+\mu^2]}\left\{\exp\left[-(1+a)^2(m\pi)^2\frac{\tau}{4a\mu^2}\right]\right\}$$

$$\times \left[\cos\frac{m\pi(x-1)}{X-1}+\frac{\mu}{m\pi}\sin\frac{m\pi(m-1)}{X-1}\right]$$
(2.261)

where

$$\mu = \left(\frac{\lambda}{2}\right)(1+a)(X-1)$$

$$\nu = \left(\frac{\lambda}{2}\right)(1-a)(X-1)$$
(2.262)

$$\lambda = \frac{1}{\epsilon'} = \frac{s\omega^2 r_1^2}{2D}$$
(2.263)

$$a = \frac{2\epsilon'}{1+X}$$
(2.264)

Although this solution is quite like the solution of Yphantis and Waugh, equation 2.241, both differ from one another in their details. For example, in equation 2.241, the space variable appears in the form $(r-r_1)/(r_2-r_1)$, while, in equation 2.261, it appears in the form $(r^2-r_1^2)/(r_2^2-r_1^2)$. The numerical behavior of equation 2.261 is considered in Section IV.C.

Nazarian[57] proposed a series expansion method in which a dimensionless quantity β defined by

$$\beta = \frac{r_2 - r_1}{r_2 + r_1} \tag{2.265}$$

was treated as the perturbation parameter. This is an ingenious idea, because this parameter is as small as 0.06 to 0.08, for the cells employed in usual sedimentation velocity experiments. His final expression for $c(r,t)$ associated with the initial condition for the conventional cell is

$$\frac{c}{c_0} = \frac{\exp(y/\alpha)}{\alpha[\exp(1/\alpha)-1]} + \exp\left[\frac{y}{2\alpha} + \frac{\beta}{4\alpha}(2y-1)^2\right]$$

$$\times \sum_{m=1}^{\infty} A_m U_m(y)\exp(-\epsilon_m\gamma\tau) \tag{2.266}$$

where

$$A_m = A_m^{(0)} + \beta A_m^{(1)} + \beta^2 A_m^{(2)} + \cdots \tag{2.267}$$

$$U_m = U_m^{(0)} + \beta U_m^{(1)} + \beta^2 U_m^{(2)} + \cdots \tag{2.268}$$

$$\epsilon_m = \epsilon_m^{(0)} + \beta^2 \epsilon_m^{(2)} + \cdots \tag{2.269}$$

and the other notation is the same as defined in Section IV.B.1. Equation 2.266 reduces to the Yphantis-Waugh solution, equation 2.241, in the limit of $\beta = 0$. Thus the various coefficients multiplied by β in equations 2.267 through 2.269 represent first-order effects due to the sectorial shape of the actual cell and the radially varying field of centrifugal force. The reader is referred to Nazarian's paper for their actual expressions.

C. FEATURES OF ARCHIBALD-TYPE SOLUTIONS

The practical value of the exact as well as approximate solutions of the Archibald type depends on the speed of convergence of the infinite series constituting each of the solutions. Taking, for example, the Yphantis-Waugh solution, the convergence of the series is primarily determined by the magnitude of $\pi^2\alpha\gamma\tau$ appearing in the exponential time factor for $E_m(\tau)$.

This quantity may be written $\pi^2 r_1^2 \epsilon' \tau / 4(r_2 - r_1)^2$. For ordinary centrifuge cells in which r_1 and r_2 are about 6 and 7 cm, respectively, the factor $\pi^2 r_1^2 / 4(r_2 - r_1)^2$ is of the order of 100. Thus the series would converge at reasonably fast rates for values of $\epsilon' \tau$ larger than about 0.01. A similar consideration on Fujita's solution as well as Nazarian's solution leads to a similar lower limit of $\epsilon' \tau$ for the rapid convergence of the series involved.

In such a respect, Archibald-type solutions are sharply contrasted with Faxén-type solutions which hold more accurately as $\epsilon' \tau$ becomes smaller. However, in certain favorable cases of parameters, the two solutions may be continued at a suitable τ, and allow a quick computation of the solute distribution over the entire period of centrifugation.

Some theoretical concentration distributions and concentration gradient distributions calculated from equation 2.261 for a relatively large ϵ' are depicted in Figs. 2.25a and 2.25b. It is seen that at relatively early stages of a centrifugation there appears a maximum gradient, though quite diffuse, close to the air-liquid meniscus, but with increasing time the peak is appreciable flattened and eventually disappears. At this stage, the distribution curves are already not very much different from that at sedimentation equilibrium. Except for the one at $\tau = 0.01$, the gradient curves do not

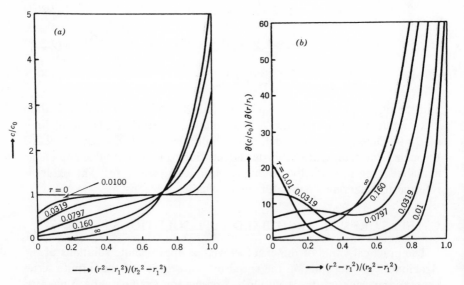

Fig. 2.25. (a) Theoretical concentration gradient curves for a system characterized by a relatively large value of ϵ' ($= 0.058211$); the following values are assigned in the computation: $r_1 = 6.067$ cm, $r_2 = 7.003$ cm, $\omega^2 = 3.919 \times 10^7$ rad^2/sec^2, $s = 0.565 \times 10^{-13}$ sec, $D = 2.372 \times 10^{-6}$ cm^2/sec. (b) Theoretical concentration gradient distributions corresponding to (a).

return to the base line ahead of the maxima. In other words, in this system, the plateau region does not persist long enough after initiation of centrifugation. Another point of interest in Fig. 2.25a is that the concentration distribution curves for different τ appear to rotate about a single point corresponding to the initial concentration. This fact was first noticed by Archibald[26] and later confirmed by Waugh and Yphantis.[50] Archibald showed that the location of the point in question, which is often called the *hinge point*, provides a measure of the molecular weight of the solute, and Ginsburg et al.[58] verified it experimentally with the sedimentation of sucrose.

V. PRESSURE EFFECTS

A. INTRODUCTION

At speeds of rotation usually employed for sedimentation velocity measurements, a large pressure difference, which amounts to several hundred atmospheres, is set up for a short distance of about 1 cm between the air-liquid meniscus and the cell bottom. Since the viscosity and density of the solution (actually those of pure solvent under the usual experimental conditions, in which dilute solutions are examined) as well as the partial specific volumes of the components may vary with pressure, the rate of sedimentation in such a field of high-pressure gradient should differ from that in a field of uniform pressure. In order to attain high precision in the determination of sedimentation coefficient s, it is necessary to investigate the magnitude of this difference and the steps that should be taken to correct the apparent values of s obtained by the conventional peak method.

This problem was first considered by Mosimann and Signer,[59] and examined more specifically by several subsequent investigators. It is generally assumed that the pressure effect may be ignored for aqueous solutions of macromolecules, because water is far less compressible than organic solvents. It is therefore not surprising that this problem has received a great deal of attention from investigators in polymer chemistry.

Pressure effects on sedimentation boundary were first investigated by Fujita,[60] who used the Lamm equation with no diffusion term after correcting the sedimentation coefficient for a pressure dependence. Subsequently, Wales[61] and Billick[62] discussed Fujita's equation from different points of view, but a study of the equation involving the diffusion term had to wait recent papers by Dishon and collaborators.[63-65] This section describes some important results from these studies, with emphasis on the ways in which the sedimentation rates in pressure-dependent binary solutions are corrected to the state of 1 atm. Before moving ahead it is important to note that, as mentioned in Chapter 1, the validity of the Lamm equation, on which all the reported theories of pressure-dependent sedimentation are based, is debatable for compressible systems.

B. BASIC EQUATIONS

Let us denote the sedimentation coefficient at a pressure P (measured from 1 atm) by s^P and, according to Oth and Desreux,[66] write it in the form*

$$s^P = s^0(1 - \gamma P) \qquad (2.270)$$

where s^0 denotes the value of s^P at 1 atm, and γ is a parameter independent of P. The quantity s^0 is the one usually sought. In general, it must be treated as a function of the solute concentration. Thus

$$s^0 = s_0^0 f(c) \qquad (2.271)$$

where s_0^0 is the limiting sedimentation coefficient reduced to 1 atm. As will be shown later, the parameter γ can be expressed approximately in terms of the compressibility of the solvent, that of the solute, and the pressure derivative of the solvent viscosity. Except for unusual systems, it may be assumed to be very small.

For a pressure-dependent system the continuity equation, the Lamm equation in the present case, must be coupled with an equation which governs the distribution of pressure in the system. For the ultracentrifuge the appropriate pressure equation is equation 1.69, that is,

$$\frac{\partial P}{\partial r} = r\omega^2 \rho \qquad (2.272)$$

where ρ, the density of the solution at a particular radial position r, may be approximated by the density of the pure solvent, ρ_0, at the same position for very dilute solutions with which we are concerned below. A reasonable first approximation to ρ_0 as a function of P is

$$\rho_0 = \rho_0^0(1 + \beta P) \qquad (2.273)$$

where ρ_0^0 is the value at 1 atm, that is, the one at the air-liquid meniscus of the solution column, and β is the isothermal compressibility of the solvent

---

*This representation of s^P may be considered as a first approximation to a more accurate series expression as $s^P = s^0(1 - \gamma_1 P + \gamma_2 P^2 + \cdots)$.

(at 1 atm). Substitution of equation 2.273 into equation 2.272, followed by integration, gives

$$P = \frac{1}{\beta}\{\exp[\nu(x-1)]-1\} \tag{2.274}$$

where

$$x = \left(\frac{r}{r_1}\right)^2 \tag{2.275}$$

$$\nu = \tfrac{1}{2}\beta\rho_0{}^0\omega^2 r_1{}^2 \tag{2.276}$$

For this typical case in which $\omega^2 = 4 \times 10^7$ (rad/sec)^2, $r_1 = 6.0$ cm, $r_2 = 7.0$ cm, $\rho_0{}^0 = 0.792$ g/ml (an organic solvent), and $\beta = 82 \times 10^{-6}$ atm^{-1}, equation 2.274 predicts a pressure of about 200 atm at the cell bottom. In general, ν is so small that this equation may be replaced, with sufficient accuracy, by

$$P = \tfrac{1}{2}\omega^2 r_1{}^2 \rho_0{}^0(x-1) \tag{2.277}$$

Introduction into equation 2.270 yields

$$s^P = s_0{}^0 f(c)[1 - m(x-1)] \tag{2.278}$$

where equation 2.271 has been inserted, and m is a dimensionless parameter defined by

$$m = \tfrac{1}{2}\gamma\omega^2 r_1{}^2 \rho_0{}^0 \tag{2.279}$$

Equation 2.278 has been the basis for most of the theoretical investigations undertaken so far concerning the effects of hydrostatic pressure on sedimentation rates. It is to be noted that the sedimentation coefficient is now an explicit function of the cell coordinate. This fact makes the integration of the ultracentrifuge equation more difficult than otherwise.

Let us consider a boundary sedimentation experiment subject to the initial condition for the conventional cell. It is assumed that the diffusion coefficient D may be treated as independent of both solute concentration c and pressure P. If equation 2.278 is substituted for s and the reduced variables defined by

$$\theta = \frac{c}{c_0}, \qquad x = \left(\frac{r}{r_1}\right)^2, \qquad \tau = 2s_0{}^0\omega^2 t$$

$$\epsilon = \frac{2D}{s_0{}^0\omega^2 r_1{}^2}, \qquad F(\theta) = f(c_0\theta) \tag{2.280}$$

are used, then the Lamm differential equation assumes the form

$$\frac{\partial \theta}{\partial \tau} = \frac{\partial}{\partial x}\left\{ x\left[\epsilon \frac{\partial \theta}{\partial x} - (1 + m - mx)F(\theta)\theta \right] \right\} \qquad (2.281)$$

When $D = 0$, that is, $\epsilon = 0$, this reduces to

$$\frac{\partial \theta}{\partial \tau} = - \frac{\partial}{\partial x}[x(1 + m - mx)F(\theta)\theta] \qquad (2.282)$$

which is the equation first treated by Fujita[60] in a study of pressure-dependent sedimentation. So far the following two cases of $f(c)$ have been investigated[60-63]:

$$f(c) = 1 \qquad \text{and} \qquad f(c) = \frac{1}{1 + k_s c} \qquad (k_s > 0) \qquad (2.283)$$

which correspond, respectively, to

$$F(\theta) = 1 \qquad \text{and} \qquad F(\theta) = \frac{1}{1 + \alpha\theta} \qquad (\alpha = k_s c_0 > 0) \qquad (2.284)$$

C. IMPORTANT FEATURES OF SEDIMENTATION BOUNDARY

Equation 2.282 can be integrated by use of the method of characteristics. The solution for the case $F(\theta) = 1$ subject to the initial condition for the conventional cell is given by

$$\theta(x, \tau) = \frac{(1 + m)^2 e^{-(1+m)\tau}}{\left[1 + m - mx(1 - e^{-(1+m)\tau}) \right]^2}$$

$$\times H\left(\frac{xe^{-(1+m)\tau} - (1 + m - mx)}{1 + m - mx(1 - x^{-(1+m)\tau})} \right) \qquad (2.285)$$

where $H(y)$ is a step function defined such that $H(y) = 0$ for $y < 0$ and $H(y) = 1$ for $y > 0$. From this it follows that the reduced concentration θ undergoes a discontinuous jump from zero to a value

$$\theta_* = e^{(1+m)\tau}\left(\frac{m + e^{-(1+m)\tau}}{1 + m} \right)^2 \qquad (2.286)$$

at a position on the x-axis, x_*, given by

$$x_* = \left(\frac{r_*}{r_1}\right)^2 = \frac{m+1}{m+e^{-(1+m)\tau}} \qquad (2.287)$$

It is evident that θ_* ultimately increases with τ. However, it can be verified that for $m < 1$ this value initially decreases as a function of τ until a time defined by $\tau = (1+m)^{-1}\ln(1/m)$, after which it begins to increase. Another important feature is that, differing from the concentration profiles in pressure-independent systems (see Section II), the concentrations in the region ahead of the discontinuous boundary between solvent and solution are not constant but increase with distance. Thus there exists no plateau region. However, this feature may not be taken as characteristic of a pressure-dependent system, as has been shown by Dishon et al.[63]

The concentration profile for the case $F(\theta) = 1/(1+\alpha\theta)$ is also represented by a step function. It has been shown by Billick[62] that the position of the step on the x-axis, x_*, is given as a function of τ by

$$\frac{\ln x_*}{\tau} = \frac{1}{1+\alpha} - \frac{(2\alpha+1)m-\alpha}{2(\alpha+1)^3}\tau + \frac{1}{6(\alpha+1)^5}$$

$$\times [\alpha(2\alpha-1) - m(10\alpha^2+5\alpha+1) + m^2(6\alpha^2+4\alpha+1)]\tau^2 + \cdots \qquad (2.288)$$

Although Billick originally regarded this as an approximation, Dishon et al.[63] have proved that it is actually exact. Fujita[60] derived an exact relation from which $\theta(x,\tau)$ for $x > x_*$ may be computed, but which requires the solution of a complicated transcendental equation. Dishon et al.[63] used a different approach in which the desired θ was expanded in powers of τ as

$$\theta(x,\tau) = 1 + \beta_1(x)\tau + \beta_2(x)\tau^2 + \cdots \qquad (x > x_*) \qquad (2.289)$$

and then the functions $\beta_1(x), \beta_2(x), \ldots$ were determined from the basic differential equation 2.282. The expressions for the first two $\beta_i(x)$ are

$$\beta_1(x) = -\frac{1+m-2mx}{1+\alpha} \qquad (2.290)$$

$$\beta_2(x) = \frac{(1+m)^2 - 6m(1+m)x + 6m^2x^2}{2(1+\alpha)^3} \qquad (2.291)$$

In this case, the concentrations beyond the discontinuous boundary between solvent and solution are not necessarily increasing monotonically with distance.

When diffusion is present, that is, $D \neq 0$ or $\epsilon \neq 0$, integration of the basic equation 2.281 becomes exceedingly difficult. Dishon et al.[64] have obtained for the case $F(\theta) = 1$ an approximate solution of the Faxén type, but it is not shown here. Instead we diagrammatically reproduce the concentration gradient distributions calculated by Dishon et al.[63] for two cases, in one of which $F(\theta) = 1$, $m = 1$, and $\sigma = s_0^0 \omega^2 / D = 23.49$ cm^{-2} (Fig. 2.26), and in the other $F(\theta) = 1/(1 + 0.5\theta)$, $m = 0.6$, and $\sigma = 23.49$ cm^{-2} (Fig. 2.27), from numerical solutions to the basic equation subject to the initial and boundary conditions for the actual cell; the sharp upturns of the gradient curves at the extreme right indicate the effects of restricted diffusion near the cell bottom. In the graphs, each of the dashed lines shows the boundary position predicted from the diffusion-free theory. For the case $F(\theta) = 1$ its agreement with the maximum of the corresponding gradient curve is surprisingly good, whereas for the case $F(\theta) = 1/(1 + 0.5\theta)$ it shifts to the right of the maximum gradient as sedimentation proceeds, but the deviations are almost negligible until the maximum travels more than half the solution column. In Fig. 2.26, the concentration gradient beyond the peak region does not return to the base line, that is, there exists no plateau region, which conforms to the prediction from the diffusion-free theory. On the other hand, each of the gradient curves shown in Fig. 2.27 has a plateau region. This feature is attributed to a compensation of the pressure effect on s by the concentration effect on s, and is consistent with an experimental observation by Billick.[67]

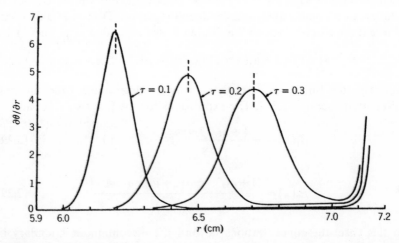

Fig. 2.26. Concentration gradient distributions in a pressure-dependent system in which $F(\theta) = 1$, $m = 1$, and $\sigma = 23.49$ cm^{-2}.

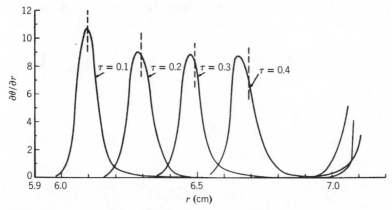

Fig. 2.27. Concentration gradient distributions in a pressure-dependent system in which $F(\theta)=1/(1+0.5\theta)$, $m=0.6$, and $\sigma=23.49$ cm^{-2}.

D. CORRECTION OF THE APPARENT SEDIMENTATION COEFFICIENT FOR PRESSURE: THE BILLICK-FUJITA METHOD

The theoretical results described above suggest that even when a given binary solution undergoes pressure effects, the position of the maximum gradient may be identified as the boundary position r_* for diffusion-free sedimentation if the gradient curve is not too asymmetric and, moreover, if the pressure dependence of s is not too pronounced. This suggestion makes it possible to determine the value of $\ln(r_*/r_1)$ as a function of time t from a series of boundary gradient curves photographed during the course of a boundary sedimentation experiment. From the data obtained a plot of s_{app} versus $[(r_*/r_1)^2-1]$ may be constructed. Here s_{app} is an apparent sedimentation coefficient defined by

$$s_{app}=\frac{\ln(r_*/r_1)}{\omega^2 t} \tag{2.292}$$

If the $f(c)$ of the system is assumed to obey the relationship $f(c)=1/(1+k_s c)$, a theoretical expression for this type of plot can be derived from equation 2.288 in the following way. First, this equation is solved for τ as a power series in x_*-1. The result reads

$$\tau=A_1(x_*-1)+A_2(x_*-1)^2+\cdots \tag{2.293}$$

where

$$A_1=1+\alpha \tag{2.294}$$

$$A_2=\tfrac{1}{2}(1+2\alpha)(m-1) \tag{2.295}$$

If equation 2.293 is substituted back into the right-hand side of equation 2.288 and the defining expression for s_{app} is considered, then

$$s_{app} = s^0(c_0)\left\{1 + B_1\left[\left(\frac{r_*}{r_1}\right)^2 - 1\right] - B_2\left[\left(\frac{r_*}{r_1}\right)^2 - 1\right]^2 + \cdots\right\} \quad (2.296)$$

where

$$B_1 = \frac{k_s c_0 - m(1 + 2k_s c_0)}{2(1 + k_s c_0)} \quad (2.297)$$

$$B_2 = \frac{1}{12(1 + k_s c_0)^2}\left\{k_s c_0(5 + 2k_s c_0)\right.$$

$$\left. + m\left[2(k_s c_0)^2 - 5k_s c_0 - 1\right] + m^2(1 + 4k_s c_0)\right\} \quad (2.298)$$

Equation 2.296 is the desired expression for the plot in question. When there is no concentration dependence of s^0, this reduces to

$$s_{app} = s_0^0\left\{1 - \frac{m}{2}\left[\left(\frac{r_*}{r_1}\right)^2 - 1\right] + \left(\frac{m - m^2}{12}\right)\left[\left(\frac{r_*}{r_1}\right)^2 - 1\right]^2 + \cdots\right\} \quad (2.299)$$

From equation 2.296 it follows that for given c_0 and ω, a plot of s_{app} versus $(r_*/r_1)^2 - 1$ gives $s^0(c_0)$ as its ordinate intercept. It is also seen that the initial slope of the plot can be used to find the value of the parameter B_1. A series of values of s^0 and B_1 may be obtained from these plots at differing initial concentrations and a fixed rotor speed. The s^0 data may then be analyzed to evaluate s_0^0 and k_s by plotting them in the form of $1/s_0$ versus c_0. If the assumed form of s^0 as a function of c is obeyed, this plot should be linear. Once k_s is known, the parameter m can be computed pointwise from the data obtained for B_1. The resulting values ought to be constant within experimental accuracy, and their average may be used to calculate the parameter γ from equation 2.279. This method of analysis for binary solutions was developed by Billick[62] and independently by Fujita.[68]

1. The Practical Method of Blair and Williams

In practical application of the Billick-Fujita method, correction must be made for two effects which have been ignored in the theory. One concerns the fact that a certain finite time is needed to bring the rotor from rest to the desired operational speed. The other is the restricted diffusion caused by the presence of the air-liquid meniscus. Usually, both are empirically taken into account by shifting the origin of the time variable t to some value t_0, which is often referred to as the zero time correction. Then the more appropriate definition of s_{app} would be

$$s_{app} = \frac{\ln(r_*/r_1)}{\omega^2(t - t_0)} \tag{2.300}$$

where t is measured from the moment the rotor is set in motion. It should be noted that the s_{app} defined by this relation cannot be determined directly from experiment because the zero time correction is unknown, in general.

To proceed with this s_{app} we ignore the third and higher terms inside the braces of equation 2.296 and substitute equation 2.300 for s_{app} on the left-hand side. Then

$$\frac{\ln(r_*/r_1)}{\omega^2(t - t_0)} = s^0(c_0) \left\{ 1 + B_1 \left[\left(\frac{r_*}{r_1} \right)^2 - 1 \right] \right\} \tag{2.301}$$

This equation contains three unknown quantities: t_0, s^0, and B_1. For a system in which the terms neglected are of minor importance the choice of t_0 may be made in such a way that plots of $[\ln(r_*/r_1)]/\omega^2(t - t_0)$ versus $(r_*/r_1)^2 - 1$ become linear over as wide a range of the abscissa as possible. Among several methods proposed, it appears that the "best fit" one of Blair and Williams[69] is the most appropriate. This method aims at determining t_0 so that the standard deviation σ, defined by

$$\sigma^2 = n^{-1} \sum_{i=1}^{n} R_i^2 \tag{2.302}$$

is minimized, where

$$R_i = s^0(c_0)\omega^2(t_i - t_0) \left\{ 1 + B_1 \left[\frac{(r_*)_i^2}{r_1^2} - 1 \right] \right\} - \ln \left[\frac{(r_*)_i}{r_1} \right] \tag{2.303}$$

and where n is the number of observations taken in a single sedimentation velocity run. The quantity $(r_*)_i$ is the position of the maximum gradient for the boundary curve measured at $t = t_i$.

The actual operation may be made in the following way. Choose an appropriate value for t_0, calculate the quantity on the left-hand side of equation 2.301 from primary experimental data for r_* as a function of t, and then plot the resulting values against $(r_*/r_1)^2 - 1$. The plots obtained may not follow a straight line, but there will be a linear portion in the region of relatively large values of the abscissa. Determine $s^0(c_0)$ and B_1 by applying the right-hand side of equation 2.301 to this linear portion, and compute σ by substituting the resulting values and the assumed t_0 value into equation 2.302. Repeat similar operations for other t_0 values chosen suitably and graph the resultant σ values against the corresponding t_0. The desired t_0 may then be found by inspection of a curve of σ versus t_0 thus obtained. A high-speed computer should be of effective use for a performance of the whole process.[70]

Figure 2.28 illustrates the Blair-Williams method applied to data of Abe et al.[71] for a narrow-distribution sample of poly-α-methylstyrene in cyclohexane at 39°C. In the insert are shown the values of σ for four chosen t_0; the open circle indicates the best-fit value of t_0 for this system. The corresponding values of $s_{app} = [\ln(r_*/r_1)]/\omega^2(t - t_0)$, also shown by open circles, are seen to vary linearly with $(r_*/r_1)^2 - 1$. As can be observed, the slope of this type of plot is quite sensitive to the choice of t_0, which implies

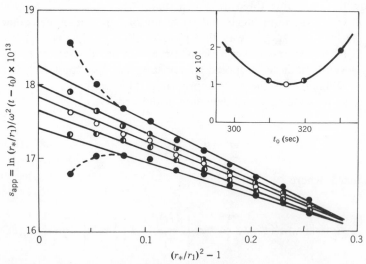

Fig. 2.28. Application of the Blair-Williams procedure to a poly-α-methylstyrene sample in cyclohexane at 39°C.[71] The five lines indicated correspond to different choices of t_0 value. The insert shows variation of the standard deviation with the assumed values of t_0.

that the determination of B_1 suffers from lack of precision. Abe et al., after averaging 64 values of the parameter γ which resulted from similar analyses of the data for five samples of widely different molecular weight at various rotor speeds and initial concentrations, obtained for poly-α-methylstyrene in cyclohexane at 39°C

$$\gamma = (1.6 \pm 0.2) \times 10^{-9} \ cm^2/dyne$$

They state that this value is in close agreement with those reported on polystyrene in cyclohexane by Billick,[67] Wales and Rehfeld,[72] Blair and Williams,[69] and Cowie and Bywater.[73] However, it disagrees with an average value of 2.3×10^{-9} cm^2/dyne obtained by Noda et al.[70] for the system poly-α-methylstyrene-cyclohexane at a comparable temperature. This disparity proves the difficulty of attaining high accuracy in the determination of B_1.

2. Remarks

1. In previous sections, when concentration-dependent systems were discussed, it was stated that a method useful for accurate determination of the initial tangent of a plot for $\ln r_*$ or $\ln r_M$ versus $\omega^2 t$ would be deferred to this section. The procedure described in Section V.D.1 is indeed the desired method, because extrapolating s_{app} to $(r_*/r_1)^2 - 1 = 0$ is equivalent to determining the desired initial slope.

2. The Blair-Williams method is restricted to the case in which terms higher than the first power of $(r_*/r_1)^2 - 1$ can be ignored from equation 2.296. This neglect is presumably permissible for most cases of practical interest, because the $(r_*/r_1)^2$ are quite close to unity at all points of the solution column in actual centrifuge cells. For a system in which the pressure dependence of s is so pronounced that the parameter m becomes quite large the B_2 term at least will have to be added to equation 2.301. However, it is doubtful whether equation 2.270 still gives a good approximation for such a system.

3. If one assumes for the friction coefficient f_0 and partial specific volume $\bar{v}_1$ of the solute at infinite dilution

$$f_0 = f_0^0 (1 + \lambda P) \tag{2.304}$$

$$\bar{v}_1 = \bar{v}_1^0 (1 - \varphi P) \tag{2.305}$$

substitutes these, with $s_0 = s_0^0 (1 - \gamma P)$ and $\rho_0 = \rho_0^0 (1 + \beta P)$, into equation 2.8 (with $\bar{v}_1^0$ in the equation being taken as $\bar{v}_1$), and expands the resulting expression in powers of P, there results from a comparison of the terms of the first power in P[60]

$$\gamma = \lambda + \frac{\bar{v}_1^{\,0}\rho_0^{\,0}}{1 - \bar{v}_1^{\,0}\rho_0^{\,0}}(\beta - \varphi) \qquad (2.306)$$

This relation permits an estimate of γ when the data for λ, β, and φ are available.

By analogy with Stokes' law it may be assumed that f_0 is approximately proportional to the viscosity of the solvent η. Then the pressure coefficient of f_0, that is, λ, may be computed from

$$\lambda = \left(\frac{1}{\eta^0}\frac{d\eta}{dP}\right)_{P=0} = \frac{(\eta/\eta^0) - 1}{P} \qquad (2.307)$$

where η^0 is the value of η at 1 atm. The coefficient φ may be approximated by the reciprocal bulk modulus of the solute, but, in general, it is much smaller than the coefficient β and may be ignored in equation 2.306. Values of β and λ for a number of typical organic liquids and water were calculated by Baldwin and Van Holde.[74] Billick[67] computed γ for polystyrene in cyclohexane and Noda et al.[70] that for poly-α-methylstyrene in cyclohexane from equation 2.306. The results were 2 and 1.96 in 10^{-9} $cm^2/dyne$, respectively. While these values are subject to uncertainty because λ is known only roughly for cyclohexane, the agreement with the above-mentioned values from sedimentation data may be considered quite satisfactory.

4. If it is assumed that the rotor speed is accelerated linearly from rest to the desired operational value, it can be shown for diffusion-free sedimentation that the zero time correction t_0 is given by[75]

$$t_0 = \tfrac{2}{3}t_f \qquad (2.308)$$

where t_f denotes the time until the rotor reaches full speed. Customarily, this t_0 is called Trautman's zero time correction or the time for two-thirds acceleration. In one experiment with a sample of polystyrene in cyclohexane (final rotor speed 59,780 rev/min, $c_0 = 0.313$ g/dl, temperature 34.2°C), Blair and Williams[69] found $t_0 = 306$ sec from the best-fit method, and obtained $t_0 = 234$ sec from equation 2.308. The difference may be accounted for, if not all, by the restricted diffusion near the air-liquid meniscus. It might also be due to the fact that the acceleration rate of the rotor was not constant.

REFERENCES

1. H. Faxén, *Arkiv Mat. Astron. Fysik* **21B**, 1 (1929).
2. W. J. Archibald, *Phys. Rev.* **53**, 746 (1938).
3. W. J. Archibald, *Phys. Rev.* **54**, 371 (1938).
4. W. J. Archibald, *Ann. N. Y. Acad. Sci.* **43**, 211 (1942).
5. G. Kegeles, *J. Am. Chem. Soc.* **74**, 5532 (1952).
6. E. G. Pickels, W. F. Harrington, and H. K. Schachman, *Proc. Natl. Acad. Sci., U.S.* **38**, 943 (1952).
7. H. K. Schachman and W. F. Harrington, *J. Polymer Sci.* **12**, 379 (1954).
8. R. T. Hersh and H. K. Schachman, *J. Am. Chem. Soc.* **77**, 5228 (1955).
9. H. E. Daniels, *J. Inst. Math. Appl.* **3**, 406 (1967).
10. For the concentration dependence of sedimentation coefficients, see, for example, H. K. Schachman, *Ultracentrifugation in Biochemistry*, Academic Press, New York, 1959.
11. J. Vinograd and R. Bruner, *Biopolymers* **4**, 131 (1966).
12. H. A. Ende, in *Polymer Handbook* (J. Brandrup and E. H. Immergut, Eds.), Interscience, New York, 1966, IV-1.
13. For example, C. Tanford, *Physical Chemistry of Macromolecules*, Wiley, New York, 1961, Chap. 6; H. Yamakawa, *Modern Theory of Polymer Solutions*, Harper & Row, New York, 1971, Chap. 6.
14. H. C. Thomas, *J. Am. Chem. Soc.* **66**, 1664 (1944); *Ann. N. Y. Acad. Sci.* **49**, 161 (1949).
15. S. Goldstein, *Proc. Roy. Soc. (London)* **A219**, 151 (1953).
16. N. K. Hiester and T. Vermeulen, *Chem. Eng. Prog.* **48**, 505 (1952).
17. A. Opler and N. K. Hiester, *Tables for Predicting the Performance of Fixed Bed Ion Exchange and Similar Mass Transfer Processes*, Stanford Research Inst., Stanford, California, 1954.
18. K. Kawahara, *Biochemistry* **8**, 2551 (1969).
19. L. J. Gosting, *J. Am. Chem. Soc.* **74**, 1548 (1952).
20. O. Lamm, *Z. Phys. Chem.* **A143**, 177 (1929).
21. H. Fujita, *J. Chem. Phys.* **24**, 1084 (1956).
22. For example, H. S. Carslaw and J. C. Jaeger, *Heat Conduction in Solids*, Oxford University Press, London and New York, 1947.
23. H. Faxén, *Arkiv Mat. Astron. Fysik* **25B**, 1 (1936).
24. H. Fujita and V. J. MacCosham, *J. Chem. Phys.* **30**, 291 (1959).
25. D. A. Yphantis, *J. Phys. Chem.* **63**, 1742 (1959).
26. W. J. Archibald, *J. Phys. Colloid Chem.* **51**, 1204 (1947).
27. M. K. Brakke, *Arch. Biochem. Biophys.* **45**, 275 (1953). For a simple account of the band sedimentation experiment with a preparative centrifuge see, for example, K. E. Van Holde, *Physical Biochemistry*, Prentice-Hall, Englewood Cliffs, N.J., 1971, pp. 114–116.
28. J. Vinograd, R. Bruner, R. Kent, and J. Weigle, *Proc. Natl. Acad. Sci., U.S.* **49**, 902 (1963).
29. J. Vinograd, R. Rabloff, and R. Bruner, *Biopolymers* **3**, 481 (1965).
30. J. Rosenbloom and V. N. Schumaker, *Biochemistry* **2**, 1206 (1963).

31. J. Vinograd and R. Bruner, *Biopolymers* **4**, 157 (1966).
32. S. Hanlon, K. Lamers, G. Lauterback, R. Johnson, and H. K. Schachman, *Arch. Biochem. Biophys.* **99**, 157 (1962); K. Lamers, F. Putney, I. Z. Steinberg, and H. K. Schachman, *ibid.* **103**, 379 (1963); H. K. Schachman and S. J. Edelstein, *Biochemistry* **5**, 2681 (1966).
33. M. Gehatia and E. Katchalski, *J. Chem. Phys.* **30**, 1334 (1959).
34. M. M. Rubin and A. Katchalsky, *Biopolymers* **4**, 579 (1966).
35. V. N. Schumaker and J. Rosenbloom, *Biochemistry* **4**, 1005 (1965).
36. J. Crank, *The Mathematics of Diffusion*, Oxford University Press, London and New York, 1956.
37. L. J. Gosting and H. Fujita, *J. Am. Chem. Soc.* **79**, 1359 (1957); see also L. J. Gosting, *Adv. Protein Chem.* **11**, 429 (1956).
38. I. H. Billick and G. H. Weiss, *J. Res. Natl. Bur. Stand.*, **70A**, 17 (1966); see also, M. Dishon and G. H. Weiss, *ibid.* **70B**, 95 (1966).
39. G. H. Weiss and D. A. Yphantis, *J. Chem. Phys.* **42**, 2117 (1965).
40. M. Dishon, G. H. Weiss, and D. A. Yphantis, *Biopolymers* **4**, 449 (1966).
41. M. Dishon, G. H. Weiss, and D. A. Yphantis, *Biopolymers* **4**, 457 (1966).
42. M. Dishon, G. H. Weiss, and D. A. Yphantis, *Biopolymers* **5**, 697 (1967).
43. R. L. Baldwin, *Biochem. J.* **65**, 503 (1957); see also H. Fujita, *J. Phys. Chem.* **63**, 1092 (1959); K. E. Van Holde, *ibid.* **64**, 1582 (1960).
44. J. M. Creeth, *Proc. Roy. Soc. (London)* **A282**, 403 (1964).
45. R. L. Baldwin, *J. Phys. Chem.* **58**, 1081 (1954).
46. R. J. Goldberg, *J. Phys. Chem.* **57**, 194 (1953).
47. H. Gutfreund and A. G. Ogston, *Biochem. J.* **44**, 163 (1949).
48. R. L. Baldwin, *Biochem. J.* **55**, 644 (1953).
49. D. A. Yphantis and D. F. Waugh, *J. Phys. Chem.* **60**, 623 (1956); **60**, 630 (1956).
50. D. F. Waugh and D. A. Yphantis, *J. Phys. Chem.* **57**, 312 (1953).
51. W. J. Archibald, *J. Appl. Phys.* **18**, 362 (1947).
52. M. Mason and W. Weaver, *Phys. Rev.* **23**, 412 (1924); see also W. Weaver, *Phys. Rev.* **27**, 499 (1926).
53. R. A. Pasternak, G. M. Nazarian, and J. R. Vinograd, *Nature* **179**, 92 (1957).
54. P. E. Hexner, L. E. Radford, and J. W. Beams, *Proc. Natl. Acad. Sci., U.S.* **47**, 1848 (1961).
55. S. I. Klenin, H. Fujita, and D. Albright, *J. Phys. Chem.* **70**, 946 (1966).
56. H. Fujita, upublished work.
57. G. M. Nazarian, *J. Phys. Chem.* **62**, 1607 (1958).
58. A. Ginsburg, P. Appel, and H. K. Schachman, *Arch. Biochem. Biophys.* **65**, 545 (1956).
59. H. Mosimann and R. Signer, *Helv. Chim. Acta* **27**, 1123 (1944).
60. H. Fujita, *J. Am. Chem. Soc.* **78**, 3598 (1956).
61. M. Wales, *J. Am. Chem. Soc.* **81**, 4758 (1959).
62. I. H. Billick, *J. Phys. Chem.* **66**, 565 (1962).
63. M. Dishon, G. H. Weiss, and D. A. Yphantis, *J. Polymer Sci. A-2* **8**, 2163 (1970).

64. M. Dishon, G. H. Weiss, and D. A. Yphantis, *Ann. N. Y. Acad. Sci.* **164**, 33 (1969).
65. G. H. Weiss and M. Dishon, *Biopolymers* **9**, 865 (1970).
66. J. Oth and V. Desreux, *Bull. Soc. Chim. Belges* **63**, 133 (1954).
67. I. H. Billick, *J. Phys. Chem.* **66**, 1941 (1962).
68. H. Fujita, *Mathematical Theory of Sedimentation Analysis*, Academic Press, New York, 1962, Chap. II.
69. J. E. Blair and J. W. Williams, *J. Phys. Chem.* **68**, 161 (1964).
70. I. Noda, S. Saito, T. Fujimoto, and M. Nagasawa, *J. Phys. Chem.* **71**, 4048 (1967); also A. Soda, T. Fujimoto, and M. Nagasawa, *ibid.* **71**, 4274 (1967).
71. M. Abe, K. Sakato, T. Kageyama, M. Fukatsu, and M. Kurata, *Bull. Chem. Soc. Japan* **41**, 2330 (1968).
72. M. Wales and S. J. Rehfeld, *J. Polymer Sci.* **62**, 179 (1962).
73. J. M. G. Cowie and S. Bywater, *Polymer* **6**, 197 (1965).
74. R. L. Baldwin and K. E. Van Holde, *Fortschr. Hochpolym. Forsch.* **1**, 451 (1960).
75. R. Trautman, *J. Phys. Chem.* **60**, 1211 (1956).

SEDIMENTATION TRANSPORT IN MULTICOMPONENT SYSTEMS

I. INTRODUCTION

A. THE SCOPE OF THIS CHAPTER

For theoretical discussions of solutions which contain more than one solute component it is convenient to classify an almost indefinite variety of such solutions into two categories usually referred to as *paucidisperse*[1] and *polydisperse*. In this monograph, a solution is called paucidisperse if it contains a small number of molecularly homogeneous macromolecular solutes, regardless of whether the solvent consists of a single pure liquid or contains a few low-molecular-weight solutes or is a mixture of two or more liquid compounds. If the number of such macromolecular solutes is indefinitely large, the solution is called polydisperse. Of course, this mode of classification is a mere convention, because there is a continuous spectrum between these two extremes.

In general, biochemical preparations are considered to be molecularly homogeneous or nearly so. Hence a mixture of several of them is typically paucidisperse. Proteins often tend to aggregate with each other to form oligomers or dissociate into subunits. Therefore it is to be noted that a protein solution cannot always be treated as composed of a single solute if the dissolved protein preparation is considered to be pure. Biochemists and molecular biologists are primarily interested in utilizing the ultracentrifuge for separating paucidisperse macromolecular solutions into component solutes and estimating the concentrations and sedimentation rates of the separated components. Section II is devoted to a discussion of the methods which have been developed for this type of sedimentation analysis.

For polydisperse solutions the methods given in Section II become ineffective, and we must abandon the desire of separating and characterizing individual solute components from observed boundary curves. Solutions of synthetic (homo)polymers are the best known examples of polydisperse systems. Any sample of this class of macromolecular substance is a mixture of an almost infinite number of long-chain molecules which are chemically indistinguishable but heterogeneous with respect to physical characteristics such as molecular weight, microstructure, and so forth. Hence when dissolved in a solvent, it gives a great number of polymer solutes differing, in the simplest case, only in molecular weight. The idea of utilizing the ultracentrifuge as a tool for determining the distributions of molecular weights in polymeric substances was recognized early by several pioneers in this field, but its substantial developments have occurred during the past two decades. Section III is concerned with a description of important contributions which have established current methods of ultracentrifugal analysis of polydisperse solutions.

As noted in Chapter 2, it is possible, according to Archibald, to evaluate the molecular weight of a solute from measurements of the concentration and concentration gradient at either end of the solution column. A detailed theory of this Archibald method is presented in Section IV on the basis of its recent extensions to multicomponent solutions.

Theoretical treatments of sedimentation phenomena in multicomponent systems involve a number of difficulties which are not encountered in those of binary solutions. Most of them arise from molecular interactions, either thermodynamic or hydrodynamic in origin, between different macromolecular solutes. As is often the case with biochemical preparations, when the solvent contains low-molecular-weight solutes, such as simple electrolytes, urea, guanidinium chloride, and the like, there arise additional complications from interactions of the macromolecular solutes with these added small solutes. To proceed under such a variety of difficulties we are forced to introduce various assumptions and approximations even if they appear to be quite drastic. As a consequence, the present state of our knowledge on the theory of sedimentation processes in multicomponent solutions is as yet far from satisfactory in many aspects.

B. EXPRESSIONS FOR REFRACTIVE INDEX AND ITS GRADIENT

It is the current practice to use the schlieren method for sedimentation velocity measurements and the Rayleigh interference method for sedimentation equilibrium studies, although, in biochemistry and molecular biology, there is an increased use of an improved light-absorption method for both velocity and equilibrium experiments.[2] The schlieren method enables us to measure the distribution of refractive index gradients along the solution column in the ultracentrifuge cell, while the Rayleigh interference method gives the corresponding distribution of refractive indices. Since the expressions derived from the phenomenological theory of sedimentation are written in terms of concentrations and their gradients, it is necessary to have relations which allow these optical data to convert to values of concentration or its gradient.

In general, the refractive index n, of a solution is a function of composition, pressure P, and temperature T. Thus for an incompressible binary solution we may write

$$n = n(c_1, P, T) \tag{3.1}$$

where c_1 denotes the c-scale concentration of the solute (component 1). For compressible solutions, c_1 cannot be taken as a variable independent of P and T, but the concentration m_1 on the molality scale [moles of solute per 1000 g of solvent (component 0)] would be satisfactory. The subsequent treatments are restricted to incompressible solutions unless otherwise stated.

For sufficiently dilute solutions the right-hand side of equation 3.1 may be expanded in powers of c_1 to give

$$n = n(0, P, T) + R_1 c_1 + O(c_1{}^2) \tag{3.2}$$

where $n(0, P, T)$ represents the refractive index of the pure solvent at pressure P and temperature T, and R_1 is a quantity defined by

$$R_1 = \lim_{c_1 \to 0} \left(\frac{\partial n}{\partial c_1} \right)_{P, T} \tag{3.3}$$

The R_1 thus defined, which is still a function of P and T,* is called the *specific refractive index increment* of the solute (at infinite dilution of the solution). Its value as a function of P and T is characteristic of a given solute-solvent combination. Equation 3.2 is limited to nonelectrolyte solutions. The corresponding expression for an electrolyte solution contains terms in $c_1^{3/2}$, $c_1^{5/2}$, and so on, in addition to the integral powers in c_1. Thus for very dilute solutions the following relation applies, regardless of whether the solute is neutral or ionizable in a given solvent:

$$n_c \equiv n(c_1, P, T) - n(0, P, T) = R_1 c_1 \tag{3.4}$$

The quantity n_c is often termed the *excess* refractive index of the solution over the solvent, and can be measured as a function of c_1 at given P and T by use of an instrument called the differential refractometer. The value of R_1 is determined from the initial slope of the plot of n_c versus c_1 so obtained.

We now apply equation 3.4 to an isothermal, binary solution which is being centrifuged in an ultracentrifuge. In this case, both c_1 and P must be treated as functions of radial distance r and time t. It is to be noted that R_1 may also change with r because it depends on P. With these facts in mind,

* For incompressible systems, R_1 may be nearly independent of P.

differentiation of equation 3.4 with respect to r (at constant t) gives

$$\frac{\partial n_c}{\partial r} = \frac{\partial n}{\partial r} - \frac{\partial n_0}{\partial r} = R_1 \frac{\partial c_1}{\partial r} + c_1 \left(\frac{\partial R_1}{\partial P} \right)_T \frac{\partial P}{\partial r} \qquad (3.5)$$

where

$$n_0 = n(0, P, T) \qquad (3.6)$$

For incompressible solutions the derivative $(\partial R_1 / \partial P)_T$ is supposed to be quite small. Hence, for dilute solutions the second term on the right-hand side of equation 3.5 may be ignored to give

$$\frac{\partial n_c}{\partial r} = R_1 \frac{\partial c_1}{\partial r} \qquad (3.7)$$

Thus, subject to the above-mentioned approximations, the concentration gradient $\partial c_1 / \partial r$ could be determined if we were able to measure $\partial n / \partial r$ and $\partial n_0 / \partial r$. The schlieren optical system provided in the ultracentrifuge indeed allows $\partial n / \partial r$ to be measured as a function of r. It might be supposed that the required data for $\partial n_0 / \partial r$ could be derived from a similar measurement with the solvent alone, that is, the "blank" experiment in which only the solvent is centrifuged at the same temperature and the same rotor speed as in the "solution" experiment. This supposition, however, is invalid. The reason comes from the fact that the two derivatives on the left-hand side of equation 3.5 must refer to the same pressure distribution $P(r)$. The values of $\partial n_0 / \partial r$ must be those that would be measured if the blank experiment were performed at a rotor speed that could give rise to exactly the same pressure distribution as that obtained in the solution experiment. It is impossible to design a blank experiment in such a way that this requirement is rigorously met. For this reason it is usual in experimental work to evaluate $\partial n_c / \partial r$ by resort to the approximation

$$\frac{\partial n_c}{\partial r} \approx \frac{\partial n}{\partial r} - \left(\frac{\partial n_0}{\partial r} \right)_b \qquad (3.8)$$

where the subscript b refers to the blank experiment.

The error, Δ, introduced by this approximation, that is,

$$\Delta = \frac{\partial n_0}{\partial r} - \left(\frac{\partial n_0}{\partial r} \right)_b \qquad (3.9)$$

can be represented by

$$\Delta = \left[\frac{\partial P}{\partial r} - \left(\frac{\partial P}{\partial r} \right)_b \right] \left(\frac{\partial n_0}{\partial P} \right)_T \qquad (3.10)$$

where $\partial P/\partial r$ and $(\partial P/\partial r)_b$ are the pressure gradients set up in the solution experiment and the corresponding blank experiment, respectively. If equation 1.69 is inserted into equation 3.10 and use is made of the relation

$$\rho = \frac{1}{\bar{v}_0} + \left(1 - \frac{\bar{v}_1}{\bar{v}_0}\right)c_1 \tag{3.11}$$

which immediately follows from the definition of the solution density ρ and equation 1.113, equation 3.10 becomes

$$\Delta = c_1\left(1 - \frac{\bar{v}_1}{\bar{v}_0}\right)\omega^2 r\left(\frac{\partial n_0}{\partial P}\right)_T \tag{3.12}$$

Since the pressure dependence of n_0 is expected to be very small unless the system is markedly compressible, it follows from this equation that except for very special cases, the error caused by the approximation indicated in equation 3.8 will be negligibly small for sufficiently dilute binary solutions.

Next we consider a multicomponent solution which contains q macromolecular solutes and p added low-molecular-weight solutes in a single solvent 0. The c-scale concentrations of the q macromolecular solutes are denoted by $c_1, c_2, \ldots, c_q$, and those of the p added small solutes by $c_{q+1}, c_{q+2}, \ldots, c_{q+p}$. By a similar argument to that given above, it can be shown that if the solution is very dilute with respect to all the macromolecular solutes, then

$$n_c \equiv n(c_1, c_2, \ldots, c_q; c_{q+1}, c_{q+2}, \ldots, c_{q+p}, P, T)$$

$$- n(0, 0, \ldots, 0; c_{q+1}, c_{q+2}, \ldots, c_{q+p}, P, T) = \sum_{i=1}^{q} R_i c_i \tag{3.13}$$

where R_i is the specific refractive index increment of solute i ($1 \leqslant i \leqslant q$) defined by

$$R_i = \lim_{c_1, c_2, \ldots, c_q \to 0} \left(\frac{\partial n}{\partial c_i}\right)_{P, T, c_{k \neq i}} \quad (1 \leqslant k \leqslant q) \tag{3.14}$$

It might be considered that differentiation of equation 3.13 with respect to r, followed by neglect of the pressure dependence of R_i's and by adoption of the approximation $\partial n_c/\partial r \approx (\partial n/\partial r) - (\partial n_0/\partial r)_b$, would lead to

$$\frac{\partial n_c}{\partial r} = \sum_{i=1}^{q} R_i \frac{\partial c_i}{\partial r} \tag{3.15}$$

However, this relation is still incorrect by an amount Δ', which is given by

$$\Delta' = \sum_{j=q+1}^{p+q} \left[\frac{\partial c_j}{\partial r} - \left(\frac{\partial c_j}{\partial r} \right)_b \right] \left(\frac{\partial n_0}{\partial c_j} \right)_{P,T,c_{i \neq j}} \tag{3.16}$$

where $(\partial c_j / \partial r)_b$ represents the concentration gradient of small solute j produced in the blank experiment. In general, there exist some interactions between macromolecular solutes and added small solutes. Therefore, redistributions of the latter should be affected by sedimentation of the former, which implies that $(\partial c_j / \partial r)_b$ may not be equal to $(\partial c_j / \partial r)$. In other words, the quantity Δ' does not vanish in general. Nevertheless, we usually adopt equation 3.15 as the basis for sedimentation analysis of multicomponent solutions because of the lack of general experimental means of measuring concentration gradients of individual small solutes. The approximation involved is equivalent to treating the solvent medium composed of a main solvent and added small solutes (or other liquid components) as if it were essentially inert to the coexisting macromolecular solutes.

The total concentration $c(r)$ of the macromolecular solutes at a radial position r in the solution column is represented by

$$c(r) = \sum_{i=1}^{q} c_i(r) \tag{3.17}$$

Differentiation with respect to r gives

$$\frac{\partial c}{\partial r} = \sum_{i=1}^{q} \frac{\partial c_i}{\partial r} \tag{3.18}$$

Comparison with equation 3.15 shows that for a solution containing more than one macromolecular solute there exists no simple correspondence between the gradient of c and the gradient of n_c, except for the case in which all R_i's have the same value, that is, $R_1 = R_2 = \cdots = R_q$. Only in this special case can the distribution of excess refractive index gradients, that is, what is usually called a schlieren boundary pattern, be converted to the distribution of total concentration gradients. In no case of multicomponent system is it possible, without any assumption, to determine the concentration distributions of the individual solutes from observed schlieren boundary patterns, because, as can be seen from equation 3.15, there may be many combinations of values of $\partial c_i / \partial r$ which give rise to the same $\partial n_c / \partial r$.

The discussion just given concerns the schlieren optical method. The Rayleigh interference method gives data for $n(r)$ and $(n_0(r))_b$. For multicomponent solutions in which the concentrations of all macromolecular solutes are sufficiently low and the pressure dependence of n_0 is negligible, the fundamental relation for use of this method is

$$n_c(r) \equiv n(r) - (n_0(r))_b = \sum_{i=1}^{q} R_i c_i(r) \qquad (3.19)$$

subject to the assumption that effects arising from interactions between macromolecular solutes and coexisting small solutes may be ignored. Equation 3.19, when equation 3.17 is referred to, indicates that only in the special case in which all R_i's $(1 \leqslant i \leqslant q)$ have the same value does the Rayleigh interference method allow the distribution of $n_c(r)$ to be converted to the distribution of total solute concentrations. Also, with no assumption, the concentration distributions of individual solutes cannot be determined by this method.

The incapability of giving information about distributions of individual solutes is common to all refractometric or interferometric methods. It is indeed one of the major difficulties which quantitative analysis of transport processes in multicomponent solutions must usually suffer. This difficulty, however, can be obviated by use of light-absorption methods if the different solutes in a given solution have characteristic absorption bands at well-separated wavelengths of light. In recent years, there has been really remarkable progress in the development of this type of optical method. The automatic photoelectric scanner designed by Schachman and collaborators[3-6] now can be regarded as an almost ideal apparatus for light-absorption measurements. It is to be noted that this new apparatus permits experiments at much lower solute concentrations than possible with the traditional schlieren and Rayleigh interference methods, for example, at 0.05 to 1 mg/ml for proteins and at 0.01 to 0.1 mg/ml for nucleic acids. Thus, in the fields of biochemistry and molecular biology, the Schachman's scanner is gradually becoming an indispensable attachment to the ultracentrifuge. However, in this treatise, we shall proceed with the premise that both refractometric and interferometric methods will continue to be valuable as the general means for studying sedimentation phenomena of macromolecular systems.

II. PAUCIDISPERSE SOLUTIONS

A. BASIC EQUATIONS AND ASSUMPTIONS

The simplest of paucidisperse solutions relevant for sedimentation analysis consists of two macromolecular solutes 1 and 2 dissolved in a single solvent 0. For simplicity, it is assumed that the partial specific volumes of all of these components are independent of pressure and composition. Then, the sedimentation processes of the solute components are governed by equations 1.124 for $q=2$, that is,

$$\frac{\partial c_1}{\partial t} = \frac{\partial}{r\partial r}\left[r\left(D_{11}\frac{\partial c_1}{\partial r} + D_{12}\frac{\partial c_2}{\partial r} \right) - r^2\omega^2 s_1 c_1 \right]$$

$$\frac{\partial c_2}{\partial t} = \frac{\partial}{r\partial r}\left[r\left(D_{21}\frac{\partial c_1}{\partial r} + D_{22}\frac{\partial c_2}{\partial r} \right) - r^2\omega^2 s_2 c_2 \right]$$

(3.20)

Here the subscript V previously affixed to s_i and D_{ij} has been omitted as in Chapter 2. In general, these transport coefficients must be treated as functions of both c_1 and c_2, but, at present, very little is known about them either theoretically or experimentally.

Equations 3.20 must be supplemented with appropriate initial and boundary conditions in order to determine their solutions uniquely. If the concentrations of solutes 1 and 2 in a solution to be placed initially in the cell cavity are denoted by $c_1{}^0$ and $c_2{}^0$, we have

$$c_1 = c_1{}^0, \qquad c_2 = c_2{}^0 \qquad (r_1 < r < r_2, \quad t=0) \tag{3.21}$$

for an experiment with a conventional cell, and

$$c_1 = 0, \qquad c_2 = 0 \qquad (r_1 < r < r_0, \quad t=0)$$

$$c_1 = c_1{}^0, \qquad c_2 = c_2{}^0 \qquad (r_0 < r < r_2, \quad t=0)$$

(3.22)

for an experiment with a synthetic boundary cell. Here r_1, r_2, and r_0 have the same geometric meaning as in Chapter 2.

The boundary conditions are obtained from $(J_1)_c = (J_2)_c = 0$ at either end of the solution column, yielding

$$D_{11}\frac{\partial c_1}{\partial r} + D_{12}\frac{\partial c_2}{\partial r} = r\omega^2 s_1 c_1 \qquad (r=r_1, r_2, \quad t>0) \tag{3.23}$$

$$D_{21}\frac{\partial c_1}{\partial r} + D_{22}\frac{\partial c_2}{\partial r} = r\omega^2 s_2 c_2 \qquad (r=r_1, r_2, \quad t>0) \tag{3.24}$$

No solution to this set of equations is as yet reported in the literature. Even when all the transport coefficients may be treated as independent of

c_1 and c_2, it is probably very difficult to integrate these equations analytically. High-speed computers may be of use for numerical solution, but too little is known about the relative magnitudes of the coefficients to undertake an actually meaningful computation. To proceed with equations 3.20 in such circumstances, therefore, it is unavoidable to introduce certain as yet unproven assumptions or fairly drastic mathematical approximations.

One such approximation is to ignore the terms associated with the cross-term diffusion coefficients in comparison with those involving the sedimentation coefficients and the main diffusion coefficients. Equations 3.20 are then simplified to

$$\frac{\partial c_1}{\partial t} = \frac{\partial}{r \partial r}\left(r D_1 \frac{\partial c_1}{\partial r} - r^2 \omega^2 s_1 c_1\right)$$

$$\frac{\partial c_2}{\partial t} = \frac{\partial}{r \partial r}\left(r D_2 \frac{\partial c_2}{\partial r} - r^2 \omega^2 s_2 c_2\right)$$

(3.25)

where D_{11} and D_{22} have been replaced by simpler symbols D_1 and D_2. In general, these two differential equations are not independent of each other because of the possible dependence of each of the coefficients s_1, s_2, D_1, and D_2 on solute concentrations c_1 and c_2. This mathematical situation corresponds to the physical state of affairs in which each solute generally cannot sediment and diffuse independently of the other solutes coexisting in the solution. The cause of such "coupling" of flows of different solute components may be either thermodynamic or hydrodynamic interactions between them.

Next, we impose restrictions on terms in equations 3.25 that the dimensionless parameters $2D_1/(s_1\omega^2 r_0^2)$ and $2D_2/(s_2\omega^2 r_0^2)$ are as small as of the order of 10^{-3} or less. As mentioned in Chapter 2, these restrictions are well met under conditions of usual sedimentation velocity experiments if solutes 1 and 2 are of high molecular weight, and are basic to the derivation of Faxén-type solutions of actual significance. For the reason discussed in Chapter 2, Section III.F, the dependence of D_i on c_1 and c_2, if any, may be ignored in solving equations 3.25. Thus, in this approximation, our essential problem becomes investigating effects of coupled sedimentation on boundary gradient curves.

B. THE CASE OF UNCOUPLED SEDIMENTATION

In this case, the sedimentation coefficient of each solute is either constant or dependent only on its own concentration, and its distribution in the solution column can be calculated by use of the theory for binary solutions. For example, if s_1 and s_2 are constant, the distributions of solutes 1 and 2 at early stages of a boundary sedimentation experiment with a synthetic boundary cell are described by

$$\frac{\partial c_1}{\partial r} = \left(\frac{c_1^0}{r_0}\right) \frac{\exp(-\tau_1)}{\left\{\pi\epsilon_1[\exp(\tau_1)-1]\right\}^{1/2}} \exp\left\{-\frac{[r/r_0 - \exp(\tau_1/2)]^2}{\epsilon_1[\exp(\tau_1)-1]}\right\} \quad (3.26)$$

$$\frac{\partial c_2}{\partial r} = \left(\frac{c_2^0}{r_0}\right) \frac{\exp(-\tau_2)}{\left\{\pi\epsilon_2[\exp(\tau_2)-1]\right\}^{1/2}} \exp\left\{-\frac{[r/r_0 - \exp(\tau_2/2)]^2}{\epsilon_2[\exp(\tau_2)-1]}\right\} \quad (3.27)$$

where

$$\epsilon_i = \frac{2D_i}{s_i\omega^2 r_0^2} \qquad (i=1,2) \tag{3.28}$$

$$\tau_i = 2s_i\omega^2 t \qquad (i=1,2) \tag{3.29}$$

Substitution of equations 3.26 and 3.27 into equation 3.15 for $q=2$, that is,

$$\frac{\partial n_c}{\partial r} = R_1 \frac{\partial c_1}{\partial r} + R_2 \frac{\partial c_2}{\partial r} \tag{3.30}$$

gives a schlieren boundary curve which is composed of two Gaussian curves. Since, in general $s_1 \neq s_2$, $D_1 \neq D_2$, and $R_1 \neq R_2$, the two component curves have different maximum heights, different peak positions, and different spreads, and move independently at different speeds as the centrifugation is continued. The equivalent boundary position, $(r_M)_i$, of the component curve for solute i changes with time in accordance with the relation

$$\ln(r_M)_i = \ln r_0 + s_i\omega^2 t \tag{3.31}$$

Under the imposed condition $\epsilon_i \ll 1$ we may replace $(r_M)_i$ by the peak position $(r_*)_i$ of the curve for solute i.

In the case when the sedimentation coefficient of each solute depends on its own concentration, the schlieren boundary becomes a sum of two non-Gaussian curves, and each component curve moves independently of the other. Equation 3.31 still holds if s_i is concentration-dependent.

1. Resolution of Schlieren Boundary Curves

Since the early days in the development of the ultracentrifuge, one of the main interests of biochemists and those in related fields has been in the ultracentrifugal separation of a mixture of macromolecular substances of biological interest. Three items to be characterized are (1) the number of solute components involved in a given solution, (2) the weight fractions of these components, and (3) the sedimentation coefficient of each of them.

The traditional procedure involves centrifuging the solution at a rotor speed high enough to cause a sufficient separation of component solutes, resolving the schlieren boundary curve obtained into Gaussian or Gaussian-like symmetric curves, and then regarding the solution as containing as many "ultracentrifugally" separable solutes as there are separated peaks. In the case where the measured curve exhibits distinctly separated peaks as illustrated in Fig. 3.1, the resolution may be made

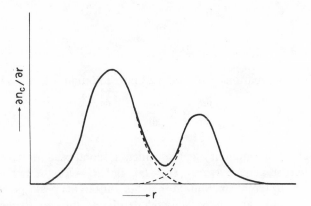

Fig. 3.1. Total and resolved schlieren curves in a multicomponent system.

without much arbitrariness. In many cases, however, observed boundary curves may not show more than one peak by the end of an experiment even at the highest speed available. Also even if two or more peaks or "shoulders" appear, the entire curve may not allow resolution into a finite number of Gaussian or even symmetric curves (Fig. 3.2). In such cases, it is usual to regard the solute as polydisperse, and to analyze the observed curves by the procedures discussed in Section III. Methods for items (2) and (3) are considered in the next section.

The above-mentioned traditional procedure is justified only under conditions in which each macromolecular solute sediments and diffuses independently and, moreover, its sedimentation coefficient is independent even of its own concentration. This latter condition is necessary, because otherwise the component curve of each solute may be neither Gaussian nor symmetric. In order to obtain these conditions the experiment will have to

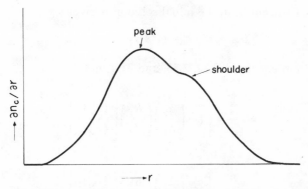

Fig. 3.2. A schlieren curve which does not allow resolution into Gaussian or Gaussian-like component curves.

be done at a sufficiently low initial concentration. In some cases, the necessary concentrations are so low that the conventional schlieren method may fail to give boundary patterns with precision. The automatic photo-electric scanning light-absorption method developed by Schachman et al.[3-6] should be of great use in such circumstances.

2. Evaluation of Weight Fractions and Sedimentation Coefficients

For a schlieren boundary curve of a ternary solution consisting of a solvent and two macromolecular solutes, the area of the component curve for solute i above the base line is designated by S_i, and the area of the corresponding concentration curve by A_i. Then

$$A_i = \frac{S_i}{R_i} \qquad (3.32)$$

If the boundary curve is freely sedimenting, the assumption of independent sedimentation of the solutes allows the concentration gradient curve of each solute to be analyzed in terms of the procedures developed for binary solutions. Thus, application of equation 2.208, with c_0 being replaced by A_i, yields

$$\frac{A_i}{c_i^{\,0}} = \frac{(r_1)^2}{(r_M)_i^{\,2}} \qquad (i = 1, 2) \qquad (3.33)$$

The total solute concentration c_0 (do not confuse with the c-scale concentration of solvent 0) in the initial solution is given by

$$c_0 = c_1{}^0 + c_2{}^0 \tag{3.34}$$

It follows from equations 3.32, 3.33, and 3.34 that

$$w_1 = \frac{c_1{}^0}{c_0} = \left\{ 1 + \left(\frac{S_2}{S_1} \right) \left(\frac{R_1}{R_2} \right) \left[\frac{(r_M)_2}{(r_M)_1} \right]^2 \right\}^{-1} \tag{3.35}$$

$$w_2 = \frac{c_2{}^0}{c_0} = \left\{ 1 + \left(\frac{S_1}{S_2} \right) \left(\frac{R_2}{R_1} \right) \left[\frac{(r_M)_1}{(r_M)_2} \right]^2 \right\}^{-1} \tag{3.35}$$

Here w_i represents the weight fraction of solute i in the original solution.

Extension of this treatment to a general case in which there are q independently sedimenting solutes gives

$$w_i = \frac{c_i{}^0}{c_0} = \left\{ \sum_{k=1}^{q} \left(\frac{S_k}{S_i} \right) \left(\frac{R_i}{R_k} \right) \left[\frac{(r_M)_k}{(r_M)_i} \right]^2 \right\}^{-1}$$

$$(i = 1, 2, \ldots, q) \tag{3.36}$$

Evaluation of the sedimentation coefficients of the solutes requires a series of schlieren boundary patterns taken at suitable intervals of time. After each of these curves is resolved into Gaussian or Gaussian-like component curves, the maximum positions, $(r_*)_i$, of the individual component curves are determined as functions of time. The desired s_i values may then be estimated by treating these data by equation 3.31, with $(r_M)_i$ being replaced by $(r_*)_i$.

C. THE CASE OF COUPLED SEDIMENTATION

Strictly speaking, mutual solute interactions would vanish only at infinite dilution of the solution ($c_0 \to 0$). Hence the methods described in the preceding section are expected to yield correct values of w_i and s_i when applied to data extrapolated to $c_0 = 0$. Dilution of the solution, however, is attended by a loss in precision of the measurements of the areas under the component curves. Moreover, there may arise a complication when some of the solutes undergo reversible association-dissociation reactions, as is often the case with proteins, since the relative concentrations of such solutes change as the total concentration is lowered (see Chapter 4).

The difficulty of analyzing sedimentation data for multicomponent solutions at finite dilutions was early recognized, and it was considered, and still is, to be one of the most serious shortcomings of the sedimentation transport methods. The best known example is a boundary anomaly usually called the *Johnston-Ogston effect*.

Let us again consider a ternary solution which contains two solutes 1 and 2 in a single solvent 0. It is assumed that solute 1 sediments faster than solute 2. Looking at Fig. 3.3, which illustrates a schlieren boundary curve

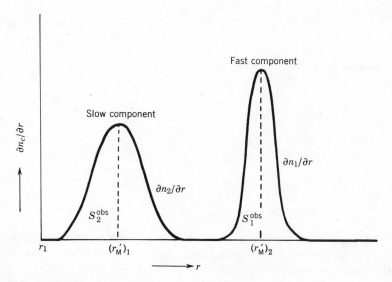

Fig. 3.3. Resolved boundary curves for fast and slow components. $(r'_M)_i$ represents the radial position defined by equation 3.41.

of such a solution, we will naturally associate the left and right component curves with solutes 2 and 1, respectively. If the contribution of the component curve i to the total excess refractive index gradient is designated by $\partial n_i / \partial r$, we have

$$\frac{\partial n_c}{\partial r} = \frac{\partial n_1}{\partial r} + \frac{\partial n_2}{\partial r} \qquad (3.37)$$

It is possible to compute from these component curves dimensionless quantities p_1 and p_2 defined by

$$p_1 = \frac{S_1^{obs}(r_M')_1{}^2}{n_1{}^0(r_1)^2}$$ (3.38)

$$p_2 = \frac{S_2^{obs}(r_M')_2{}^2}{n_2{}^0(r_1)^2}$$ (3.39)

where S_i^{obs} is the area of the component curve i above the base line, that is,

$$S_i^{obs} = \int_{r_1}^{(r_p)_i} \frac{\partial n_i}{\partial r} \, dr$$ (3.40)

$(r_M')_i$ is a quantity defined by

$$(r_M')_i = \left| \frac{\int_{r_1}^{(r_p)_i} r^2 \frac{\partial n_i}{\partial r} \, dr}{\int_{r_1}^{(r_p)_i} \frac{\partial n_i}{\partial r} \, dr} \right|^{1/2}$$ (3.41)

and $n_i{}^0$ is given by

$$n_i{}^0 = R_i c_i{}^0$$ (3.42)

In these equations, $(r_p)_i$ is the radial position of a point arbitrarily chosen in the plateau region for the component curve i. The sum of $n_i{}^0$ over all solutes gives the excess refractive index of the given solution, $n_c{}^0$:

$$n_c{}^0 = n_1{}^0 + n_2{}^0$$ (3.43)

If no interaction is operative between solutes 1 and 2, the component curve i differs from the concentration gradient curve of solute i by a constant factor R_i. In this case, S_i^{obs} should be equal to R_i times the area A_i of the concentration gradient curve of solute i, and $(r_M')_i$ should agree with the equivalent boundary position $(r_M)_i$ of the same curve. Therefore, by equation 2.208, both p_1 and p_2 become unity.

In actual cases, however, it is often observed that p_1 is smaller than unity, p_2 is larger than unity, and both change as sedimentation proceeds. In accordance with Trautman et al.,[7] the difference $\Delta_d = 1 - p_1$ is called the *builddown* of the fast component, and the difference $\Delta_u = p_2 - 1$ is called the *buildup* of the slow component. Observations of this type of boundary anomaly date back to McFarlane[8] and Pedersen[9] in the 1930s. The anomaly was found to be negligible at low concentrations and to become appreciable as the concentration was increased, but no one before Johnston and Ogston[10] was able to describe its real cause. In 1946, these authors first recognized the builddown and buildup phenomena to be due to coupled sedimentation of solutes, that is, the physical state of affairs that the rate of sedimentation of a particular solute is accelerated or retarded by the coexistence of other solutes. At present, these phenomena are usually referred to as the Johnston-Ogston effect.

No satisfactory theory is as yet available for this effect. The problem, which is concerned with handling a set of nonlinear, coupled ultracentrifuge equations, appears insuperably difficult to solve analytically without introduction of certain drastic simplifications. Numerical solution with the aid of a high-speed computer is very inviting, but, at present, available information about the concentration dependence of sedimentation coefficients of individual solutes in multicomponent solutions is too limited to undertake such work.

In what follows, the formulation due to Trautman et al.[7] is chosen as representative of the theories so far proposed and given a detailed account. The reader should consult Schachman's monograph[11] for a further study of the Johnston-Ogston effect and related topics.

1. Generalization of the Square-Dilution Rule

By application of the same theoretical consideration as in Chapter 2, Section III.H.1 to the generalized Lamm equations for multicomponent systems it can be verified that even when there are interactions between different solutes in a multicomponent solution, equation 2.208 holds exactly for each solute component. Thus for solute i in a solution containing q macromolecular solutes we have

$$c_i^0(r_1)^2 = (c_i)_p(r_M)_i^2 \qquad (3.44)$$

or

$$c_i^0 = \frac{1}{r_1^2} \int_{r_1}^{r_p} r^2 \frac{\partial c_i}{\partial r}\, dr \qquad (3.45)$$

where r_p denotes the radial position of an arbitrary point chosen in the plateau region of the observed boundary curve. It should be noted that the concentration gradient curves of individual solute components are not experimentally obtainable in general.

Multiplication of equation 3.45 by R_i, followed by summation over all solute components, gives

$$n_c{}^0 = \frac{(r'_M)^2}{r_1{}^2} S \tag{3.46}$$

where $n_c{}^0$ is the excess refractive index of the given solution, that is, by equation 3.19

$$n_c{}^0 = \sum_{i=1}^{q} R_i c_i{}^0 \tag{3.47}$$

r'_M is defined by

$$r'_M = \left(\frac{\displaystyle\int_{r_1}^{r_p} r^2 \frac{\partial n_c}{\partial r}\, dr}{\displaystyle\int_{r_1}^{r_p} \frac{\partial n_c}{\partial r}\, dr} \right)^{1/2} \tag{3.48}$$

and S is the area enclosed by the observed boundary curve above the base line, that is,

$$S = \int_{r_1}^{r_p} \frac{\partial n_c}{\partial r}\, dr \tag{3.49}$$

Equation 3.46, due originally to Trautman and Schumaker,[12] naturally reduces to equation 3.44 for two-component systems. It may be considered to be a generalization of the square-dilution rule to multicomponent systems.

Summation of equation 3.45 over all solute components gives, after rearrangement of terms,

$$c_0 = \frac{(r_M)^2}{(r_1)^2} A \tag{3.50}$$

where A and r_M are the area and the equivalent boundary position of the distribution of total concentration gradients. This equation is of little practical use, because A and r_M cannot be evaluated from observed schlieren patterns unless all solutes have the same specific refractive index increment (see Section I.B).

On the other hand, both S and r'_M can be calculated from such patterns. Hence, with the help of equation 3.46, the quantity n_c^0 can be evaluated, which means that it is possible to correct the total area of a schlieren pattern for radial dilution, regardless of whether or not different solutes in a given solution interact with each other.

If several peaks are observed on a schlieren pattern, the pattern would be resolved into as many symmetric or nearly symmetric component curves as there are peaks, and equation 3.46 would be rewritten

$$n_c^0 = \sum_{j=1}^{Q} \frac{(r'_M)_j^2}{(r_1)^2} S_j^{obs} \tag{3.51}$$

Here $(r'_M)_j$ is defined by

$$(r'_M)_j = \left(\frac{\int_{r_1}^{r_P} r^2 \frac{\partial n_j}{\partial r} \, dr}{\int_{r_1}^{r_P} \frac{\partial n_j}{\partial r} \, dr} \right)^{1/2} \tag{3.52}$$

and S_j^{obs} is the area of the component curve associated with the jth peak. A point to note is that we have no positive reason that the number of separated component curves, Q, may be equated to the number of different solutes present in the solution and that each component curve may be regarded as the distribution of refractive index gradients of a particular solute component. The essential difficulty of sedimentation analysis of multicomponent solutions is indeed associated with this point.

2. The Theory of Trautman et al. for the Johnston-Ogston Effect

For a ternary solution we presuppose the excess refractive index distributions of solutes 1 and 2 as sketched in Fig. 3.4. This picture follows the hypothesis of Johnston and Ogston[10] and uses the boundary and phase notation of moving boundary electrophoresis theory.[13, 14] The postulate here is that the solution is divided into three distinct phases: (1) an α phase, composed of pure solvent 0; (2) a β phase, containing the slow component 2 at a concentration c_2^β; and (3) a γ phase, containing the slow component 2 and the fast component 1 at the concentrations c_2^γ and c_1^γ, respectively. Separating these phases are two boundaries, the slower being termed the $\alpha\beta$-boundary and the faster the $\beta\gamma$-boundary. We define the position of the former boundary by $(r'_M)_2$ and that of the latter boundary by $(r'_M)_1$, where $(r'_M)_j$ ($j = 1$ or 2) is calculated by equation 3.52.

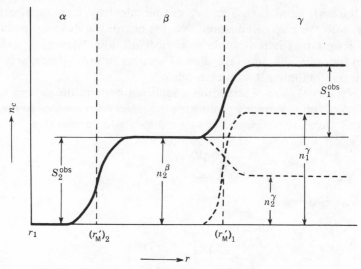

Fig. 3.4. Refractive index distributions assumed in the theory of Trautman et al. for the analysis of a system containing two solutes 1 and 2.

Inspection of Fig. 3.4 allows the following relation to be written:

$$S_1^{obs} = R_1 c_1^\gamma + R_2 c_2^\gamma - R_2 c_2^\beta = n_1^\gamma + n_2^\gamma - n_2^\beta \qquad (3.53)$$

$$S_2^{obs} = R_2 c_2^\beta = n_2^\beta \qquad (3.54)$$

where S_1^{obs} and S_2^{obs} represent the areas of the schlieren patterns at the $\beta\gamma$-boundary and the $\alpha\beta$-boundary, respectively. If equation 3.51 is applied, with substitution of equations 3.53 and 3.54 for S_1^{obs} and S_2^{obs}, and rearrangement of terms is made by use of equation 3.43, we obtain

$$n_1^0 - n_1^\gamma \left[\frac{(r_M')_1}{r_1} \right]^2 = \left[\frac{(r_M')_1}{r_1} \right]^2 (n_2^\gamma - S_2^{obs}) + \left[\frac{(r_m')_2}{r_1} \right]^2 S_2^{obs} - n_2^0 \quad (3.55)$$

To proceed further Trautman et al.[7] assumed that the square-dilution rule applies for solute 1 in the γ phase. That is, they related n_1^γ to $(r_M')_1$ by the equation

$$n_1^\gamma (r_M')_1^2 = n_1^0 (r_1)^2 \qquad (3.56)$$

Physically, this is equivalent to assuming that effects of the slow component on the sedimentation of the fast component, if any, may be ignored. However, it does not readily mean that transport of the slow component also is not affected by the presence of the fast component. With equation 3.56, the left-hand side of equation 3.55 vanishes, and rearrangement of the remaining terms gives

$$p_2 = \frac{[(r'_M)_1/r_1]^2(n_2^{\gamma}/n_2^0) - 1}{[(r'_M)_1/(r'_M)_2]^2 - 1}$$ (3.57)

This is introduced into equation 3.51 for $Q = 2$ to yield

$$p_1 = 1 - \frac{n_2^0}{n_1^0}(p_2 - 1)$$ (3.58)

Thus if a buildup occurs at the $\alpha\beta$-boundary, there is a concomitant builddown at the $\beta\gamma$-boundary, a conclusion that conforms to experimental observations.

According to Trautman et al.,[7] we introduce a new quantity ψ by the relation

$$\frac{n_2^{\gamma}}{n_2^0} = \left[\frac{r_1}{(r'_M)_1}\right]^{2\psi}$$ (3.59)

Then equation 3.57 may be written

$$p_2 = \frac{[(r'_M)_1/r_1]^{2(1-\psi)} - 1}{[(r'_M)_1/(r'_M)_2]^2 - 1}$$ (3.60)

which shows that if ψ is known in advance, the buildup p_2 can be evaluated, since both $(r'_M)_1$ and $(r'_M)_2$ are experimentally measurable. Equation 3.39 may be rewritten

$$n_2^0 = \frac{S_2^{obs}(r'_M)_2^2}{(r_1)^2 p_2}$$ (3.61)

Hence, once p_2 is known in this way, substitution of the observed values of S_2^{obs} and $(r'_M)_2$ into equation 3.61 allows n_2^0 to be calculated. Since n_c^0 may be determined by a separate refractometric measurement, it is then possible to find n_1^0 from equation 3.43. Finally, with these data, the builddown p_1 can be computed from equation 3.58. Thus the problem of analyzing sedimentation data on ternary solutions becomes one of determining the quantity ψ from experiment, provided the assumption of independent sedimentation of the fast solute component is obeyed.

If the second equation of 3.20 is applied to the plateau region in the γ phase, we obtain

$$\frac{dc_2^{\gamma}}{dt} = -2\omega^2 s_2^{\gamma} c_2^{\gamma} \tag{3.62}$$

which, upon integration, gives

$$\ln\left(\frac{n_2^{\gamma}}{n_2^{0}}\right) = -2\omega^2 \int_0^t s_2^{\gamma}\, dt \tag{3.63}$$

Similarly, we obtain for the fast component in the γ phase

$$\ln\left(\frac{n_1^{\gamma}}{n_1^{0}}\right) = -2\omega^2 \int_0^t s_1^{\gamma}\, dt \tag{3.64}$$

which, upon substitution of equation 3.56, yields

$$\ln\left[\frac{r_1}{(r_M')_1}\right] = -\omega^2 \int_0^t s_1^{\gamma}\, dt \tag{3.65}$$

In these equations, the superscript γ affixed to s_1 and s_2 refers to the concentrations in the γ phase. Insertion of equations 3.63 and 3.65 into equation 3.59 allows ψ to be expressed as

$$\psi = \frac{\displaystyle\int_0^t s_2^{\gamma}\, dt}{\displaystyle\int_0^t s_1^{\gamma}\, dt} \tag{3.66}$$

During the period of centrifugation in which a well-defined plateau region is observed beyond the $\beta\gamma$-boundary, the concentrations c_1 and c_2 may not be appreciably different from their initial values c_1^{0} and c_2^{0}. Therefore, it is reasonable to replace the s_1 and s_2 in equation 3.66 by their values corresponding to c_1^{0} and c_2^{0}. In this approximation, the quantity ψ becomes a constant given by

$$\psi = \frac{s_2^{0}}{s_1^{0}} \tag{3.67}$$

where the superscript 0 refers to the initial solution. The necessary values of s_1^{0} and s_2^{0} for the calculation of ψ by equation 3.67 may be obtained as the initial slopes of $\ln(r'_M)_1$ and $\ln(r'_M)_2$ plotted against $\omega^2 t$ (see equation 3.65).

3. Experimental Tests of the Theory of Trautman et al.

Trautman et al.[7] tested their theory with experimental data reported by Harrington and Schachman[15] for mixtures of bushy stunt virus (BSV) and tobacco mosaic virus (TMV). In these mixtures, the concentration of the slow component, BSV in this case, was held constant at $c_2^0 = 3$ mg/cc or $n_2^0 = 50 \times 10^{-5}$ in refractive index units, and the concentration of the fast component (TMV) was varied from 2 to 10 mg/cc. First, the values of p_2 were calculated from equation 3.61 by use of the known value of n_2^0, and then they were introduced into equation 3.60 to compute ψ. The results for three mixtures are given in Table 3.1. It is seen that for each mixture the

TABLE 3.1. Analysis of Mixtures of TMV (Component 1) and BSV (Component 2) by the Theory of Trautman et al.[7]

Photo. No.	$\Delta_u = p_2 - 1$	$n_2^0 \times 10^5$	ψ
	BSV, 3 mg/cc : TMV, 2 mg/cc		
5	0.24	47.6	0.713
6	0.21	46.5	0.721
7	0.23	48.3	0.710
8	0.20	46.4	0.721
9	0.18	45.4	0.724
10	0.18	45.5	0.726
	BSV, 3 mg/cc : TMV, 6 mg/cc		
5	1.30	53.1	0.682
6	1.17	50.9	0.695
7	1.16	52.7	0.684
8	1.04	49.7	0.702
9	0.98	48.8	0.704
10	0.89	46.4	0.721
	BSV, 3 mg/cc : TMV, 10 mg/cc		
8	3.02	51.3	0.692
9	3.00	54.4	0.676
10	2.84	51.4	0.692
11	2.49	49.5	0.703
12	2.49	51.1	0.694
13	2.66	50.2	0.698

computed values of ψ are almost independent of the time of centrifugation (indicated by the photograph number) and that these values are essentially the same for the three mixtures. Similar results were obtained for other mixtures examined. With a constant value of 0.7 assigned to ψ for all mixtures, Trautman et al.[7] calculated p_2 from equation 3.60 and then $n_2{}^0$ from equation 3.61 for each schlieren pattern in a sedimentation run. The resulting $n_2{}^0$ values agreed well with the known value 50×10^{-5} refractive index units (see Table 3.1).

The data shown in Table 3.1 indicate the occurrence of an enormous buildup in the $\alpha\beta$-boundary. The degree of buildup increases with the concentration of the fast component relative to that of the slow component, and for each run it decreases as sedimentation proceeds. The immediate recognition from these values of Δ_u is that if we are ignorant of mutual solute interactions, the traditional methods of sedimentation analysis will lead to gross errors in the estimate of relative amounts of the components.

For ternary solutions which contain two solutes 1 and 2 we may write

$$s_1{}^0 = \left(s_1{}^0\right)_0 f_1\left(c_1{}^0, c_2{}^0\right), \qquad s_2{}^0 = \left(s_2{}^0\right)_0 f_2\left(c_1{}^0, c_2{}^0\right) \qquad (3.68)$$

where $\left(s_1{}^0\right)_0$ and $\left(s_2{}^0\right)_0$ designate the values of $s_1{}^0$ and $s_2{}^0$ at infinite dilution of the solution, that is, $c_1{}^0 = 0$ and $c_2{}^0 = 0$. In actuality, $\left(s_i{}^0\right)_0$ may be determined from experiments in which solutions containing solute i alone are centrifuged at differing initial concentrations. The approximate constancy of ψ shown in Table 3.1 for different combinations of $n_1{}^0$ and $n_2{}^0$ may be reckoned as a consequence of the fact that the functions f_1 and f_2 for the BSV-TMV systems happened to be nearly identical. Trautman et al.[7] state that the approximation $\psi \approx \left(s_2{}^0\right)_0 / \left(s_1{}^0\right)_0$ is probably accurate enough for a determination of $n_2{}^0$ by equations 3.60 and 3.61, but it seems to be an oversimplification.

Soda et al.[16] undertook a more extensive test of the theory of Trautman et al. with data obtained for very narrow-distribution samples of poly-α-methylstyrene. The boundary anomalies of the Johnston-Ogston type were clearly observed in the two solvents examined, cyclohexane and toluene at 35°C. The values obtained for ψ are plotted against the total solute concentration c_0 in Figs. 3.5 and 3.6, where different marks indicate different values of $c_1{}^0 / c_2{}^0$. The data for cyclohexane are seen to be essentially independent of both c_0 and $c_1{}^0 / c_2{}^0$. They agreed closely with the value of $\left(s_2{}^0\right)_0 / \left(s_1{}^0\right)_0$ determined separately, in conformity to the proposal of Trautman et al. mentioned above. On the other hand, the data for toluene solutions indicate such a proposal to be generally inapplicable. Here the values of ψ depend on c_0 as well as on $c_1{}^0 / c_2{}^0$, but they tend to $\left(s_2{}^0\right)_0 / \left(s_1{}^0\right)_0$ at the limit of infinite dilution.

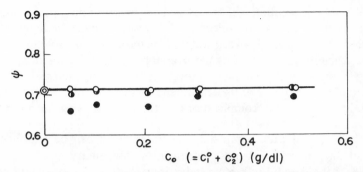

Fig. 3.5. Concentration dependence of the parameter ψ for binary mixtures of poly-α-methylstyrenes A-S (component 2, $M_w = 7.3 \times 10^5$) and A-F (component 1, $M_w = 1.52 \times 10^6$) in cyclohexane at 35°C.[16] Values of $c_1{}^0/c_2{}^0$ are 1.85, 1.00, and 0.478 for open, half-filled, and filled circles, respectively. The double circle shows the value of $(s_2{}^0)_0/(s_1{}^0)_0$.

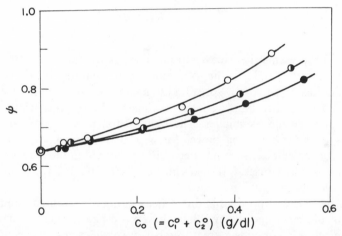

Fig. 3.6. Concentration dependence of the parameter ψ for binary mixtures of poly-α-methylstyrenes B (component 2, $M_w = 4.6 \times 10^5$) and C (component 1, $M_w = 1.38 \times 10^6$) in toluene at 35°C.[16] Values of $c_1{}^0/c_2{}^0$ are 1.96, 1.00, and 0.498 for open, half-filled, and filled circles, respectively. The double circle has the same meaning as in Fig. 3.5.

D. BAND CENTRIFUGATION

The technique of band centrifugation displays distinguished merits in its application to paucidisperse mixtures of macromolecular solutes if the sedimentation coefficients of the solutes are distinctly different from each other in the given solvent (which, of course, contains appropriate concentrations of small solutes so that suitable field and diffusion gradients may be set up during centrifugation). In this case, the initially layered thin band of a sample solution will sooner or later be split into separate bands, each consisting of macromolecules having the same sedimentation coefficient. Interactions between different macromolecular solutes, if any, become insignificant as soon as the overlapping of the bands disappears, and after that time, each band moves at a rate characteristic of the sedimentation coefficient of its constituent solute molecules, and its shape is entirely free from the complications of the type discussed in preceding sections.

The behavior of each completely separated band can be analyzed in terms of the theory of band centrifugation for binary solutions. For example, if the specific refractive index increment or the extinction coefficient is the same for all macromolecular components contained in the solution, the weight fraction of solute i may be evaluated by the equation[17]

$$w_i = \frac{\int_{r_1}^{r_2} n_i(r)r\,dr}{\sum_{j=1}^{q} \int_{r_1}^{r_2} n_j(r)r\,dr} \tag{3.69}$$

where $n_i(r)$ represents the excess refractive index distribution in the band formed by solute i, and q is the total number of separated bands.

A band has a narrower spread and hence becomes more distinctly separated from other bands as the average diffusion coefficient of its constituent molecules is smaller. The band sedimentation method, therefore, has a higher resolving power for a paucidisperse mixture of more slowly diffusing solutes, so it has been used primarily in the study of large nucleic acids and viruses.[18] It also can be of effective use for proteins of higher molecular weight.

When there is no appreciable difference among the sedimentation coefficients of the macromolecular solutes in a given paucidisperse mixture or when the solution contains a great number of solutes whose sedimentation coefficients vary almost continuously from solute to solute, the initially layered band will not separate into discrete bands but will develop into an increasingly spreading single band. In such cases, the movement and shape of the band may be complicated by interactions between different solutes. For the analysis of bands of this type the reader is referred to an article by Vinograd and Bruner.[17]

III. POLYDISPERSE SOLUTIONS

A. DISTRIBUTION FUNCTIONS

It is convenient for the ensuing discussion to call a multicomponent solution *ultracentrifugally ideal* if all of its cross-term diffusion coefficients are negligible and the sedimentation coefficient and main diffusion coefficient of each solute can be treated as constant. In such a solution, each solute sediments and diffuses independently of other solutes and forms a sedimentation boundary characteristic of its s and D values. Thus the solute molecules having the same s and D values are ultracentrifugally indistinguishable, regardless of whether or not they are chemically identical. In other words, the solute composition in an ultracentrifugally ideal solution can be defined in terms of paired values of s and D. When the solution is polydisperse, the number of ultracentrifugally classified solute components may be correspondingly large, and the associated s and D values may vary almost continuously from component to component. We begin this section with a discussion of standard mathematical methods for use in expressing heterogeneity of a substance with respect to s or D or both.

It is possible to sort out the molecules composing a substance into a series of groups in such a way that those in each group have the same s value, regardless of their D values. If the sample is polydisperse in s, the number of such groups is indefinitely large. We consider a differential fraction of such a sample which contains groups whose s values are between s and $s + ds$. Let its weight relative to the total weight of the sample be denoted by dw_s. Since dw_s should be proportional to ds, we may write

$$dw_s = f(s)\,ds \qquad (3.70)$$

The function $f(s)$ defined by this relation is called the *differential distribution* (function), or simply the distribution, of s in the sample. It may be recognized readily that the broader the curve of f versus s, the more heterogeneous the sample is with respect to s. It follows from equation 3.70 that

$$\int_0^\infty f(s)\,ds = 1 \qquad (3.71)$$

provided that, without much loss of generality, we confine ourselves to the system in which all molecules have positive s.

A new function $F(s)$ defined by

$$F(s) = \int_0^s f(\xi)\,d\xi \qquad (3.72)$$

represents the weight fraction of that portion of the sample having sedimentation coefficients whose values are less than or equal to a specified value of s. Thus $F(\infty)$ should be equal to unity. The function $F(s)$ is called the *integral distribution* (function) of s. It follows from equation 3.72 that

$$f(s) = \frac{dF(s)}{ds} \tag{3.73}$$

Thus once $F(s)$ is obtained experimentally or theoretically, the function $f(s)$ can be determined by its graphical or numerical differentiation.

For a sample which is homogeneous with respect to s, $F(s)$ is represented by a step function having a discontinuous point at $s = s'$, where s' is the s value of molecules composing the sample (Fig. 3.7). As $F(s)$ deviates more from this extreme type the sample is more heterogeneous in s (Fig. 3.8). The forms of $f(s)$ corresponding to these $F(s)$ are as shown in Figs.

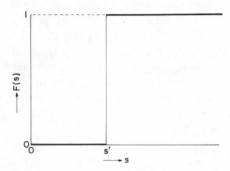

Fig. 3.7. Integral distribution of s in a sample homogeneous with respect to s.

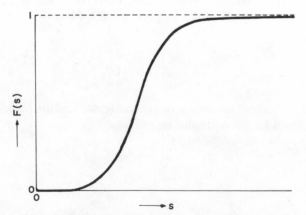

Fig. 3.8.   Integral distribution of s in a sample heterogeneous with respect to s.

3.9 and 3.10. The $f(s)$ of a sample homogeneous in s is represented by an infinitely sharp line, which is often called the line spectrum. The $f(s)$ for a sample paucidisperse in s consists of an array of line spectra.

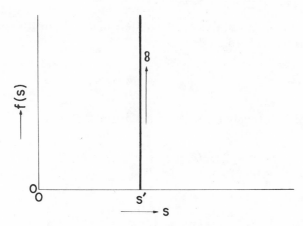

Fig. 3.9. Differential distribution of s corresponding to Fig. 3.7.

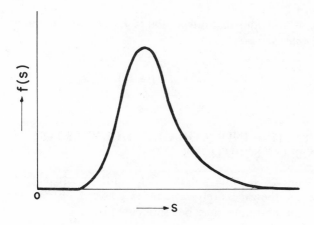

Fig. 3.10. Differential distribution of s corresponding to Fig. 3.8.

Next we sort out the molecules composing a substance in terms of s and D, and denote by $dw_{s,D}$ the weight fraction of a differential portion of the sample whose s and D are between s and $s + ds$ and between D and $D + dD$, respectively. Again we confine ourselves to the system in which all molecules have positive s values in a given solvent. Note that only positive values are physically meaningful for D. Since the value of $dw_{s,D}$ should be proportional to $ds\,dD$, we may write

$$dw_{s,D} = f(s,D)\,ds\,dD \tag{3.74}$$

The function $f(s,D)$ is termed the differential s–D distribution in the sample. It is normalized so that

$$\int_0^\infty \int_0^\infty f(s,D)\,ds\,dD = 1 \tag{3.75}$$

The integral s–D function $F(s,D)$ is defined by

$$F(s,D) = \int_0^s \int_0^D f(\xi,\eta)\,d\xi\,d\eta \tag{3.76}$$

Thus

$$f(s,D) = \frac{\partial^2 F(s,D)}{\partial s\,\partial D} \tag{3.77}$$

Both $f(s,D)$ and $F(s,D)$ are graphically represented by surfaces above the s–D plane.

Finally, it is obvious that integration of $f(s,D)$ over the range $0 < D < \infty$ gives $f(s)$, that is,

$$f(s) = \int_0^\infty f(s,D)\,dD \tag{3.78}$$

B. REFRACTIVE INDEX GRADIENT CURVES FOR POLYDISPERSE SOLUTIONS

For a schlieren boundary curve of an ultracentrifugally ideal polydisperse solution, the excess refractive index gradient, $\partial n_c / \partial r$, at a particular radial position and time, may be represented by

$$\frac{\partial n_c}{\partial r} = \int_0^\infty R_s \frac{\partial c_s}{\partial r}\,ds \tag{3.79}$$

where R_s is the specific refractive index increment of solute molecules having a sedimentation coefficient s, and $(\partial c_s / \partial r)\,ds$ is the concentration gradient, at the position and time considered, of solutes having s values between s and $s + ds$. Equation 3.79 follows from equation 3.15 if the running index i is replaced by a continuous variable s.

The total concentration of the original solution, c_0, is represented by

$$c_0 = \int_0^\infty c_s^0 \, ds \tag{3.80}$$

where $c_s^0 \, ds$ is the concentration in the original solution of solutes having s values between s and $s + ds$. The excess refractive index of the original solution, n_c^0, is expressed by

$$n_c^0 = \int_0^\infty R_s c_s^0 \, ds \tag{3.81}$$

By using equations 3.80 and 3.81, equation 3.79 may be rewritten

$$\frac{\partial n_c}{\partial r} = n_c^0 \int_0^\infty \alpha(s) \frac{\partial(c_s/c_s^0)}{\partial r} \, ds \tag{3.82}$$

where $\alpha(s)$ stands for

$$\alpha(s) = \frac{n_s^0}{n_c^0} = \frac{R_s c_s^0}{n_c^0} \tag{3.83}$$

The quantity $\alpha(s)\,ds$ represents the fractional amount, measured in terms of the excess refractive index, of solutes in the original solution whose s values are between s and $s + ds$. Thus the function $\alpha(s)$ is called the *differential distribution of s on a refractive index increment basis*. The function $f(s)$ defined previously is then termed the differential distribution of s on a *weight basis*. Sometimes these functions are simply called the refractive distribution of s and the weight distribution of s, respectively.

Obviously, $f(s)$ may be expressed as

$$f(s) = \frac{c_s^0}{c_0} \tag{3.84}$$

Equation 3.83 thus may be written

$$\alpha(s) = \frac{R_s c_0}{n_c^0} f(s) \tag{3.85}$$

Hence only when R_s is independent of the index s, $\alpha(s)$ coincides with $f(s)$, and equation 3.82 then becomes

$$\frac{\partial n_c}{\partial r} = n_c^0 \int_0^\infty f(s) \frac{\partial(c_s/c_s^0)}{\partial r} \, ds \tag{3.86}$$

Analogous relations can be derived for a sample which contains a continuous s–D distribution. For example, the one corresponding to equation 3.82 is

$$\frac{\partial n_c}{\partial r} = n_c^0 \int_0^\infty \int_0^\infty \alpha(s,D) \frac{\partial(c_{s,D}/c_{s,D}^0)}{\partial r} \, ds \, dD \tag{3.87}$$

where $\alpha(s,D)$ is defined by

$$\alpha(s,D) = \frac{n_{s,D}^0}{n_c^0} = \frac{R_{s,D} c_{s,D}^0}{n_c^0} \tag{3.88}$$

with $R_{s,D}$ being the specific refractive index increment of the solute having sedimentation coefficient s and diffusion coefficient D. The quantity n_c^0 has the same physical meaning as defined above, but this time it may be expressed by

$$n_c^0 = \int_0^\infty \int_0^\infty R_{s,D} c_{s,D}^0 \, ds \, dD \tag{3.89}$$

The differential $c_{s,D} \, ds \, dD$ represents the concentration, at a particular radial position and time, of solute molecules having s values between s and $s + ds$ and simultaneously having D values between D and $D + dD$, and $c_{s,D}^0$ is the value of $c_{s,D}$ in the original solution. The function $\alpha(s,D)$ may be termed the differential s–D distribution on a refractive index increment basis or the refractive distribution of s and D. It is related to $f(s,D)$ by

$$\alpha(s,D) = \frac{R_{s,D} c_0}{n_c^0} f(s,D) \tag{3.90}$$

Also it is obvious that

$$\alpha(s) = \int_0^\infty \alpha(s,D) \, dD \tag{3.91}$$

The formulation given above refers to a continuous s–D distribution. For a more general case in which the number of solutes having different paired values of s and D is finite, say q, the function $\alpha(s, D)$ is represented by

$$\alpha(s, D) = \frac{n_i^0}{n_c^0} \delta(s - s_i) \delta(D - D_i) \qquad (i = 1, 2, \ldots, q) \qquad (3.92)$$

where s_i and D_i are the values of s and D for solute i ($i = 1, 2, \ldots, q$), n_i^0 is the excess refractive index of solute i in the original solution, and δ denotes a normalized delta function. If the sample is polydisperse only with respect to s, its $\alpha(s, D)$ is given by

$$\alpha(s, D) = \alpha(s) \delta(D - D') \qquad (3.93)$$

where D' is the D value of the sample.

For $\alpha(s, D)$ given by equations 3.92 and 3.93, equation 3.87 is still applicable, yielding

$$\frac{\partial n_c}{\partial r} = \sum_{i=1}^{q} n_i^0 \frac{\partial(c_i / c_i^0)}{\partial r} \qquad (3.94)$$

and

$$\frac{\partial n_c}{\partial r} = n_c^0 \int_0^\infty \alpha(s) \frac{\partial(c_{s, D'} / c_{s, D'}^0)}{\partial r} ds \qquad (3.95)$$

A subtle difference between equation 3.95 and equation 3.82 should be noted.

C. THE BOUNDARY SPREADING EQUATION

For ultracentrifugally ideal solutions the expression for $\partial c_{s, D} / \partial r$ to be inserted into equation 3.87 is obtained from a solution to the Lamm equation with constant s and D, because in these solutions each solute sediments and diffuses as if it alone were in a single solvent. The mathematical solution to be chosen depends on the system and the stage of centrifugation. Thus if, as is generally the case for macromolecular systems under conditions of usual sedimentation velocity experiments, the parameter $\epsilon = 2D / (s\omega^2 r_0^2)$ for each solute is sufficiently small compared to unity and if we are concerned with the period of centrifugation in which freely sedimenting boundary curves appear in the cell, equation 2.106 may be substituted for $\partial c_{s,D}/\partial r$. Equation 3.87 then becomes

$$
\frac{\partial n_c}{\partial r} = n_c^{0} \int_0^{\infty} \int_0^{\infty} \frac{(s\omega^2)^{1/2} \exp(-2s\omega^2 t)}{\left\{2\pi D\left[\exp(2s\omega^2 t)-1\right]\right\}^{1/2}}
$$

$$
\times \exp\left\{ -\frac{s\omega^2 \left[r - r_0 \exp(s\omega^2 t)\right]^2}{2D\left[\exp(2s\omega^2 t)-1\right]} \right\} \alpha(s,D)\, ds\, dD \qquad (3.96)
$$

This integral equation for the function $\alpha(s,D)$ is called the *boundary spreading equation* for ultracentrifugally ideal polydisperse solutions, and forms the basis of sedimentation analysis of polydisperse systems.

When the number of solutes is finite, equation 3.94 must be used for deriving the boundary spreading equation. After equation 2.106 has been inserted, it yields

$$
\frac{\partial n_c}{\partial r} = \sum_{i=1}^{q} R_i c_i^{0} \frac{(s_i\omega^2)^{1/2} \exp(-2s_i\omega^2 t)}{\left\{2\pi D_i\left[\exp(2s_i\omega^2 t)-1\right]\right\}^{1/2}}
$$

$$
\times \exp\left\{ -\frac{s\omega^2\left[r - r_0\exp(s_i\omega^2 t)\right]^2}{2D_i\left[\exp(2s_i\omega^2 t)-1\right]} \right\} \qquad (3.97)
$$

D. THE BRIDGMAN EQUATION

The idea that schlieren boundary curves observed in sedimentation velocity experiments may be converted to some form of distribution of s was conceived by earlier investigators.[19-21] However, it was Bridgman[22] in 1942 who derived for the first time a mathematical relation which permits this conversion.* In his equation is found the beginning of an extensive literature on the use of the sedimentation velocity method for investigation of polydispersity of polymeric substances.

Bridgman's equation is derived from the special case of the boundary spreading equation in which all solutes have zero diffusion coefficient. In this case, there is a distribution of s only, so that equation 3.95 may be applied. When $D=0$, Case I in Chapter 2, Section II.B shows that the concentration distribution of a solute having a sedimentation coefficient s

---

* It appears to be the prevailing misunderstanding that a form equivalent to Bridgman's equation had been reported earlier by Signer and Gross.[21]

is represented by a step function as*

$$c_{s,0} = 0 \qquad [r_1 \leqslant r < (r_s)_*, \quad t > 0] \tag{3.98}$$

$$c_{s,0} = \left[\frac{r_1^2}{(r_s)_*^2} \right] c_{s,0}^0 \qquad [(r_s)_* < r < \infty, \quad t > 0] \tag{3.99}$$

Here $(r_s)_*$, the radial position of the discontinuous step, is given as a function of t and s by the equation

$$(r_s)_* = r_1 \exp(s\omega^2 t) \tag{3.100}$$

It follows from equations 3.98 and 3.99 that

$$\frac{\partial(c_{s,0}/c_{s,0}^0)}{\partial r} = \left[\frac{r_1}{(r_s)_*} \right]^2 \delta(r - (r_s)_*) \tag{3.101}$$

or, substituting equation 3.100,

$$\frac{\partial(c_{s,0}/c_{s,0}^0)}{\partial r} = \exp(-2s\omega^2 t)\delta(r - r_1 \exp(s\omega^2 t)) \tag{3.102}$$

Here $\delta(r)$ denotes a delta function normalized in the range $r_1 < r < \infty$. Thus

$$\int_{r_1}^{\infty} \delta(r - r')\,dr = 1 \qquad (\text{for } r_1 < r' < \infty) \tag{3.103}$$

and

$$\int_{r_1}^{\infty} p(r)\delta(r - r')\,dr = \int_{r_1}^{\infty} p(r)\delta(r' - r)\,dr = p(r') \text{ (for } r_1 < r' < \infty) \tag{3.104}$$

where $p(r)$ is an arbitrary function of r. Introduction of equation 3.101 into equation 3.95, followed by use of equation 3.100, gives

$$\frac{\partial n_c}{\partial r} = n_c^0 \int_{r_1}^{\infty} \alpha\left(\frac{\ln[(r_s)_*/r_1]}{\omega^2 t} \right) \frac{r_1^2}{(r_s)_*^3 \omega^2 t} \delta(r - (r_s)_*)\,d(r_s)_*$$

* $c_{s,0}$ denotes the concentration of a solute which has sedimentation coefficient s and zero diffusion coefficient.

which, with the aid of equation 3.104, reduces to

$$\frac{\partial n_c}{\partial r} = n_c^0 \frac{r_1^2}{r^3 \omega^2 t} \alpha\left(\frac{\ln(r/r_1)}{\omega^2 t}\right)$$
(3.105)

This equation may be rewritten to give Bridgman's formula

$$\alpha(s) = \frac{\omega^2 t [r^3 (\partial n_c / \partial r)]_{r = r_1 \exp(s\omega^2 t)}}{n_c^0 r_1^2}$$
(3.106)

Thus it is found that the refractive distribution of s is obtained by a simple transformation of an observed schlieren pattern. The actual procedure to carry out this transformation is as follows.

From the schlieren curve observed at a time t, plot the quantity $\omega^2 t r^3 (\partial n_c / \partial r)/(n_c^0 r_1^2)$ against r; the necessary value of n_c^0 may be determined from a separate refractometric measurement. Values of the ordinate at suitable intervals of r are then plotted against the corresponding values of s defined by $(1/\omega^2 t)\ln(r/r_1)$. The curve so obtained gives the required $\alpha(s)$. Since $\alpha(s)$ is a function of s only, the right-hand side of equation 3.106 should be independent of time t. In other words, the curves of $\alpha(s)$ at different times should merge on a single curve, provided the assumptions underlying Bridgman's equation are obeyed, that is, the diffusion coefficient of each solute is zero and the solution is ultracentrifugally ideal. In actual systems, however, these conditions are not met, and formal application of Bridgman's equation to schlieren patterns observed at different times in a sedimentation run may yield "apparent" distributions of $s, \alpha^*(s)$, which do not superimpose as sketched in Fig. 3.11. Our task is

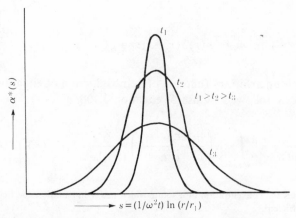

Fig. 3.11. Change in the apparent distribution of s with time (schematic).

then to correct such distribution curves for diffusion and deviations from ultracentrifugal ideality.

E.   BALDWIN'S APPROACH BY THE METHOD OF MOMENTS

First we consider the correction for diffusion, assuming ultracentrifugally ideal solutions. This problem has been investigated by many authors since the beginning of the 1950 s, and it has been shown to be essential for the desired correction that a series of apparent distributions at various times be extrapolated to infinite time. In this and next sections, we describe the typical contributions that served to establish the conclusion.

Baldwin[23] treated the case in which there are q discrete solutes, each having characteristic paired values of s and D in a given solvent, and in which freely sedimenting boundary curves are observed. His theory was actually a refinement of the earlier theories of Baldwin and Williams[24] and Williams et al.[25]

Let us denote the reduced first moment of a freely sedimenting boundary curve about the center of rotation by r_m, and the reduced second moment of the curve about it by σ_2:

$$r_m = \frac{\displaystyle\int_{r_1}^{r_p} r \frac{\partial n_c}{\partial r}\, dr}{\displaystyle\int_{r_1}^{r_p} \frac{\partial n_c}{\partial r}\, dr} \qquad (3.107)$$

$$\sigma_2 = \frac{\displaystyle\int_{r_1}^{r_p} (r - r_m)^2 \frac{\partial n_c}{\partial r}\, dr}{\displaystyle\int_{r_1}^{r_p} \frac{\partial n_c}{\partial r}\, dr} \qquad (3.108)$$

where r_p denotes the position of an arbitrarily chosen point in the plateau region beyond the maximum gradient(s). Equation 3.107 is substituted into equation 3.108 to give

$$\sigma_2 = \frac{\displaystyle\int_{r_1}^{r_p} r^2 \frac{\partial n_c}{\partial r}\, dr}{\displaystyle\int_{r_1}^{r_p} \frac{\partial n_c}{\partial r}\, dr} - \left(\frac{\displaystyle\int_{r_1}^{r_p} r \frac{\partial n_c}{\partial r}\, dr}{\displaystyle\int_{r_1}^{r_p} \frac{\partial n_c}{\partial r}\, dr} \right)^2 \qquad (3.109)$$

For the case considered here equation 3.97 is the relevant boundary spreading equation. With this equation, Baldwin evaluated σ_2, obtaining

$$\sigma_2 = p_2(\omega^2 r_m t)^2 \left\{ 1 - \left(\frac{p_3}{p_2}\right)\omega^2 t + \left[\tfrac{7}{12}\left(\frac{p_4}{p_2}\right) - \tfrac{5}{4}p_2 \right]\omega^4 t^2 + \cdots \right\}$$

$$+ 2\overline{D}t \left\{ 1 + \bar{s}\omega^2 t + 2(\bar{s} - \bar{s}')\omega^2 t \right.$$

$$\left. + \left[\tfrac{9}{2}(\bar{s})^2 + \tfrac{13}{6}(\bar{s}')^2 - 6\bar{s}\bar{s}' + \tfrac{13}{6}p_2' - \tfrac{5}{2}p_2 \right]\omega^4 t^2 + \cdots \right\} \qquad (3.110)$$

where

$$\bar{s} = \sum_{i=1}^{q} \nu_i s_i \qquad (3.111)$$

$$\overline{D} = \sum_{i=1}^{q} \nu_i D_i \qquad (3.112)$$

$$\bar{s}' = \frac{\displaystyle\sum_{i=1}^{q} \nu_i s_i D_i}{\displaystyle\sum_{i=1}^{q} \nu_i D_i} \qquad (3.113)$$

$$p_j = \sum_{i=1}^{q} \nu_i (s_i - \bar{s})^j \qquad (j = 2, 3, \ldots) \qquad (3.114)$$

$$p_2' = \frac{\displaystyle\sum_{i=1}^{q} \nu_i (s_i - \bar{s})^2 D_i}{\displaystyle\sum_{i=1}^{q} \nu_i D_i} \qquad (3.115)$$

$$\nu_i = \frac{n_i^0}{n_c^0} = \frac{R_i c_i^0}{n_c^0} \qquad (3.116)$$

If all solutes have the same specific refractive index increment, ν_i reduces to w_i, the weight fraction of solute i in the original solution, and $\bar{s}$ and $\overline{D}$ become, respectively, the *weight-average* sedimentation coefficient and the *weight-average* diffusion coefficient. The quantity p_j has the same physical meaning as the jth moment of the refractive distribution of s about its mean $\bar{s}$.

Equation 3.110 provides additional terms to the equation previously derived by Baldwin and Williams[24] and Williams et al.,[25] which reads

$$\sigma_2 = p_2(\omega^2 r_m t)^2(1 + \cdots) + 2\bar{D}t(1 + \bar{s}\omega^2 t + \cdots) \tag{3.117}$$

Baldwin[23] states that these additional terms will rarely amount to 2% of σ_2 (unless the distribution of s is too asymmetric). Thus for nearly symmetric distributions of s, a good approximation to σ_2 would be

$$\sigma_2 = p_2(\omega^2 r_m t)^2 + 2\bar{D}t(1 + \bar{s}\omega^2 t) \tag{3.118}$$

if the observations are restricted to early stages of a centrifugation in which terms higher than $O(\bar{s}\omega^2 t)$ may be ignored in comparison with unity.

According to equation 3.118, σ_2, a measure of the boundary spread, is given by the sum of two terms, one being associated with the heterogeneity of the sample with respect to s [the term $p_2(\omega^2 r_m t)^2$] and the other being the effect due to diffusion [the term $2\bar{D}t(1 + \bar{s}\omega^2 t)$]. The most important is the fact that these two terms increase in different powers in t: The first term is roughly proportional to t^2 (note that r_m varies with t), and the second term changes almost linearly with t. Thus, the boundary spreading due to diffusion becomes smaller compared with that due to heterogeneity in s as sedimentation proceeds. Hence, in principle, the correct form of $\alpha(s)$ could be determined by extrapolation of a series of apparent distributions of s at various times of centrifugation to infinite time. The functional form for use in the extrapolation was investigated in detail by Gosting.[26] Fujita[27] has recently approached this problem by use of a different method.

As shown in Appendix A, when the distribution of s is not too asymmetric and $\bar{D}$ is not large, r_m is related to $\bar{s}$ by the equation

$$r_m = r_1(1 + \bar{s}\omega^2 t) \tag{3.119}$$

provided the solution is ultracentrifugally ideal and terms higher than $O(\bar{s}\omega^2 t)$ are neglected compared with unity (these conditions are incorporated in the derivation of equation 3.118). Combining equation 3.119 with equation 3.118, we obtain

$$\frac{\sigma_2(r_1/r_m)}{2t} = \bar{D} + \tfrac{1}{2}(p_2\omega^4 r_1)(r_m t) \tag{3.120}$$

Relations slightly different in form from but essentially equivalent to this equation were originally given by Baldwin and Williams[24] and then by Williams et al.[25] Equation 3.120 shows that plots of $\sigma_2(r_1/r_m)/2t$ versus $r_m t$ should be linear and that $\overline{D}$ and p_2 can be evaluated from the intercept and slope of the resulting straight line, respectively. For a system homogeneous with respect to s, p_2 vanishes so that the measurements of $\sigma_2(r_1/r_m)/2t$ should yield a constant value equal to $\overline{D}$. In other words, the dependence of $\sigma_2(r_1/r_m)/2t$ on time may be used as a means of judging whether or not a given system is heterogeneous in s.*

F. GOSTING'S SOLUTIONS TO THE BOUNDARY SPREADING EQUATION

Methods to use for extrapolating the correct form of $\alpha(s)$ from apparent refractive distributions $\alpha^*(s)$ could be derived by investigating the asymptotic behavior of the boundary spreading equation for very large values of time. Gosting[26] carried out such an investigation for both electrophoresis and velocity ultracentrifugation. Here his principal results are presented, with a description of the details of derivation omitted.

For the case in which there is only a continuous distribution of s (so that all solutes are assumed to have the same diffusion coefficient, say D), the basis is equation 3.95. For the term $\partial c_{s,D}/\partial r$ in this equation Gosting[26] substituted the exact asymptotic expansion of the Faxén solution, rather than its approximate form 2.106. Therefore his result refers to sedimentation in synthetic boundary cells. However, here we modify it to suit the experiment with a conventional cell, by replacing his x_0 (our r_0) with r_1. This replacement is permissible as long as freely sedimenting sedimentation boundaries are considered.

His expression for $\alpha^*(s)$ thus modified reads

$$\alpha^*(s) = \left[1 + \tfrac{5}{6}q_2 Z^2 + O(Z^4)\right]\alpha(s)$$

$$-\left[q_3 + O(Z^2)\right]\frac{Z^2}{\omega^2 t}\frac{d\alpha}{ds}$$

$$+\left[1 + 3q_4 Z^2 + O(Z^4)\right]\frac{Z^2 q_1}{4(\omega^2 t)^2}\frac{d^2\alpha}{ds^2}$$

$$-\left[q_5 + O(Z^2)\right]\frac{Z^4 q_1}{4(\omega^2 t)^3}\frac{d^3\alpha}{ds^3} + \cdots \qquad (3.121)$$

---

* This method would be particularly useful when the observed schlieren patterns are so close to Gaussian curves that we might be tempted to conclude that the system under study is homogeneous with respect to s.

where

$$Z^2 = \frac{4Dt}{r^2}$$ (3.122)*

and

$$q_1 = \frac{\exp(2s\omega^2 t) - 1}{2s\omega^2 t}$$

$$q_2 = 1 + \tfrac{3}{5}(s\omega^2 t) + \tfrac{7}{25}(s\omega^2 t)^2 + \cdots$$

$$q_3 = 1 + \tfrac{5}{6}(s\omega^2 t) + \tfrac{1}{2}(s\omega^2 t)^2 + \cdots$$ (3.123)

$$q_4 = 1 + \tfrac{1}{3}(s\omega^2 t) + \tfrac{7}{90}(s\omega^2 t)^2 + \cdots$$

$$q_5 = 1 + \tfrac{2}{3}(s\omega^2 t) + \cdots$$

In order that Faxén's solution may represent a freely sedimenting boundary it is required that $\epsilon' = 2D/(s\omega^2 r_1^2)$ and $\tau = 2s\omega^2 t$ be sufficiently small in comparison with unity. Therefore, equation 3.121 would be of little practical value if at least those solute components contributing to the principal portion of $\alpha(s)$ did not satisfy these conditions. Thus equation 3.121 may be approximated by

$$\alpha^*(s) = \alpha(s) - \frac{4Dq_3}{\omega^2 r^2} \frac{d\alpha}{ds} + \left(\frac{D}{\omega^4}\right)\left(\frac{q_1}{r^2 t}\right)\left(\frac{d^2\alpha}{ds^2} - \frac{4Dq_5}{\omega^2 r^2} \frac{d^3\alpha}{ds^3}\right)$$

$$+ \frac{1}{2}\left(\frac{D}{\omega^4}\right)^2\left(\frac{q_1}{r^2 t}\right)^2 \frac{d^4\alpha}{ds^4} + \text{higher terms}$$ (3.124)

According to Gosting, the terms $(4Dq_3/\omega^2 r^2)d\alpha/ds$ and $(4Dq_5/\omega^2 r^2)$ $\times d^3\alpha/ds^3$ are seldom more than a few percent of the maximum values of $\alpha(s)$ and $d^2\alpha/ds^2$, respectively, so they may be neglected in most experi-

---

* Z^2 is written in terms of ϵ' and τ as $Z^2 = \epsilon'\tau(r_1/r)^2$. Since $r \approx r_1$ throughout the cell, the conditions that $\epsilon' \ll 1$ and $\tau \ll 1$ give $Z^2 \ll 1$. Thus terms of $O(Z^2)$ and of higher orders of Z may be dropped from the series in the brackets in equation 3.121.

ments. Then equation 3.124 reduces to

$$\alpha^*(s) = \alpha(s) + \frac{D}{\omega^4 r_1^2} \frac{1 - \exp(-2s\omega^2 t)}{2s\omega^2 t} \frac{1}{t} \frac{d^2\alpha}{ds^2}$$

$$+ \frac{1}{2}\left(\frac{D}{\omega^4 r_1^2}\right)^2 \left[\frac{1 - \exp(-2s\omega^2 t)}{2s\omega^2 t}\right]^2 \left(\frac{1}{t}\right)^2 \frac{d^4\alpha}{ds^4} + \cdots \qquad (3.125)$$

where use has been made of the fact that r in equation 3.124 is related to s and t by the equation $r = r_1 \exp(s\omega^2 t)$.

Equation 3.125 clearly indicates that $\alpha^*(s)$ is not a function of s only but changes with time, so that its more appropriate notation would be $\alpha^*(s,t)$. It is not evident from equation 3.125 that plots of $\alpha^*(s)$ versus $1/t$, or more precisely, $[1 - \exp(-2s\omega^2 t)]/(2s\omega^2 t^2)$, will extrapolate linearly to $\alpha(s)$ at very long times, because this equation was originally derived under the condition $\tau = 2s\omega^2 t \ll 1$. In other words, because of this condition, the variable t in equation 3.125 is not allowed to increase indefinitely. However, if the factor $D/(\omega^4 r_1^2)$ is sufficiently small, there would be an intermediate range of times in which t is so large that the third and higher terms in equation 3.125 may be neglected but in which $2s\omega^2 t$ is still fairly small compared with unity for values of s corresponding to the main portion of $\alpha^*(s)$. Plots of $\alpha^*(s)$ for a fixed s against $1/t$, or more precisely, $[1 - \exp(-2s\omega^2 t)]/(2s\omega^2 t^2)$, should then be linear over such a region of t, and the desired $\alpha(s)$ may be obtained as the intercept of this limiting linear region at $1/t = 0$. The problem of evaluating $\alpha(s)$ (subject to the conditions that the solution is ultracentrifugally ideal and the sample is heterogeneous only in s) thus becomes one of reaching the range of time for which Gosting's limiting law will hold. However, there is a basic limitation on the time for which a sedimentation experiment may be continued. The experiment must be stopped before the leading edge of the sedimentation boundary between solvent and solution reaches the cell bottom. Thus it may occur that by the end of an experiment we do not enter the range of t for Gosting's limiting law, in the case where the factor $D/(\omega^4 r_1^2)$ is so large that the asymptotic series on the right-hand side of equation 3.125 converges slowly.

With a Gaussian form assumed for $\alpha(s)$ Baldwin[23] examined how small the average diffusion coefficient $\bar{D}$ should be in order that the domain of Gosting's limiting law may appear in the range of time accessible to experiment. His theory is described in Section III.F.2.

Gosting[26] carried out a similar calculation for the case in which D is not the same for all solute molecules, and obtained an expression for $\alpha^*(s)$

identical to equation 3.121, except that the products $D^j(d^k\alpha/ds^k)$ $(j,k = 0, 1, 2, \ldots)$ have to be replaced by $d^k(\overline{D}^j\alpha)/ds^k$, in which

$$\overline{D}^j = \frac{\int_0^\infty D^j\alpha(s, D)\,dD}{\int_0^\infty \alpha(s, D)\,dD} \tag{3.126}$$

Thus, Gosting's limiting law is still useful when there is a distribution of D among the solute molecules.

1. The Case of Electrophoresis

Analogous to the case of ultracentrifugation, an electrophoretically ideal solution is defined as one in which the electrophoretic mobility u and the diffusion coefficient D of each solute can be treated as constant. The concentration gradient distribution of such a solute during an electrophoresis experiment, in which a solution initially separated from the solvent by an infinitely sharp boundary is subject to an electric field of constant strength E, is exactly represented by[28, 29]

$$\frac{\partial c_{u,D}}{\partial x} = \frac{c_{u,D}{}^0}{2\sqrt{\pi Dt}} \exp\left[-\frac{(x - uEt)^2}{4Dt} \right] \tag{3.127}$$

Here x denotes the distance measured along the cell (assumed to have a uniform cross section) from the initial boundary between solution and solvent. A similar argument to that in Section III.B shows that the excess refractive index gradient $\partial n_c/\partial x$ of an electrophoretically ideal polydisperse solution is expressed by

$$\frac{\partial n_c}{\partial x} = n_c{}^0 \int_{-\infty}^\infty \int_0^\infty \beta(u, D)\frac{\partial(c_{u,D}/c_{u,D}{}^0)}{\partial x}\,du\,dD \tag{3.128}$$

where $\beta(u, D)$ denotes the differential distribution of u and D on a refractive index increment basis.* Introduction of equation 3.127 into equation 3.128 gives the electrophoretic boundary spreading equation[28, 29]:

$$\frac{\partial n_c}{\partial x} = \frac{n_c{}^0}{2\sqrt{\pi t}} \int_{-\infty}^\infty \int_0^\infty \frac{\beta(u, D)}{\sqrt{D}} \exp\left[-\frac{(x - uEt)^2}{4Dt} \right] du\,dD \tag{3.129}$$

---

* Note that u ranges from $-\infty$ to $+\infty$.

Gosting[26] deduced from this equation that the apparent refractive distribution of u, designated here by $\beta^*(u)$, is asymptotically represented by

$$\beta^*(u) = \frac{Et}{n_c^{\,0}} \left(\frac{\partial n_c}{\partial x} \right)_{x=uEt} = \sum_{k=0}^{\infty} \frac{1}{k!} \left(\frac{1}{E^2 t} \right)^k \frac{d^{2k}[D^k \beta(u)]}{du^{2k}} \qquad (3.130)$$

where

$$\beta(u) = \int_0^{\infty} \beta(u, D)\, dD \qquad (3.131)$$

and

$$\overline{D}^k = \frac{\displaystyle\int_0^{\infty} D^k \beta(u, D)\, dD}{\displaystyle\int_0^{\infty} \beta(u, D)\, dD} \qquad (3.132)$$

When D is the same for all solute molecules, equation 3.130 reduces to

$$\beta^*(u) = \sum_{k=0}^{\infty} \frac{1}{k!} \left(\frac{D}{E^2 t} \right)^k \frac{d^{2k}\beta(u)}{du^{2k}} \qquad (3.133)$$

which may be compared to equation 3.121 for the case of ultracentrifugal sedimentation. Either equation 3.130 or 3.133 indicates that a plot of $\beta^*(u)$ versus $1/t$ becomes linear at sufficiently large values of t. Thus, as in the case of sedimentation, the problem of determing $\beta(u)$ from the measurement of $\beta^*(u)$ becomes one of reaching, within the period accessible to experiment, the range of time in which this limiting law holds.

2. Baldwin's Analyses with a Gaussian Distribution

Baldwin[23] examined the applicability of Gosting's limiting law, assuming for $\alpha(s)$ a Gaussian form represented by

$$\alpha(s) = \frac{1}{\sqrt{2\pi p_2}} \exp\left[-\frac{(s-\bar{s})^2}{2p_2} \right] \qquad (3.134)$$

where $\bar{s}$ and p_2 have the same significance as defined in Section III.E.1. This $\alpha(s)$ does not satisfy the required normalization condition

$$\int_0^{\infty} \alpha(s)\, ds = 1$$

but gives for this integral the value $\frac{1}{2}[1 + \Phi(\frac{\bar{s}}{\sqrt{2p_2}})]$, where, as before, Φ is the error function. Therefore, it is necessary to impose the conditions that $\bar{s}$ is sufficiently removed from zero and p_2 is so small that the ratio $\bar{s}/\sqrt{2p_2}$ is greater than about 1.5, for which $\Phi \approx 1$.

Baldwin introduced equation 3.134, together with the exact asymptotic expansion of Faxén's solution for $\partial c_{s,D}/\partial r$, into equation 3.95 (therefore he treated the case in which all solute molecules have the same D) and, after complicated calculations, arrived at

$$\alpha^*(s) = (\tfrac{3}{2}e^{2b} - \tfrac{1}{2}e^{3b})\frac{1}{\sqrt{2\pi p_2^*}} \exp\left[-\frac{(s-\bar{s})^2}{2p_2^*}\right] \qquad (3.135)^*$$

Here

$$b = \frac{\omega^2 t(s - \bar{s})}{\left[1 + p_2\left(\frac{\omega^4 r r_1 t}{2D}\right)\right]} \qquad (3.136)$$

$$p_2^* = p_2 + \frac{2D}{\omega^4 r r_1 t} \qquad (3.137)$$

and terms giving very small contributions (less than 0.2%, according to Baldwin) to $\alpha^*(s)$ have been omitted. Also, as in Section III.F, r_0 in the Faxén solution has been replaced by r_1 to make equation 3.135 applicable to experiments with a conventional cell.

Figure 3.12 shows the change with time of $\alpha^*(s)$ calculated from equation 3.135, with the following typical values assigned to the parameters involved: $\bar{s} = 4 \times 10^{-13}$ sec, $\bar{D} = 5 \times 10^{-7}$ cm^2/sec, $p_2 = 1 \times 10^{-26}$ sec^2, $\omega^2 = 3.9 \times 10^7$ rad^2/sec^2, and $r_1 = 5.8$ cm. It is seen that the effects of diffusion are appreciable near the beginning of the experiment ($t = 1 \times 10^3$ sec) and still quite important at $t = 5 \times 10^3$ sec, which is roughly the length of time for which a usual sedimentation velocity experiment is continued. Figure 3.13 indicates, in comparison with the true distribution, the extrapolated distribution obtained by placing a straight line through two values of $\alpha^*(s)$ plotted against $1/t$, at $t = 1 \times 10^3$ and 5×10^3 sec. According to the theory

* This is an extension to the ultracentrifugal sedimentation of the result derived by Alberty[29] in his study of boundary spreading in electrophoresis.

in Section III.F, the extrapolated curve should give the true distribution of s if these two times were in the range of Gosting's limiting law. It is observed that the extrapolated values are low at the center of the curve, where the error in extrapolation is the most serious, and high at the sides. Consequently, the area of the extrapolated curve is close to unity, thus satisfying the normalization condition required for the true $\alpha(s)$. If at $t = 5 \times 10^3$ sec the boundary gradient curve is no longer freely sedimenting, the curves for smaller values of t must be used in order to extrapolate $\alpha(s)$. The error in the extrapolated curve would then be greater than in Fig. 3.13.

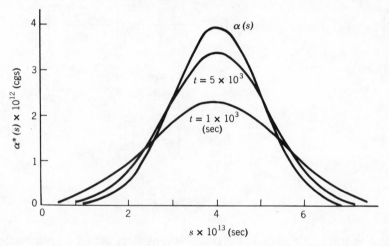

Fig. 3.12. Change in $\alpha^*(s)$ with time of centrifugation when the true distribution $\alpha(s)$ is Gaussian.

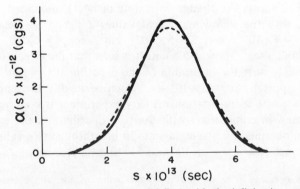

Fig. 3.13. Comparison of the true $\alpha(s)$ (solid line) with the infinite-time values of $\alpha^*(s)$ (dashed line) extrapolated from the range of time between 1×10^3 and 5×10^3 sec.

Equation 3.120 may be put in the form

$$\sigma_2 = 2t\left(\frac{r_m}{r_1}\right)\overline{D}[1 + R(t)] \qquad (3.138)$$

where

$$R(t) = \frac{p_2\omega^4 r_1 r_m t}{2\overline{D}} \qquad (3.139)$$

The quantity $R(t)$ is a measure of the boundary spread due to heterogeneity in s relative to that due to diffusion. Baldwin[23] showed that for the Gaussian $\alpha(s)$ considered here this quantity may be used to estimate the extent by which the values of $\alpha^*(s)$ at the end of the experiment, t_f, deviate from those of the true $\alpha(s)$.

Combination of equations 3.134 and 3.135 gives

$$\frac{\alpha^*(\bar{s})}{\alpha(\bar{s})} \approx \left[\frac{p_2}{(p_2^*)_{s=\bar{s}}}\right]^{1/2} \qquad (3.140)$$

where

$$(p_2^*)_{s=\bar{s}} = p_2 + \frac{2\overline{D}}{\omega^4 r_1 r_m t} \qquad (3.141)$$

In terms of $R(t)$ equation 3.140 may be written

$$\frac{\alpha^*(\bar{s})}{\alpha(\bar{s})} = \left[1 + \frac{1}{R(t)}\right]^{-1/2} \qquad (3.142)$$

or

$$\frac{\alpha^*(\bar{s})}{\alpha(\bar{s})} = \left[1 + \frac{1}{R(t_f)}\left(\frac{t_f}{t}\right)\right]^{-1/2} \qquad (3.143)$$

In Table 3.2 are listed the values of $\alpha^*(\bar{s})/\alpha(\bar{s})$ calculated from equation 3.143 at $t/t_f = 0.2$ and 1 for various fixed values of $R(t_f)$. The fourth column of the table gives the values of $\alpha^*(\bar{s})/\alpha(\bar{s})$ extrapolated to infinite time by placing a straight line through values of this ratio plotted against $1/t$ for the two time $t/t_f = 0.2$ and 1. It is probably safe to infer from this table that if the distribution of s is symmetric or nearly so and if $R(t_f)$ is smaller than 3, that is, the boundary spread due to heterogeneity in s is roughly less than three times that due to diffusion at the end of the experiment, Gosting's limiting law will cease to be effective for extrapolating the correct form of $\alpha(s)$.

TABLE 3.2. The Ratio of $\alpha^*(\bar{s})$ to $\alpha(\bar{s})$ for the Case when $\alpha(s)$ is Gaussian

$R(t_f)$	$\alpha^*(\bar{s})/\alpha(\bar{s})$ $t/t_f=0.2$	$t/t_f=1$	$\alpha^*(\bar{s})/\alpha(\bar{s})$ extrapolated
100	0.974	0.995	1.000
10	0.806	0.954	0.996
5	0.696	0.913	0.981
3	0.601	0.866	0.956
2	0.523	0.817	0.922
1	0.398	0.707	0.828

The question then arises: What other method can be used to find $\alpha(s)$ if $R(t_f)$ is smaller than 3 or so? Such cases may be encountered when relatively small globular proteins are studied, since these substances are generally quite homogeneous in s and have larger diffusion coefficients. No general approach to this problem is as yet known. Here we outline a method proposed by Baldwin[30] for the special case in which $\alpha(s)$ is Gaussian and D is the same for all solutes constituting the sample.

If r in the expressions for p_2^* and b is approximated by r_m and terms of the order of $\omega^2 t(s-\bar{s})$ are ignored in comparison with unity, it follows from equation 3.135 that

$$\frac{\alpha^*(s)}{\alpha^*(\bar{s})} = \exp\left[-\frac{(s-\bar{s})^2}{2p_2(1+B/r_m t)} \right] \qquad (3.144)$$

where

$$B = \frac{2D}{p_2\omega^4 r_1} \qquad (3.145)$$

Equation 3.144 may be solved for $(s-\bar{s})^2$ to give

$$(s-\bar{s})^2 = 2p_2\left(\ln\frac{\alpha^*(\bar{s})}{\alpha^*(s)} \right)\left(1+\frac{B}{r_m t} \right) \qquad (3.146)$$

Thus if the values of s found at a fixed value of $\alpha^*(s)0\alpha^*(\bar{s})$ are plotted in the form $(s - \bar{s})^2$ versus $1/(r_m t)$, there should be obtained a straight line, and the quantity p_2 can be determined from its ordinate intercept. The value of $\bar{s}$ necessary to construct this plot may be found from the position at which $\alpha^*(s)$ becomes maximum. Substitution of the values of p_2 and $\bar{s}$ so obtained into equation 3.134 allows the shape of the desired $\alpha(s)$ to be determined. This method is strictly applicable to Gaussian distributions of s, but it involves no limitation on the magnitude of R, that is, the boundary spread due to diffusion relative to that due to heterogeneity.* Figure 3.14 illustrates plots of $(s - \bar{s})^2$ versus $1/(r_m t)$ at four fixed values of $\alpha^*(s)/\alpha^*(\bar{s})$ on the trailing side of the boundary for a preparation of thyroglobulin in a $M/15$ phosphate buffer (the protein concentration is 0.553 g/dl).[30] The lines are slightly curved and, as far as one can tell, they all pass through the origin, suggesting that the protein preparation studied (at least the portion associated with the $\bar{s}$ value in this graph) was essentially homogeneous in s.

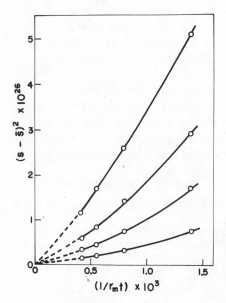

Fig. 3.14. Plots of $(s - \bar{s})^2$ versus $1/(r_m t)$ at fixed values of $\alpha^*(s)/\alpha^*(\bar{s})$ (0.2, 0.4, 0.6, 0.8 from top) on the trailing edge of the sedimentation boundary for a thyroglobulin preparation.[30] The initial concentration in this experiment was 0.55$_3$ g/dl.

Eriksson[31] reported a method similar in principle to Baldwin's described above, along with two other procedures of extrapolation.

---

* Note that D is assumed to be the same for all solute molecules. When D is allowed to vary with s, the conclusion given here is no longer valid.[27]

G. THE CASE OF INTERRELATED s AND D

In the considerations of the boundary spreading equations just given, the sedimentation coefficient s and the (main) diffusion coefficient D have been treated as independent variables. In one extreme case, it was assumed that a given sample had a distribution of s but all molecules constituting it had the same D. However, in many cases of practical interest with macromolecular solutes, there is a unique relation between s and D for a molecule of given chemical structure, so that if there is a distribution of s in a given sample, there must be a corresponding distribution of D. For example, the values of s and D of a linear polymer molecule in a theta solvent (see Section IV.E.1 for the definition of this term) depend quite accurately on the molecular weight M in accordance with the empirical relations[33]

$$s = kM^{1/2}, \qquad D = k'M^{-1/2}$$

where k and k' are constants independent of M and characteristic of the given polymer-solvent pair. Thus for a homologous series of linear polymers in a theta solvent both s and D are related uniquely by the condition $sD = \text{constant}$. Also, the theory[32] for rigid spherical molecules, such as globular proteins in aqueous media, predicts that their s and D are related by $(s)^{1/2}D = \text{constant}$.

For polydisperse solutions in which s and D of each solute molecule are interrelated in such a way, a new treatment of the boundary spreading equation is required. Recently, Fujita[27] has investigated this problem on the basis of a simplified boundary spreading equation, and has derived asymptotic expansion of the apparent refractive distributions of s for two cases in which $sD = \text{constant}$ and $(s)^{1/2}D = \text{constant}$.

The fundamental equation of Fujita's analysis is derived from equation 3.96 by neglecting terms of $O(s\omega^2 t)$ and higher orders in comparison with unity. Thus it reads

$$\frac{\partial n_c}{\partial x} = \frac{n_c^0}{2\sqrt{\pi t}} \int_0^\infty \int_0^\infty \frac{\alpha(s,D)}{\sqrt{D}} \exp\left[-\frac{(x - \omega^2 s r_1 t)^2}{4Dt} \right] ds\, dD \quad (3.147)$$

where x denotes

$$x = r - r_1 \qquad (3.148)$$

Here, unlike the original article of Fujita,[27] we have replaced r_0 in equation

3.96 by r_1, the position of the meniscus, so that the results may fit to experiments initiated with the condition for the conventional cell. Thus the equations given below are restricted to early stages of centrifugation in which freely sedimenting boundaries are observed. Equation 3.147 may also be derived directly from equation 3.129 if u, E, and $\beta(u, D)$ are replaced by s, $\omega^2 r_1$, and $\alpha(s, D)$, respectively. This fact implies that equation 3.147 indeed refers to an ultracentrifugally ideal solution in a rectangular cell in which there is a force field of constant strength equal to $\omega^2 r_1$. Because of this relationship between equations 3.147 and 3.129, the apparent refractive distribution of s relevant for the present case is expressed by equation 3.130 if u and E in the latter are replaced by s and $\omega^2 r_1$, that is,

$$\alpha^*(s) = \frac{\omega^2 r_1 t}{n_c^{\,0}} \left(\frac{\partial n_c}{\partial x} \right)_{x = s\omega^2 r_1 t} \tag{3.149}$$

In the case when s and D are related by $sD = \text{constant} = K$, the function $\alpha(s, D)$ is represented by $\alpha(s) \cdot \delta(D - K/s)$, so that equation 3.147 becomes

$$\frac{\partial n_c}{\partial x} = \frac{n_c^{\,0}}{2\sqrt{\pi K t}} \int_0^\infty \sqrt{S}\, \alpha(S) \exp\left[-\frac{S}{4Kt}(x - S\omega^2 r_1 t)^2 \right] dS \tag{3.150}$$

where the integration variable s in equation 3.147 has been replaced by S in order to avoid confusion with s which appeared in equation 3.149. Introduction of equation 3.150 into equation 3.149 yields

$$\alpha^*(s) = \frac{\omega^2 r_1 \sqrt{t}}{2\sqrt{\pi K}} \int_0^\infty \sqrt{S}\, \alpha(S) \exp\left[-\left(\frac{\omega^4 r_1^2 t}{4K} \right) S(s - S)^2 \right] dS \tag{3.151}$$

The corresponding equation for the case in which s and D are related by $(s)^{1/2} D = \text{constant} = K$ is

$$\alpha^*(s) = \frac{\omega^2 r_1 \sqrt{t}}{2\sqrt{\pi K}} \int_0^\infty (S)^{1/4} \alpha(S) \exp\left[-\frac{\omega^4 r_1^2 t}{4K} \sqrt{S}\,(s - S)^2 \right] dS \tag{3.152}$$

Fujita derived from these equations the asymptotic expansions for $A*(s)$, the apparent integral distribution of s defined by

$$A*(s) = \int_0^s \alpha*(s)\,ds \qquad (3.153)$$

His results are summarized as follows:

CASE I.   Where $sD = K$:

$$A*(s) = A(s) - \frac{1}{4\tau}\left[\frac{\alpha(s)}{s^2} - \frac{\alpha'(s)}{s}\right]$$

$$-\frac{3}{16\tau^2}\left[\frac{4\alpha(s)}{s^5} - \frac{3\alpha'(s)}{s^4} + \frac{\alpha''(s)}{s^3} - \frac{\alpha'''(s)}{6s^2}\right] + O\left(\frac{1}{\tau^3}\right) \qquad (3.154)$$

CASE II.   Where $(s)^{1/2}D = K$:

$$A*(s) = A(s) - \frac{1}{8\tau}\left[\frac{\alpha(s)}{s^{3/2}} - \frac{2\alpha'(s)}{s^{1/2}}\right]$$

$$-\frac{1}{32\tau^2}\left[\frac{6\alpha(s)}{s^4} - \frac{6\alpha'(s)}{s^3} + \frac{3\alpha''(s)}{s^2} - \frac{\alpha'''(s)}{s}\right] + O\left(\frac{1}{\tau^3}\right) \qquad (3.155)$$

Here $A(s)$ is the true integral distribution of s defined by

$$A(s) = \int_0^s \alpha(s)\,ds \qquad (3.156)$$

τ is a new time variable defined by

$$\tau = \frac{\omega^4 r_1^2 t}{4K} \qquad (3.157)$$

and the primes attached to $\alpha(s)$ denote the derivatives of $\alpha(s)$ with respective to s; for example,

$$\alpha''(s) = \frac{d^2\alpha(s)}{ds^2}$$

All these expansions are subject to the following restrictions: (1) that $\alpha(0) = 0$ and $\alpha'(0) = 0$, which means that the solute contains no detectable amount of small molecules, and (2) that $\exp(-4s^2\tau/27) \ll 1$ for Case I and $\exp[-16s^{5/2}\tau/25(5)^{1/2}] \ll 1$ for Case II, which implies that these expansions do not apply for the distribution function in the region of small s.

Fujita[27] also derived the asymptotic expansion of $A^*(s)$ for the case when there is no distribution of D. The result was

$$A^*(s) = A(s) + \frac{\alpha'(s)}{4\tau} + \frac{\alpha'''(s)}{32\tau^2} + O\left(\frac{1}{\tau^4}\right) \qquad (3.158)$$

where

$$\tau = \frac{\omega^4 r_1^2 t}{4D} \qquad (3.159)$$

and the restriction that $\exp(-s^2\tau) \ll 1$ had to be assigned to s and τ. Equation 3.158 agrees with the result obtained by integrating equation 3.133 with respect to s after u, E, and $\beta(u)$ have been replaced by s, $\omega^2 r_1$, and $\alpha(s)$, respectively.

These results indicate that even if there is a unique relation between s and D, either $\alpha^*(s)$ or $A^*(s)$ varies linearly with $1/t$ at very large values of t, and that the intercept of the limiting straight line gives $\alpha(s)$ or $A(s)$. Thus Gosting's limiting law holds, regardless of whether s and D are dependent or independent of each other. The difference between the two cases appears in the slope of the limiting straight line and in the range of $1/t$ for which the linear relation holds.

H. EXTRAPOLATION TO INFINITE DILUTION

The condition of "ultracentrifugal ideality" is generally not met by actual solutions of macromolecules. In these solutions, s and D depend not only on the concentration of a particular solute but also on the concentrations of coexisting solutes, as has been repeatedly noted in earlier parts of this chapter. Therefore, in general, the distributions of s calculated by Bridgman's formula or other related formulas are still apparent even after having been corrected for diffusion by extrapolation to infinite time. Various empirical procedures have been proposed to correct such apparent distributions for concentration effects, but none of them has as yet enjoyed theoretical justification. At present, it seems by no means easy to find a theoretical guideline relevant for this purpose if one recalls that for ternary solutions there was a formidable difficulty in the theoretical treatment of coupled sedimentation.

Baldwin[34,35] sought a method which allows the correction for concentration effects to be made by calculation, assuming that s and D of each constituent solute depend on the excess total refractive index n_c as follows:

$$s = s_0 - a_1 n_c - a_2 n_c^2 - \cdots$$

$$D = D_0 - b_1 n_c$$

$\qquad\qquad\qquad\qquad\qquad\qquad\qquad\qquad$ (3.160)

where $a_1, a_2, \ldots,$ and b_1 are treated as independent not only of n_c but also of the solute species. His theory, however, has received scant attention from subsequent investigators. Its pertinent summary can be found in Schachman's monograph.[11]

In what follows, three empirical procedures are dictated, together with their applications to actual systems. These procedures, due, respectively, to Williams et al.,[37] Baldwin et al.,[38] and Gralén and Lagermalm,[39] all purport to extrapolate a series of $\alpha^*(s, c_0)$ or $A^*(s, c_0)$ to infinite dilution ($c_0 = 0$). Here $\alpha^*(s, c_0)$ denotes the diffusion-corrected apparent distribution of s at an initial concentration c_0, and $A^*(s, c_0)$ denotes the corresponding apparent integral distribution of s, that is,

$$A^*(s, c_0) = \int_0^s \alpha^*(s', c_0)\, ds'$$

$\qquad\qquad\qquad\qquad\qquad\qquad\qquad\qquad$ (3.161)

1. The Procedure of Williams et al.

Values of $\alpha^*(s, c_0)$ corresponding to different fixed values of s, for example, those equally spaced, are plotted against c_0, and the resulting plots are extrapolated graphically to infinite dilution. It is the proposal of Williams et al.[37] to regard the extrapolated values plotted against the chosen s values as the desired $\alpha(s)$. This method encounters a profound difficulty in strongly concentration-dependent systems, in which there occur a marked flattening and a concomitant appreciable shift of $\alpha^*(s, c_0)$ curves as c_0 is lowered. It is rarely employed in current sedimentation analyses.

2. The Procedure of Baldwin et al.

Values of s or $1/s$ which are found at a series of specified values (e.g., 0.1, 0.2,...,0.9, 1.0) of $u = \alpha(s, c_0)/[\alpha^*(s, c_0)]_{\max}$ are plotted against c_0, and the resulting curves are graphically extrapolated to infinite dilution. The chosen values of u are then plotted against the corresponding s values (say s_0) at the limit of $c_0 = 0$. Baldwin et al.[38] proposed to regard a u versus s_0 curve thus obtained as equivalent to $\alpha(s)/[\alpha(s)]_{max}$ plotted against s. If the area under this curve is denoted by S, it follows from the normalization condition for $\alpha(s)$ that $[\alpha(s)]_{max} = 1/S$. Thus the curve of u versus s_0 can be converted to the desired $\alpha(s)$.

3. The Procedure of Gralén and Lagermalm

In this procedure, we first calculate $A^*(s)$ defined by equation 3.153 for a series of $\alpha^*(s)$ determined as a function of t at a given initial concentration c_0. It is important to check whether each of the calculated $A^*(s)$ approaches unity as s is increased. Values of s or $1/s$ which correspond to a specified value of $W = A^*(s)$ are plotted against $1/t$, and the resulting plot is graphically extrapolated to $1/t = 0$ to find the limiting value of s or $1/s$ (designated here by s_∞ or $1/s_\infty$). Repetition of similar operations at many different W values between 0 and 1 allows a relation between W and s_∞ to be established. Gralén and Lagermalm[39] proposed to take this relation as equivalent to $A^*(s, c_0)$. The theory described in Section III.G suggests that this proposal is correct at least for ultracentrifugally ideal solutions.

After determination of $A^*(s, c_0)$ for several initial concentrations, values of s or $1/s$ which correspond to a particular value of $U = A^*(s, c_0)$ are plotted against c_0, and the resulting plot is graphically extrapolated to $c_0 = 0$. The limiting value of s or $1/s$ so obtained is designated by s_0 or $1/s_0$. If similar operations are made at many other fixed values of U between 0 and 1, it is then possible to construct a curve of U versus s_0. Gralén and Lagermalm[39] predicted that this curve should be a graphical representation of the desired $A(s)$.

I. APPLICATIONS TO ACTUAL SYSTEMS

Among the various procedures so far proposed for the determination of $\alpha(s)$ the Gralén-Lagermalm one appears to have been in most widespread use for heterogeneity analyses of synthetic polymers and proteins. In fact, it has often been reported that application of this method to typical synthetic polymers yielded distributions of molecular weights which compared very favorably to those derived from precise fractionation experiments. Examples of such comparisons are presented in Section III.I.2.

1. Gelatin Solutions

There are few systems for which results of different procedures of extrapolation have been compared. An example of one such system is a gelatin sample in an aqueous buffer investigated by Williams and his collaborators.[37] Solid curves in Fig. 3.15 depict their $f^*(s,c_0)$ data for three different c_0 values. It was assumed that the specific refractive index increment was the same for all gelatin molecules, so $\alpha^*(s,c_0)$ was taken equal to $f^*(s,c_0)$, the apparent distribution of s on a weight basis. It is seen from Fig. 3.15 that with increasing c_0 the curve shifts toward the region of smaller s and, at the same time, it is markedly sharpened. These features may be attributed to a normal decrease of the sedimentation coefficient of each gelatin molecule with the concentration.

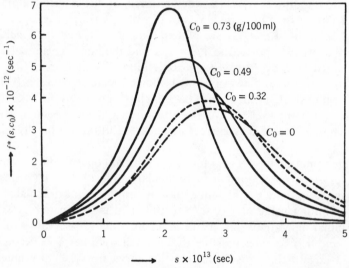

Fig. 3.15. Apparent and true distributions of s in a sample of gelatin.[37] Solid lines, apparent distributions at various initial concentrations c_0 extrapolated to infinite time. Dashed line, true distribution extrapolated by the Baldwin procedure. Chain line, true distribution extrapolated by the Gralén-Lagermalm procedure.

Figure 3.16 shows the application of the Baldwin procedure to the data shown in Fig. 3.15. Here are depicted the plots of s versus c_0 for various fixed values of $u = f^*(s,c_0)/[f^*(s,c_0)]_{max}$. The distribution function $f(s)$ obtained by this method is indicated by a dashed line in Fig. 3.15.

The application of the Gralén-Lagermalm procedure to the present data is illustrated in Fig. 3.17, where the values of s at fixed values of $U = F^*(s,c_0)$ are plotted against c_0 and extrapolated to infinite dilution. The chain line in Fig. 3.15, showing the $f(s)$ curve obtained by this method, agrees reasonably well with the dashed line deduced by the Baldwin

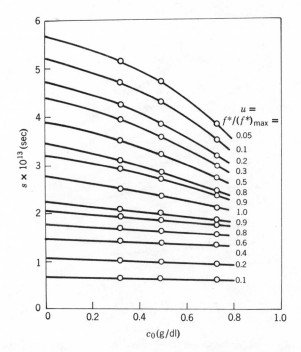

Fig. 3.16. Plots of s versus c_0 for various fixed values of u. Data are taken from Fig. 3.15.

method. It is important to observe that these distributions at infinite dilution markedly differ from the apparent distributions at finite dilutions. Also it should be observed that the plots of s versus c_0 by the Gralén-Lagermalm procedure are linear for all fixed values of $F^*(s,c_0)$ examined, whereas those by the Baldwin procedure exhibit an appreciable curvature for fixed values of $f^*(s,c_0)/[f^*(s,c_0)]_{max}$ corresponding to the region beyond the maximum of the distribution curve. This difference tells something of the greater popularity of the Gralén-Lagermalm method in current sedimentation analyses.

2. Synthetic Polymers

A survey of published applications of the Gralén-Lagermalm method to synthetic polymers would find its pertinent place in the literature on polymer chemistry. To illustrate, we quote here a recent study by Yamamoto et al.[40] on binary mixtures of narrow-distribution samples of poly-α-methylstyrene (PAMS) in cyclohexane at 35°C, because the results show quite clearly the power and limitations of the method.

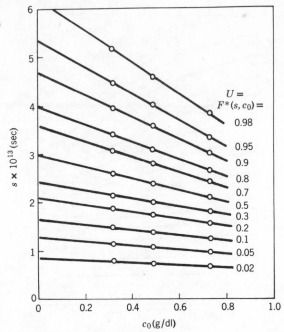

Fig. 3.17. Plots of s versus c_0 for various fixed values of U. Data are taken from Fig. 3.15 after they are corrected so that $F^*(s, c_0)$ for each c_0 gives unity in the limit of $s = \infty$.

Figure 3.18 shows plots of s versus $1/t$ for various fixed values of $W = F^*(s, t)$ derived from the data for sample No. 11 at $c_0 = 0.254$ g/dl. This sample was a 50:50 (by weight) mixture of nearly monodisperse PAMS samples whose number-average molecular weights were 34.2×10^4 and 50.0×10^4, respectively. Except for the two on the bottom, each set of data points is fitted accurately by a straight line and can be extrapolated to $1/t = 0$ with no ambiguity. Figure 3.19 compares the integral distributions of molecular weights M obtained for sample No. 11 by three independent methods: sedimentation velocity (crosses), gel permeation chromatography (filled circles), and precipitation chromatography (open circles). The procedure used for converting the distribution of s to that of M is discussed in Section III.J.1 Figure 3.20 displays the corresponding results for sample No. 9, which was a 50:50 (by weight) mixture of nearly monodisperse PAMS samples whose number-average molecular weights were 14.2×10^4 and 34.2×10^4, respectively. In each of these figures, the chain line indicates the ideal integral distribution of M which would be obtained if the components were perfectly homogeneous in M. A further comparison of the three methods made for sample No. 4, the component common to the mixtures No. 9 and No. 11, is illustrated in Fig. 3.21.

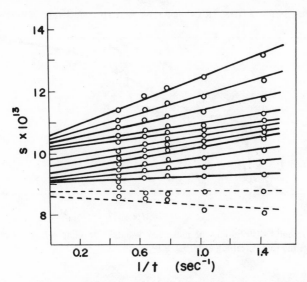

Fig. 3.18. Plots of s versus $1/t$ for various fixed values of W (0.975, 0.900, 0.800, 0.700, 0.650, 0.600, 0.550, 0.500, 0.400, 0.300, 0.200, 0.100, 0.025 from top) for a binary mixture (50:50 by weight) of monodisperse poly-α-methylstyrene samples in cyclohexane at 35°C.[40] $c_0 = 0.254$ g/dl.

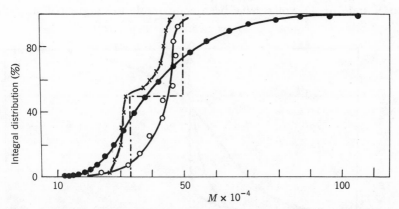

Fig. 3.19. Integral distributions of molecular weights for sample No. 11 (see text) determined by sedimentation velocity (crosses), gel permeation chromatography (filled circles), and precipitation chromatography (open circles).[40]

199

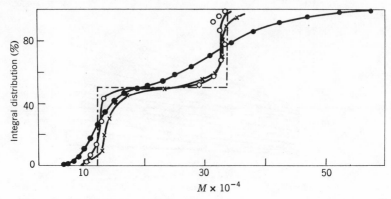

Fig. 3.20. Integral distributions of molecular weights for sample No. 9 (see text).[40] The symbols have the same meaning as in Fig. 3.19.

Except for sample No. 11, both sedimentation velocity and precipitation chromatograpy methods are seen to yield almost identical distributions of molecular weights. The curves for sample No. 4 are characteristic of a monodisperse sample, and are consistent with the result that the weight-average molecular weight relative to the number-average one of this sample (determined from separate light-scattering and osmotic pressure measurements) was unity within the limits of experimental error. It is to be observed that the sedimentation velocity method is indeed capable of detecting molecular weight heterogeneity of as homogeneous a polymer sample as this. The heterogeneities of the other components in mixtures No. 9 and No. 11 would have been comparable to sample No. 4. If this

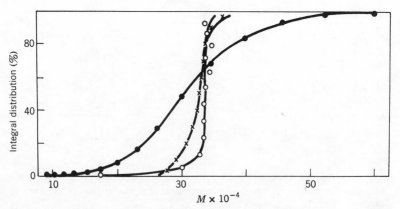

Fig. 3.21. Integral distributions of molecular weights for sample No. 4 (see text).[40] The symbols have the same meaning as in Fig. 3.19.

were the case, the chain lines in Figs. 3.19 and 3.20 would be close approximations to the true distributions of M in samples No. 9 and No. 11. It is seen that for the former sample both sedimentation velocity and precipitation chromatography provide distribution curves quite close to the chain line. For the latter sample the situation is different: Precipitation chromatography fails to disclose the existence of two components having different molecular weights, while the sedimentation velocity method is sensitive enough to detect it.

Gel permeation chromatography gives too broad a distribution curve for any of the three samples examined, especially for the almost monodisperse sample No. 4, suggesting that this method is of little use as a quantitative tool for investigating relatively homogeneous samples or their paucidisperse mixtures. Kotaka and Donkai[41] reached a similar conclusion from their recent measurements on polystyrene solutions. On the other hand, the sedimentation velocity method may be less effective for heterogeneity analysis of broad-distribution samples of polymer, because components with larger s will be swept down to the cell bottom at relatively early stages of centrifugation, whereas components with smaller s will remain close to the air-liquid meniscus so that freely sedimenting boundary curves may not be observed during the entire period of a sedimentation run.[41]

For synthetic polymers in organic solvents the distributions of s determined by any of the empirical procedures described above will have to be corrected further for the dependence of s on pressure. In fact, it has been pointed out that when the experiment is performed under theta conditions, this correction may become more important than that for concentration effects.[42,43] Some empirical procedures to effect it have been proposed,[41-45] all on the basis of the theory for binary solutions (see Section V in Chapter 2).

J. MOLECULAR WEIGHT DISTRIBUTIONS AND AVERAGE MOLECULAR WEIGHTS

Any sample of synthetic polymers prepared by the polymerization techniques now available is a mixture of an almost infinite number of chain molecules, either linear or branched, whose molecular weights vary from zero to infinity in an essentially continuous fashion. In such a sample, the weight fraction of molecules having a specified molecular weight M is effectively zero, but that of molecules whose molecular weights lie between M and $M + dM$ may be a nonzero differential dw_M. The magnitude of dw_M must be proportional to dM if dM is sufficiently small. Thus we may write

$$dw_M = g(M)\,dM \qquad (3.162)$$

The proportionality factor $g(M)$ represents, on a weight basis, the way in which the polymer molecules of different M distribute in the sample, that is, heterogeneity of the sample with respect to M. It is called the *differential distribution of molecular weights* or simply the *molecular weight distribution*. From the definition it immediately follows that

$$\int_0^\infty g(M)\,dM = 1 \qquad (3.163)$$

Next, we define a function $G(M)$ by

$$G(M) = \int_0^M g(\xi)\,d\xi \qquad (3.164)$$

and call it the *integral distribution of molecular weights*. This function is another means of characterizing molecular weight heterogeneity of a polymer sample.

The form of $g(M)$ or $G(M)$ is characteristic of a given sample of polymer, being independent of the external conditions under which it is determined experimentally. Sometimes, from kinetic or statistical considerations on the kinetics of polymerization,[46] it is possible to deduce in advance what particular form either $g(M)$ or $G(M)$ of a polymeric material assumes, but usually we must resort to suitable experimental means for its determination.

In Appendix B are listed some of the typical forms of $g(M)$ which have been derived from appropriate theoretical or experimental studies.* These forms are often useful in theoretical calculations of the effects of molecular weight heterogeneity on various physical properties of polymeric substances in solution or in bulk. In fact, such calculations form an important part of polymer physical chemistry.[32,33]

In place of either $g(M)$ or $G(M)$, average molecular weights defined in terms of $g(M)$ are very often used for characterization of molecular weight heterogeneity of polymeric substances, although information obtained from them is more restrictive than that from $g(M)$ or $G(M)$. Typical of them are the number-average molecular weight M_n, the weight-average molecular weight M_w, and the z-average molecular weight M_z, which are

---

*An elegant description of many possible molecular weight distributions can be found in L. H. Peebles, Jr., *Molecular Weight Distributions in Polymers*, J. Wiley, New York, 1971.

defined, respectively, as follows:

$$M_n = \frac{1}{\int_0^\infty [g(M)/M]dM} \tag{3.165}$$

$$M_w = \int_0^\infty Mg(M)dM \tag{3.166}$$

$$M_z = \frac{\int_0^\infty M^2 g(M)dM}{\int_0^\infty Mg(M)dM} \tag{3.167}$$

In general, the $(z+p)$-average molecular weight M_{z+p} is defined by

$$M_{z+p} = \frac{\int_0^\infty M^{p+2}g(M)dM}{\int_0^\infty M^{p+1}g(M)dM} \qquad (p=1, 2, \ldots) \tag{3.168}$$

Inasmuch as the function $g(M)$ [and $G(M)$ as well] is positive or zero for any value of $M(\geqslant 0)$, all these average molecular weights are positive, and, as is verified in Appendix C, they are in the sequence

$$M_n \leqslant M_w \leqslant M_z \leqslant M_{z+1} \leqslant M_{z+2} \leqslant \cdots \tag{3.169}$$

Here the equality holds only for a perfectly monodisperse sample. In general, there is no definite relation between any pairs of members in this sequence of average molecular weights.

The formulas above refer to a sample with a continuous distribution of M. The corresponding formulation can be developed for a sample consisting of a finite number (say q) of discrete components, each of which is homogeneous with respect to M. In this case, the function $g(M)$ is represented by

$$g(M) = g_i \delta(M - M_i) \qquad (i=1, 2, \ldots, q) \tag{3.170}$$

where M_i and g_i are the molecular weight and weight fraction of component i, and δ is a normalized delta function. The corresponding $G(M)$ is given by

$$G(M) = \sum_{i=1}^{j} g_i \qquad (M_j \leqslant M < M_{j+1}; \; j=1, 2, \ldots, q) \tag{3.171}$$

The average molecular weights M_n, M_w, and M_z are expressed as follows:

$$M_n = \frac{1}{\sum\limits_{i=1}^{q} (g_i/M_i)} \qquad (3.172)$$

$$M_w = \sum\limits_{i=1}^{q} M_i g_i \qquad (3.173)$$

$$M_{z+1} = \frac{\sum\limits_{i=1}^{q} M_i^2 g_i}{\sum\limits_{i=1}^{q} M_i g_i} \qquad (3.174)$$

The inequalities 3.169 apply for this case, too.

1. Relationship between Distribution Functions of s and M

If there exists a unique relationship between s (precisely, the limiting sedimentation coefficient) and M, experimental determination of $f(s)$ provides information about $g(M)$. In fact, it is well known experimentally and theoretically that such a relationship is available for a homologous series of linear homopolymers.[33] Empirically, it is represented quite accurately by a simple equation of the form

$$s = kM^\nu \qquad (3.175)$$

where k and ν are constants which depend on the kinds of polymer and solvent as well as on the temperature of the system. For polymer molecules which assume randomly coiled conformations in solution the values of ν lie between 0.5 and 0.4.[33] Especially under theta conditions* ν becomes 0.5.

For a substance which consists of molecules whose s and M are uniquely related, it can be shown that there holds a relation

$$g(M) dM = f(s) ds \qquad (3.176)$$

Integration gives

$$G(M) = F(s)|_{s \to M} \qquad (3.177)$$

where $s \to M$ stands for transforming the argument s in $F(s)$ to M by means of the functional interrelation between s and M: $s = s(M)$. Thus for this substance its integral distribution of M can be found from sedimentation transport measurements if information about $s = s(M)$ is available from other sources. It is usual for the determination of this relation to

*The definition of this term is given in Section IV.E.1.

prepare a series of narrow-distribution samples by either careful fractionation or a special polymerization technique and to evaluate the M of each
of the samples by osmotic pressure or light-scattering or sedimentation
equilibrium experiment. However, when theta conditions convenient for
sedimentation velocity experiments are available for the polymer under
study, it is sufficient if one prepares samples of arbitrary heterogeneity
which differ in M_w over a wide range. This statement can be verified as
follows.

Under theta conditions equation 3.175 assumes the form

$$s = kM^{1/2} \tag{3.178}$$

If this relation and equation 3.176 are combined with the defining expression for M_w, equation 3.166, it follows that

$$\langle s^2 \rangle^{1/2} = kM_w^{1/2} \tag{3.179}$$

where $\langle s^2 \rangle$ denotes the mean-square sedimentation coefficient defined by

$$\langle s^2 \rangle = \int_0^\infty s^2 f(s)\, ds \tag{3.180}$$

According to equation 3.179, a straight line of slope 1.0 should be obtained
if the values of $\langle s^2 \rangle$ for the given samples are calculated from their $f(s)$
determined experimentally under theta conditions and are plotted double
logarithmically against the corresponding M_w. The k value required to
establish the relation 3.178 should be found from its intercept at $M_w = 1$.

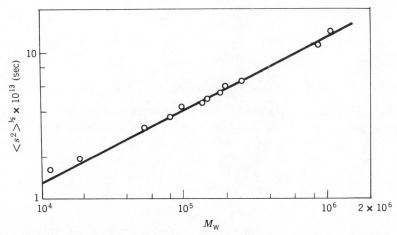

Fig. 3.22. Log-log plots of $\langle s^2 \rangle^{1/2}$ versus M_w for polystyrene fractions in cyclohexane at
35°C.[47] The line has been drawn with slope 0.5.

Figure 3.22 tests the validity of equation 3.179 with McCormick's data[47] for polystyrene samples in cyclohexane at 35°C, a theta temperature for this polymer-solvent pair (see Section IV.E.1). Except for the two on the extreme left, the data points are seen to follow closely a straight line of the expected slope, yielding a k value of 1.32×10^{-15} sec (mole/g)$^{1/2}$. McCormick's data were not corrected for the pressure effect, which has been found by subsequent investigators [41-44] to be rather appreciable in cyclohexane. The pressure-corrected values of k for the polystyrene-cyclohexane system at 35°C scatter around 1.45×10^{-15} sec (mole/g)$^{1/2}$.

IV. THE ARCHIBALD METHOD

A. INTRODUCTION

The traditional sedimentation-diffusion $(s-D)$ method for molecular weight determination resorts to separate measurements of the limiting sedimentation coefficient s_0 and the limiting diffusion coefficient D_0, but these measurements demand a considerable amount of experimental work, although for binary solutions the D_0 may be estimated from sedimentation boundary curves, as explained in Chapter 2. Because only the ratio of s_0 to D_0 appears in the Svedberg relation 1.121, it would be advantageous if we were able to determine this ratio directly. This idea was embodied by Archibald,[48] who found that the ratio s/D for binary solutions can be calculated from measurements of c and $\partial c/\partial r$ at either end of a solution column in the ultracentrifuge cell. In fact, from the boundary conditions, equations 2.4,

$$\frac{s}{D} = \frac{(\partial c/\partial r)_{r=r_1}}{r_1\omega^2(c)_{r=r_1}} \tag{3.181}$$

$$\frac{s}{D} = \frac{(\partial c/\partial r)_{r=r_2}}{r_2\omega^2(c)_{r=r_2}} \tag{3.182}$$

For dilute binary solutions in which the density of the solvent and the partial specific volume of the solute may be approximated by their infinite-dilution values we have from equations 1.120 and 1.121

$$\frac{s}{D} = \frac{s_0}{D_0}\left[1 + c\left(\frac{\partial \ln y_1}{\partial c}\right)_{T,P}\right]^{-1} \tag{3.183}$$

The second term in the brackets is generally nonzero. Accordingly, the s/D ratios calculated from equations 3.181 and 3.182 may not be substituted directly into the Svedberg relation. They may also change with time in different ways at the ends, because the concentration at the air-liquid meniscus decreases and that at the cell bottom increases as sedimentation proceeds. However, if extrapolated back to zero time, they should converge to an identical value $(s_0/D_0)[1 + c_0(\partial \ln y_1/\partial c)_{T,P,c=c_0}]^{-1}$. Here c_0 denotes the concentration in the initial uniform solution. If data for $\ln y_1$ as a function of c are available, this zero time value of s/D can be converted to the desired s_0/D_0. When, as is generally the case with macromolecular solutions, such data are not available, it is necessary to determine s/D at zero time for several different initial concentrations and then to extrapolate the results to infinite dilution. Thus unless the activity coefficient y_1 is constant, data for s/D must be obtained at a number of times and initial concentrations so that the extrapolations to $t = 0$ and $c_0 = 0$ may be made with certainty.

For dilute multicomponent solutions it was shown by Archibald[48] that even if the activity coefficient of each solute is constant, the values of s/D at either end of the solution column vary with time, but the value extrapolated back to $t = 0$ yields, when substituted for s_0/D_0 in the Svedberg relation, the weight-average molecular weight of the total solute, provided the partial specific volumes of all solute components are identical. Kegeles et al.[49] attempted extending this Archibald theory to a more general case in which the activity coefficients of the solutes depend on the composition. Later, Fujita et al.[50] refined the treatment of Kegeles et al. by use of the formulation of nonequilibrium thermodynamics.

From the experimental point of view, the most crucial problem is the measurements of concentration gradients at the ends of the solution column. Two problems are involved. One is the choice of experimental conditions, and the other is the methods to be used for extrapolation of observed gradient curves to the ends. These are discussed in Sections IV.D and IV.D.2.

Historically, the Archibald method once acquired a greater popularity as a promising substitute for the traditional s–D method. However, as it became evident that reliable determinations of the concentration gradients at the ends were not always an easy task, the investigators have kept this method at a distance, and have been more and more attracted by the sedimentation equilibrium method, which now enables thermodynamic information about dilute macromolecular solutions to be obtained at speeds comparable to and with an accuracy higher than the Archibald method.

B. FORMULATION OF THE METHOD

Let us consider an incompressible solution which contains q different nonelectrolyte solutes in a single solvent. As before, the solutes are numbered 1, 2,...,q. If the physical requirement, basic to the Archibald method, that the flow of each solute relative to the cell must vanish for all times at either end of the solution column is applied to equations 1.102, the following set of q equations is derived:

$$(1 - \bar{v}_i\rho)\omega^2 r = \sum_{k=1}^{q} \mu_{ik} \frac{\partial c_k}{\partial r} \qquad (i = 1, 2, ..., q; \quad r = r_1 \text{ or } r_2) \quad (3.184)$$

As will be seen in Chapter 5, Section I.C, the equations descriptive of sedimentation equilibrium of an incompressible multicomponent solution in the ultracentrifuge cell are formally identical to this set of equations. For this reason there is often a misunderstanding that the theory of the Archibald method is essentially the same as that of sedimentation equilibrium. A clear distinction must be made, however, that the term $\partial c_k / \partial r$ in equations 3.184 is not a function of r but one of t, whereas the corresponding term in the equilibrium equations 5.1 is a function of r and not one of t. Thus equations 3.184 are not differential equations but algebraic relations for q concentration gradients at either $r = r_1$ or $r = r_2$. No information can be derived from them about second and higher derivatives of c_i with respect to r. In other words, they may not be differentiated with respect to r.

According to thermodynamics, we may write μ_i in the form

$$\mu_i = (\mu_i^0)_c + \frac{RT}{M_i} \ln(y_i c_i) \tag{3.185}$$

provided solute i may be treated as a nonelectrolyte. Here $(\mu_i^0)_c$ is the reference chemical potential *per gram* of solute i appropriate to the c-concentration scale, M_i and y_i are the molecular weight and the practical activity coefficient on the c-scale of solute i, and R is the gas constant. Thus the activity coefficient y_i has the limiting property:

$$\lim_{c_1, c_2, ..., c_q \to 0} y_i = 1 \qquad (i = 1, 2, ..., q) \tag{3.186}$$

Note that, in general, y_i depends not only on c_i but also on the concentrations of coexisting solutes, $c_1, c_2, ..., c_{i-1}, c_{i+1}, ..., c_q$. Substitution of equation 3.185 into equation 3.184 gives

$$\frac{M_i(1 - \bar{v}_i\rho)r\omega^2 c_i}{RT} = \frac{\partial c_i}{\partial r} + c_i \sum_{k=1}^{q} \left(\frac{\partial \ln y_i}{\partial c_k}\right)_{T, P, c_m} \frac{\partial c_k}{\partial r}$$

$$(i = 1, 2, ..., q; \quad r = r_1 \text{ or } r_2) \tag{3.187}$$

Multiplication of equations 3.187 by R_i, the specific refractive index increment of solute i in the given solvent, followed by summation over all solute components and use of equation 3.15, yields

$$\frac{\omega^2 r}{RT} \sum_{i=1}^{q} R_i (1 - \bar{v}_i \rho) M_i c_i = \frac{\partial n_c}{\partial r} + \sum_{i=1}^{q} \sum_{k=1}^{q} R_i c_i \left(\frac{\partial \ln y_i}{\partial c_k} \right)_{T,P,c_m} \frac{\partial c_k}{\partial r}$$

$$(r = r_1 \text{ or } r_2) \tag{3.188}$$

The definition of y_i allows $\ln y_i$ to expand in Taylor's series as

$$\ln y_i = M_i \sum_{k=1}^{q} B_{ik} c_k + M_i^2 \sum_{k=1}^{q} \sum_{m=1}^{q} B_{ikm} c_k c_m$$

$$+ \text{ higher terms in } c_1, c_2, \ldots, c_q \tag{3.189}$$

provided the solution is sufficiently dilute with respect to all solute components. The subsequent development will confine itself to the range of concentrations in which terms higher than $O(c_k)$ need not be included. The expansion coefficients B_{ik}, B_{ikm}, and so on refer to infinite dilution of the solution, that is, $c_1 = c_2 = \cdots = c_q = 0$. Hence they are functions of temperature T and pressure P.

Now we substitute equation 3.189 into equations 3.187 and solve the resulting equation for $\partial c_i / \partial r$ to obtain

$$\frac{\partial c_i}{\partial r} = \frac{\omega^2 r}{RT} (1 - \bar{v}_i \rho) M_i c_i + \text{ higher terms in } c_1, c_2, \ldots, c_q$$

$$(i = 1, 2, \ldots, q; \quad r = r_1 \text{ or } r_2) \tag{3.190}$$

If these and equation 3.189 are introduced into equations 3.188 and if all solute components are assumed to have the same partial specific volume $\bar{v}$ and the same specific refractive index increment $\bar{R}$, it follows, after some rearrangement of terms, that

$$\frac{RT(\partial n_c / \partial r)}{\omega^2 r (1 - \bar{v} \rho)} = \bar{R} \sum_{i=1}^{q} M_i c_i - \bar{R} \sum_{i=1}^{q} \sum_{k=1}^{q} M_i M_k B_{ik} c_i c_k$$

$$+ \text{ higher terms in } c_1, c_2, \ldots, c_q \tag{3.191}$$

$$(r = r_1 \text{ or } r_2)$$

To proceed further we introduce still another assumption that $\bar{v}$ is independent of the composition of the solution.* Then it can be shown (by use of the Gibbs-Duhem relation) that the partial specific volume of the solvent may be equated to $1/\rho_0$, where ρ_0 is the density of the pure solvent. The density of the solution, ρ, is then represented by

$$\rho = \rho_0 + (1 - \bar{v}\rho_0)c \tag{3.192}$$

where c is the total solute concentration, that is,

$$c = \sum_{i=1}^{q} c_i \tag{3.193}$$

Also, if the above-mentioned assumption for the specific refractive index increment holds, it follows from equation 3.19 that

$$n_c = \bar{R}c \tag{3.194}$$

If equations 3.192 and 3.194 are used and rearrangement of terms are made, equation 3.191 may be transformed to

$$M_{\text{app}}(t) = \frac{\displaystyle\sum_{i=1}^{q} c_i M_i}{c} - \frac{\displaystyle\sum_{i=1}^{q} \sum_{k=1}^{q} M_i M_k [B_{ik} + (\bar{v}/M_k)]c_i c_k}{c}$$
$$+ \text{higher terms in } c \qquad (r = r_1 \text{ or } r_2) \tag{3.195}$$

where $M_{\text{app}}(t)$, a quantity having dimensions of molecular weight, is defined by

$$M_{\text{app}}(t) = \frac{RT(\partial n_c/\partial r)}{\omega^2 r (1 - \bar{v}\rho_0)n_c} \qquad (r = r_1 \text{ or } r_2) \tag{3.196}$$

Since n_c and $\partial n_c/\partial r$ at the ends of the solution column vary with time as sedimentation proceeds, the right-hand side of equation 3.196 generally depends on time t, as indicated by the notation $M_{\text{app}}(t)$.

In the limit when t tends to zero, equation 3.195 becomes

$$M_{\text{app}} \equiv M_{\text{app}}(0) = M_w \left[1 - M_w B_{AB} c_0 + O(c_0^2) \right] \tag{3.197}$$

where c_0 is the initial concentration of the solution, M_w is the weight-

---

*It is shown in Chapter 5, Section II.D that this assumption is equivalent to neglecting the dependence of the activity coefficient of each solute on pressure.

average molecular weight of the solute, that is,

$$M_w = \sum_{i=1}^{q} M_i g_i \qquad \left(g_i = \frac{c_i^0}{c_0} \right) \tag{3.198}$$

and B_{AB} is given by (AB means Archibald)

$$B_{AB} = \frac{\displaystyle\sum_{i=1}^{q} \sum_{k=1}^{q} M_i M_k [B_{ik} + (\bar{v}/M_k)] g_i g_k}{(M_w)^2} \tag{3.199}$$

The right-hand side of equation 3.197 contains no term which depends on r. Hence, the values of M_{app} at both ends of the solution column should agree with each other. Mathematically, equation 3.197 is equivalent to

$$\frac{1}{M_{app}} = \frac{1}{M_w} + B_{AB}c_0 + O(c_0^2) \tag{3.200}$$

which indicates that a plot of $1/M_{app}$ against c_0 allows evaluation of M_w from its ordinate intercept and of B_{AB} from its *initial* slope. Equation 3.200 has exactly the same form as the c_0 expansion for the reciprocal apparent molecular weight obtained from the theory of light scattering[51] for an incompressible multicomponent solution of the type considered here, that is, one in which the partial specific volume and the specific refractive index increment are the same for all solute components (see equation 5.176). The coefficient B_{AB} is identical to the quantity usually referred to as the light-scattering second virial coefficient, which, in this monograph, is denoted by B_{LS}. Thus the Archibald method affords the same thermodynamic information as does the light-scattering photometry, provided all solutes in the solution have the same specific refractive index increment and the same partial specific volume. Recently, Kotaka et al.[52] have broadened the range of applicability of this method by working out a theory for binary copolymers whose constituent units have different specific refractive index increments and different partial specific volumes.

Prior to Fujita et al.,[50] Kegeles et al.[49] derived an expansion similar in form to equation 3.197. However, the quantity M_{app} which appeared in their theory did not refer to zero time, but was essentially equivalent to $M_{app}(t)$ in the development given above. Since, as had early been pointed out by Archibald,[48] it was essential for the study of polydisperse solutes to use $M_{app}(t)$ extrapolated to zero time, Kegeles et al. should have elaborated their theory a step further. The coefficient for c_0^2 in equation 3.200 is not as yet available in the literature.

C. THE TRAUTMAN METHOD

Equation 3.196 may be recast in the form

$$Y = M_{app}(t)\left[(n_c - n_c^0) + n_c^0\right] \qquad (r = r_1 \text{ or } r_2) \qquad (3.201)$$

where Y stands for

$$Y = \frac{RT(\partial n_c/\partial r)}{(1 - \bar{v}\rho_0)\omega^2 r} \qquad (r = r_1 \text{ or } r_2) \qquad (3.202)$$

and n_c^0 denotes, as before, the excess refractive index of the original solution.

As is shown in Section IV.D.1, the values of $Z_1 = (n_c)_{r=r_1} - n_c^0$ and $Z_2 = (n_c)_{r=r_2} - n_c^0$ can be calculated from observed schlieren patterns, without separate determination of n_c^0. Equation 3.201 indicates that if the values of $(Y)_{r=r_1}$ and $(Y)_{r=r_2}$ determined at various times are plotted against the corresponding values of Z_1 and Z_2, then there will be obtained a single curve as schematically shown in Fig. 3.23. Proceeding downward and to the left from the point E on the vertical axis, the curve which appears in the region of negative abscissa represents the variation, as sedimentation proceeds, of $(Y)_{r=r_1}$ with Z_1. Similarly, the curve, starting also at the point E and proceeding upward and to the right, shows the increase of $(Y)_{r=r_2}$ with Z_2. The curve of $(Y)_{r=r_1}$ versus Z_1 stops at the point D, which corresponds to $(n_c)_{r=r_1} = 0$, that is, to the time when the trailing edge of the boundary gradient curve just clears the meniscus. On the other hand, the curve of $(Y)_{r=r_2}$ versus Z_2 extends up to the point which corresponds to the equilibrium value of $(n_c)_{r=r_2}$. This type of plot was first proposed by Trautman[53] and then discussed in greater detail by Trautman and Crampton.[54] It is usually referred to as *Trautman's plot*.

Experimentally, paired values of Y and Z ($= Z_1$ and Z_2) needed for the preparation of Trautman's plot can be obtained not only from the measurements at different times during a sedimentation run but also from the experiments at different rotor speeds. One of the merits of the Trautman method is that data for different rotor speeds can be incorporated into the plot. Another merit of the method is that no separate determination of n_c^0 is needed. This quantity, if desired, can be evaluated by finding the position at which the Trautman plot intersects the horizontal axis.

Now, according to equation 3.201, if $M_{app}(t)$ is independent of time, the Trautman plot should be linear and yield M_{app} from its slope and $-n_c^0$ from its intercept on the abscissa axis. In the case when $M_{app}(t)$ varies with time, the Trautman plot exhibits curvature, and the slope of the line connecting the point D (see Fig. 3.23) to a point, say A, on the curve gives

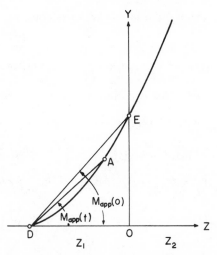

Fig. 3.23. Trautman's plot for Archibald-type ultracentrifugation. The ordinate Y is defined by equation 3.202, while $Z_1 = (n_c)_{r=r_1} - n_c^0$ and $Z_2 = (n_c)_{r=r_2} - n_c^0$.

the value of $M_{app}(t)$ at the time when Z_1 or Z_2 equals the value corresponding to the point A at the given rotor speed. Note that the value of Z_1 or Z_2 for a fixed A varies with the rotor speed used for the experiment. Since, regardless of the rotor speed, both Z_1 and Z_2 tend to zero at the start of a centrifugation, the slope of the line drawn from the point D to the point E yields $M_{app}(0)$, that is, M_{app}. In practice, one may reach the point D by conducting the experiment at a sufficiently high speed, whereas there will be a difficulty in approaching the region near the point E. Thus the point E will have to be determined by a suitable extrapolation.

As will be expounded in Section IV.E, the second and higher terms on the right-hand side of equation 3.195 are associated with the departure of a solution from thermodynamic ideality. For a solution in which these "nonideality" effects, so termed here for the convenience of description, are negligible, the quantity $M_{app}(t)$ represents the weight-average molecular weight of the solute at either the meniscus or the cell bottom as a function of time. If the solute in this solution is homogeneous with respect to molecular weight M, this quantity equals the true molecular weight of the solute, so the Trautman plot becomes linear over the entire range. If the solute is heterogeneous in M, then the value of $M_{app}(t)$ at the meniscus *decreases* with time, since the heavier solutes clear the meniscus earlier than the lighter ones. Obviously, the reverse should be the case at the cell bottom. These effects make the Trautman plot *convex downward*.

To examine nonideality effects upon the Trautman plot it is convenient to treat the case in which the solute is homogeneous. In this case, equation 3.195 reduces to

$$M_{app}(t) = M_1 - B_{AB}c_1 + \cdots \qquad (r = r_1 \text{ or } r_2) \qquad (3.203)$$

At the limit of zero time, this equation becomes

$$M_{app} = M_1 - B_{AB}c_0 + \cdots \qquad (r = r_1 \text{ or } r_2) \qquad (3.204)$$

From these equations it follows that

$$M_{app}(t) = M_{app} + B_{AB}(\Delta c_1)_m + \cdots \qquad (r = r_1) \qquad (3.205)$$

$$M_{app}(t) = M_{app} - B_{AB}(\Delta c_1)_b + \cdots \qquad (r = r_2) \qquad (3.206)$$

where

$$(\Delta c_1)_m = c_0 - (c_1)_{r=r_1}, \qquad (\Delta c_1)_b = (c_1)_{r=r_2} - c_0 \qquad (3.207)$$

Physically, it is obvious that both $(\Delta c_1)_m$ and $(\Delta c_1)_b$ increase with increasing time of centrifugation. Therefore, it follows from equations 3.205 and 3.206 that if the solute is homogeneous and, moreover, the thermodynamic factor B_{AB} is positive, the value of $M_{app}(t)$ *increases* at the meniscus and *decreases* at the cell bottom as sedimentation proceeds, provided the solution is so dilute that the contributions from the third and higher terms in these equations need not be considered. Obviously, the reverse is the case if B_{AB} is negative. Thus, nonideality effects will make the Trautman plot either convex upward or downward, depending on whether B_{AB} is positive or negative.

In sum, the curvature of Trautman's plots, and hence the time dependence of $M_{app}(t)$, is a reflection of two effects, one arising from solute heterogeneity and the other from thermodynamic nonideality of the solution. Since, in general, B_{AB} is positive, the two effects act in the opposite direction, and, under special conditions, they may exactly cancel each other to make the plot linear over a range. Therefore, a great caution must be paid in the interpretation of experimental results when a linear Trautman plot is observed. These theorectical predictions have been checked by several investigators,[55,56] using samples of typical linear homopolymers.

Figure 3.24 shows Trautman's plots for ribonuclease in acetate buffers (pH 5.5) with and without 6 M urea.[54] Here the ordinate is not Y but $(1 - \bar{v}\rho_0)Y$, while the abscissa is $(c)_{r=r_1} - c_0$. The lack of data points in the region of positive abscissa indicates that these data were all obtained at the air-liquid meniscus. The plotted points for each system contain experiments at two or three different rotor speeds: 25,980, 52,640, and 59,780

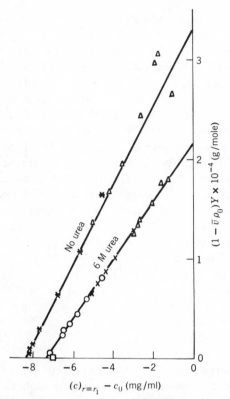

Fig. 3.24. Trautman's plots for ribonuclease in pH 5.5 acetate buffer of ionic strength 0.1 both with (right-hand line) and without (left-hand line) 6 M urea.[54] Δ, at 25,980 rev/min; *, at 52,640 rev/min; O, at 59,780 rev/min; □, average of three values at 59,780 rev/min, $c_0 = 7.07, 7.17, 7.19$ mg/ml.

rev/min. They are fitted by a straight line, which suggests that the protein preparation examined was essentially homogeneous and the nonideality effect was very small. The values of M_{app} calculated from the solid lines were found to agree closely with the results from amino acid analyses.[54]

D. EXPERIMENTAL PROBLEMS

To obtain reliable figures for $M_{app}(t)$ it is crucial to select experimental conditions so that the magnitudes of n_c and $\partial n_c/\partial r$ at the meniscus become moderately large. Apparently, this problem need not be considered for the cell bottom, where these quantities always (if the buoyancy factor is positive) become large, or even too large to be pertinent for the measurements. As shown in Chapter 2, if the parameter $\epsilon' = 2D/s\omega^2 r_1^2$ is very small, the concentration and its gradient at the meniscus diminish to zero soon after initiation of centrifugation, whereas for relatively large ϵ' these quantities remain at measurable magnitudes for a considerable interval of time during a sedimentation run. Since, in general, s/D cannot be known in advance and since r_1 is fixed to about 6 cm for ordinary ultracentrifuge cells, the rotor speed ω is the only experimentally adjustable parameter for ϵ'. For many macromolecular solutes the values of s/D are so small that relatively low speeds must be employed to obtain large ϵ', and hence accurately measurable values of n_c and $\partial n_c/\partial r$ at the meniscus. On the other hand, fairly high speeds may be used to ensure the corresponding conditions for small solutes. In practice, the best working condition for a given system will have to be sought empirically. Customarily, the sedimentation velocity experiment designed for obtaining such a condition is called *Archibald type*.

A distinguished feature of the Archibald method is that it allows the determination of molecular weights over a very wide range. By operating the ultracentrifuge at a few thousand revolutions per minute, it is possible to evaluate molecular weights up to several hundred thousand (molecular weights of usual high polymers). At the maximum operational speeds attainable in the current machines (about 60,000 rev/min), molecular weights as low as a few hundred can be determined accurately, as has been demonstrated by Klainer and Kegeles.[57]

The obvious drawback of the Archibald method is that it resorts to data at the ends of the solution column, where optical phenomena cause deterioration of schlieren patterns. The experimental determination of n_c at the ends of the solution column involves no technical difficulty, as is shown in the next section; data over the entire solution column can be used for this purpose. On the other hand, the desired $\partial n_c/\partial r$ must be found by extrapolation, with no generally accepted guiding principle being available. In this point is the real difficulty one must face in the application of the Archibald method to actual systems, and it is the principal reason that this method has become less inviting to experimentalists.

1. Evaluation of n_c at the Ends of the Solution Column

Again, let us consider ultracentrifugation of an incompressible solution in a conventional cell. The solution consists of q different nonelectrolyte solutes and a single solvent. The generalized Lamm equation 1.124 for solute i yields the following relation valid for the period of centrifugation in which there exists a plateau region:

$$\left(\frac{dc_i}{dt}\right)_{r=r_p} = -2\omega^2(s_ic_i)_{r=r_p} \tag{3.208}$$

As before, r_p denotes the radial position of an arbitrary point in the plateau region. Next, after multiplication by r, equation 1.124 is integrated from $r = r_1$ to $r = r_p$ and the condition that $(J_i)_c$ at $r = r_1$ vanishes for all times is used. Then we obtain

$$\frac{d}{dt} \int_{r_1}^{r_p} rc_i \, dr = -r_p^2 \omega^2 (s_i c_i)_{r=r_p} \qquad (3.209)$$

Combination of equations 3.208 and 3.209 gives

$$\left(\frac{dc_i}{dt} \right)_{r=r_p} = \frac{2}{r_p^2} \frac{d}{dt} \int_{r_1}^{r_p} rc_i \, dr$$

which, upon integration with respect to t and use of the condition that c_i and $(c_i)_{r=r_p}$ tend to c_i^0 at $t = 0$, yields

$$(c_i)_{r=r_p} - c_i^0 = \frac{2}{r_p^2} \int_{r_1}^{r_p} r(c_i - c_i^0) \, dr \qquad (3.210)$$

This is further transformed to give

$$(c_i)_{r=r_1} = c_i^0 - \frac{1}{r_1^2} \int_{r_1}^{r_p} r^2 \frac{\partial c_i}{\partial r} \, dr \qquad (3.211)$$

Both sides are multiplied by R_i and then summed over all solute components. Then*

$$(n_c)_{r=r_1} = n_c^0 - \frac{1}{r_1^2} \int_{r_1}^{r_p} r^2 \frac{\partial n_c}{\partial r} \, dr \qquad (3.212)$$

which indicates that the required value of n_c at the air-liquid meniscus can be calculated from the data for n_c^0 and the distribution of $\partial n_c / \partial r$ between the meniscus and the plateau region. A similar consideration yields the corresponding expression for n_c at the cell bottom, that is,

$$(n_c)_{r=r_2} = n_c^0 + \frac{1}{r_2^2} \int_{r_p}^{r_2} r^2 \frac{\partial n_c}{\partial r} \, dr \qquad (3.213)$$

---

*Equations 3.212 and 3.213 show that the quantities Z_1 and Z_2 needed for the application of Trautman's method (Section IV.C) can be determined without separate measurement of n_c^0.

These important formulas are due originally to Klainer and Kegeles.[57] It can be shown that equation 3.213 is also applicable to experiments with a synthetic boundary cell.

If equation 3.212 is applied to a freely sedimenting boundary curve, in which $(n_c)_{r=r_1} = 0$, then

$$n_c^0 r_1^2 = \int_{r_1}^{r_p} r^2 \frac{\partial n_c}{\partial r} \, dr$$

which is equivalent to equation 3.46. Both equations 3.212 and 3.213 cease to hold after the plateau region disappears. Expressions for $(n_c)_{r=r_1}$ and $(n_c)_{r=r_2}$ applicable at all times have been derived by Ginsburg et al.[58] For example, their expression for $(n_c)_{r=r_1}$ is

$$(n_c)_{r=r_1} = n_c^0 - \frac{1}{r_2^2 - r_1^2} \int_{r_1}^{r_2} (r_2^2 - r^2) \frac{\partial n_c}{\partial r} \, dr \qquad (3.214)$$

provided the initial condition of the experiment is for the conventional cell.

2. Evaluation of $\partial n_c / \partial r$ at the Ends of the Solution Column

The accuracy of $M_{app}(t)$ depends predominantly on how correctly the refractive index gradients at the ends of the solution column can be extrapolated from observed schlieren diagrams. The guideline for this operation ought to be found from a detailed investigation of the theoretical behavior of concentration gradients in the vicinity of the ends. Unfortunately, such behavior is presently known only for binary solutions, and even for these solutions no generally accepted method of extrapolation is as yet established.

Archibald[48] proposed an empirical procedure, in which values of $(1/rn_c)$ $(\partial n_c / \partial r)$ are plotted against r and extrapolated graphically to the ends of the solution column.* Smith et al.[59] applied this method to lysozyme and apurinic acid. Fujita et al.,[50] in their study on polystyrene in methyl ethyl ketone, performed the extrapolation on a plot of $\log[(1/rn_c)(\partial n_c / \partial r)]$ versus r. Still another method, used by Ginsburg et al.[58] and several others,[60] was to extrapolate a plot of $\partial n_c / \partial r$ versus r. This last one was

---

*The following relation may be used to calculate $n_c(r)$ at an arbitrary position r necessary for the application of this empirical method:

$$n_c(r) = (n_c)_{r=r_1} + \int_{r_1}^{r} \frac{\partial n_c}{\partial r} \, dr$$

where equation 3.212 is to be substituted for $(n_c)_{r=r_1}$.

subjected to a detailed theoretical investigation by LaBar.[61]

An entirely different approach to the evaluation of $\partial n_c/\partial r$ at the meniscus was introduced by Ehrenberg.[62] His method assumes that if the experiment is performed at a relatively high speed, the curve of $\partial n_c/\partial r$ versus r will become horizontal near the meniscus at a short time after initiation of the experiment. The height of the curve in this horizontal region is equated to the desired $\partial n_c/\partial r$ at the meniscus. Thus no extrapolation is involved in this method.

Peterson and Mazo[63] were probably the first to investigate theoretical conditions which will guide the extrapolation to the meniscus. However, they were unable to establish such conditions, and recommended that the extrapolation be performed with a computer by successive iteration using constants from the Fujita-MacCosham solution to the Lamm equation (Chapter 2, Section III.D). Recently, similar studies, also using the Fujita-MacCosham solution, have been reported by LaBar[61] and Paetkau.[64]

LaBar[61] examined how the slope of the concentration gradient curve for a binary solution (in which s and D are constant) varies with distance r in the regions near the ends of the solution column, in the hope that such a study would be of use for establishing conditions under which a linear extrapolation of the gradient curve is valid. From the Fujita-MacCosham theory he derived an approximate expression for $\partial^2 c/\partial r^2$ near the meniscus, omitting terms which, for most experimental conditions, contribute less than 2%. His expression reads

$$\frac{\partial^2 c}{\partial r^2} = \frac{c_0(1-2\delta/r_1)\exp(-\epsilon' T)}{(\epsilon')^2 r_1^2}\left\{(6-2\epsilon')-\frac{2\epsilon' r_1}{\sqrt{\pi D t}}\left[1+\frac{(3-\epsilon')\delta}{\epsilon' r_1}\right]\right\}$$

$$(3.215)$$

where, as in Chapter 2, Section III.D,

$$\epsilon' = \frac{2D}{s\omega^2 r_1^2}, \qquad T = \frac{4Dt}{(\epsilon')^2 r_1^2} \qquad (3.216)$$

and δ is the distance from the meniscus, that is,

$$\delta = r - r_1 \qquad (3.217)$$

LaBar states that the term containing $(Dt)^{-1/2}$ always exceeds the constant term $6-2\epsilon'$, as long as the plateau region exists in the cell. Thus, the gradient curve will have a negative slope at the meniscus, and the steepness of this slope will decrease with time, as is observed routinely in experiments.

In practice, the direct reading of data can be made reliably to within

0.015 cm of the ends, and the data used for the extrapolation are usually taken from 0.015 to 0.050–0.10 cm of the ends. Accordingly, if the term $1+(3-\epsilon')(\delta/\epsilon'r_1)$ is constrained to vary no greater than 5% for $0<\delta<0.075$ cm and $r_1=6$ cm, then the gradient curve will be linear in r near the meniscus, provided $\epsilon'>0.5$. LaBar estimated the time t^* required to achieve this linearity for practical work by imposing the condition that the gradient at $\delta=0.075$ cm be one-half its value at the meniscus (this implies a moderate slope of the gradient curve), and obtained

$$Dt^*\approx0.56\times10^{-2}(cm^2) \tag{3.218}$$

Thus the desired concentration gradient at the meniscus may be determined correctly by linear extrapolation of the gradient curve between $\delta=0.015$ and 0.10 cm if $\epsilon'>0.5$ and data at $t^*\approx10^{-2}/D$ are used.

LaBar showed that the same conclusion also holds for the behavior near the cell bottom, that is, if $\epsilon'>0.5$ and $t^*\approx10^{-2}/D$, the gradient curve near the cell bottom should be linear.

The validity of these predictions was checked with numerical data calculated from the Yphantis-Waugh series solution to the Lamm equation (see Chapter 2, Section IV.B.1).[61] Figure 3.25 illustrates the typical results. In general, it appears that more reliable figures for the concentration gradient can be obtained at the meniscus than at the cell bottom by linear extrapolation of the gradient curve.

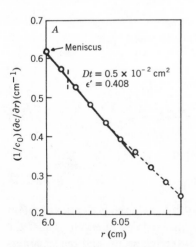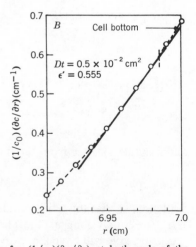

Fig. 3.25. Linear extrapolation of computer data for $(1/c_0)(\partial c/\partial r)$ at both ends of the solution column. The solid line is the best straight line for extrapolation through the points near the ends, and the dashed curve shows actual behavior of the data when the points depart from a line. In experimental situations data within 0.015 cm cannot be obtained accurately; the vertical dash marks define this interval.

Paetkau[64] proposed to calculate the following two "recip" functions from observed schlieren curves:

$$X_1 = \frac{rc}{(\partial c / \partial r)} \left(\frac{S4}{S4 + S5} \right) \frac{(1 - \bar{v}\rho_0)\omega^2}{RT}$$

(r near the meniscus)   (3.219)

$$X_2 = \frac{rc}{(\partial c / \partial r)} \left(\frac{S6}{S6 + S7} \right) \frac{(1 - \bar{v}\rho_0)\omega^2}{RT}$$

(r near the cell bottom)   (3.220)

The terms $S4$, $S5$, $S6$, and $S7$ are complex functions of r, t, s, D, and ω, which can be derived from the Fujita-MacCosham solution to the Lamm equation. The reader is referred to Paetkau's paper for the actual expressions of these functions.

The procedure offered by Paetkau consists of plotting X_1 and X_2 against r^2 [such that $(r^2 / r_1^2) > 1.012$ near the meniscus and $(r^2 / r_2^2) < 0.988$ near the cell bottom] and taking the extrapolated values at $r = r_1$ and $r = r_2$ as the desired values for $1 / M_{app}(t)$ corresponding to the particular sedimentation pattern examined. In order to perform these operations it becomes necessary to know the values of s and D for the system. Paetkau proposed the use of Baldwin's method (Chapter 2, Section III.I) for the evaluation of s. This idea is relevant because a series of schlieren patterns taken in an experiment of the Archibald type should be just suitable for the application of Baldwin's method. On the other hand, no such direct method is available for the evaluation of D. Paetkau set forth the following iteration method.

First, the concentration gradient at the meniscus is extrapolated from a given schlieren diagram by use of a suitable method, for example, the one discussed by LaBar.[61] Introduction of the result, together with the concentration at the meniscus computed from the same diagram by use of equation 3.211, into equation 3.196 yields a first approximation value to the $M_{app}(t)$ at the meniscus. With this $M_{app}(t)$ and the s value obtained by Baldwin's method, a first approximation to D may be computed from the Svedberg relation. Similarly, the corresponding D value for the cell bottom may be determined. These D values allow the various S factors in the recip functions to be computed, and then the initial Paetkau plots for the system can be constructed. The values of $M_{app}(t)$ derived from these plots may be used to find second approximations to D at the ends of the solution column. Similar operations are repeated until convergent values are obtained for $M_{app}(t)$ and D. For a systematic performance of this iteration process we may utilize a computer program worked out by Paetkau.[64]

Figures 3.26 and 3.27 are the final Paetkau plots for a solution of β-lactoglobulin B in 0.05 M Tris(phosphate) buffer at pH 7.1.[64] Note that the abscissa here is not r^2 but r. Although either X_1 and X_2 is explicitly a linear function of r^2, it will also be essentially linear in r over short distances near the ends of the solution column. For comparison, the corresponding plots of $\partial c / \partial r$ versus r are also shown in these graphs. It is seen that the recip function method of Paetkau gives an essentially horizontal curve, and allows a more reliable determination of its intercept at either end of the solution column. It should be noted that small errors in the determination of the position of meniscus or cell bottom do not seriously affect the intercept value. This feature distinguishes the Paetkau method from all other empirical procedures of extrapolation so far proposed.

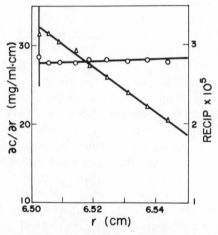

Fig. 3.26. Values of $\partial c / \partial r$ (Δ, left-hand ordinate) and the recip function (O, right-hand ordinate) for β-lactoglobulin B near the air-liquid meniscus, which is indicated by the line at $r = 6.5025$ cm. Data were obtained from a photograph taken 55 min after the rotor reached full speed.[64]

Inasmuch as the theories described above refer to binary solutions with constant s and D, the conclusions derived are of limited value from the practical point of view. Further theoretical developments with inclusion of concentration dependence of s and D and also of solute heterogeneity are highly desirable. A basis for the extension to concentration-dependent systems may be obtained, for example, from a recent exact solution of

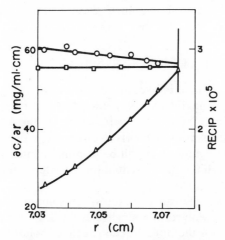

Fig. 3.27. Values of $\partial c / \partial r$ and the recip function near the cell bottom, located at $r = 7.0746$ cm, obtained from the same experimental data as in Fig. 3.26.[64] Squares show recip function values calculated from the Fujita-MacCosham equations (Chapter 2, Section III.D) for a protein of molecular weight 36,000 and a rotor speed of 12,459 rev/min.

Weiss and Yphantis[65] to the rectangular-cell approximation of the Lamm equation with $s = s_0(1 - k_s c)$.

3. Extrapolation to Zero Time

Another problem involved in actual use of the Archibald method concerns the extrapolation of $M_{app}(t)$ to zero time. This is essential for the study of pauci- or polydisperse solutions by this method. On the basis of a theoretical consideration involving the Yphantis-Waugh solution (Chapter 2, Section IV.B.1) to the Lamm equation, Yphantis[66] has recommended linear extrapolation on a plot of $M_{app}(t)$ versus $(t)^{1/2}$. This idea predicts that a plot of $M_{app}(t)$ versus t should exhibit a curvature at small values of t. However, in practice, such a curve is seldom observed, but the plot is linear in t, usually over the whole interval of an Archibald-type ultracentrifugal experiment. Thus, routinely, the extrapolation is made on a plot of $M_{app}(t)$ versus t. In this case, again, the real origin of time becomes a serious problem for the rigorous determination of $M_{app}(0)$, but no theory is as yet worked out for the necessary zero time correction. Often, Trautman's time for two-thirds acceleration (see Chapter 2, Section V.D.2) is used as a rough approximation.

For proteins or other biochemical preparations of relatively low molecular weights the observed values of $M_{app}(t)$ are essentially independent of time, and no extrapolation to zero time is required. This feature is indicative of the situation in which these macromolecular solutes are essentially homogeneous and their dilute solutions are not very nonideal.

E. NONIDEALITY COEFFICIENTS

In the theory presented above the coefficients B_{ik}, B_{ikm}, and so on have been introduced as mathematical parameters for representing the concentration dependence of activity coefficients y_i in the region of low solute concentrations. This section purports to discuss how they are related to the thermodynamics of solutions. It will be seen in Chapter 5 that comprehension of this relationship is also essential for the interpretation of sedimentation equilibrium data.

As before, we consider an incompressible solution of q different nonelectrolyte solutes dissolved in a single solvent 0. Let us denote by f_i^∞ the activity coefficient on the mole fraction scale of solute i which is defined so that it approaches unity as the concentrations of all solutes, $c_1, c_2, \ldots, c_q$ tend to zero. According to thermodynamics,[67] this solution is thermodynamically ideal if all f_i^∞ are unity at any composition of the system. Therefore, the extent to which $\ln f_i^\infty$ differs from zero may be taken as a measure of thermodynamic nonideality of the solution.

Since the solutes are assumed to be nonelectrolytes, the quantity $\ln f_i^\infty$ for small values of $c_1, c_2, \ldots, c_q$ may be expressed in Taylor's series as

$$\ln f_i^\infty = M_i \sum_{k=1}^{q} A_{ik} c_k + M_i^2 \sum_{k=1}^{q} \sum_{m=1}^{q} A_{ikm} c_k c_m + \cdots \qquad (3.221)$$

where the expansion coefficients A_{ik}, A_{ikm}, and so on refer to infinite dilution of the solution. If the solution is ideal, these A coefficients all must vanish. Hence, in the range of solute concentrations for which this expansion converges, the degree of nonideality of a solution depends on how much they deviate from zero.

It can be shown that y_i and f_i^∞ are related to each other by the equation

$$f_i^\infty = \left[1 - \sum_{k=1}^{q} \left(\bar{v}_k - \frac{M_0 \bar{v}_0}{M_k} \right) c_k \right] y_i \qquad (3.222)$$

where $\bar{v}_k$ is the partial specific volume of component k. Introduction of equations 3.189 and 3.221 into this equation, followed by expansion of the resulting relation in powers of solute concentrations, yields

$$B_{ik} = A_{ik} + \frac{\bar{v}_k}{M_i} - \frac{\bar{v}_0 M_0}{M_i M_k} \qquad (i, k = 1, 2, \ldots, q) \qquad (3.223)$$

Similar and more complex expressions are available for B_{ikm} and higher B coefficients, but they are not shown here.

Equation 3.223 indicates that B_{ik} varies with thermodynamic nonideality of the solution. However, it is not legitimate to regard the magnitude of B_{ik} as a direct measure of thermodynamic nonideality, because this coefficient does not vanish for ideal solutions. The same argument applies for the higher B coefficients. These coefficients nonetheless are often referred to as nonideality coefficients. Moreover, terms containing them, such as the factor B_{AB} defined by equation 3.199, are usually called nonideality terms. These conventions are used throughout the subsequent presentation.

1. Pseudoideality

If the partial specific volumes of all solute components are identical (say $\bar{v}$), it follows from equation 3.223 that

$$B_{ik}^* \equiv B_{ik} + \frac{\bar{v}}{M_k} = A_{ik} + \frac{\bar{v}(M_i + M_k)}{M_i M_k} - \frac{\bar{v}_0 M_0}{M_i M_k}$$

$$(i, k = 1, 2, \ldots, q) \qquad (3.224)$$

The new coefficient B_{ik}^* generally remains nonzero for a thermodynamically ideal solution. It can be shown (Appendix D) that A_{ik} is invariant for the exchange of i and k. Hence we find from equation 3.224 that B_{ik}^* also has the same property, that is,

$$B_{ik}^* = B_{ki}^* \qquad (3.225)$$

Now, equation 3.224 indicates that there will be a system in which B_{ik}^* vanishes for all pairs of i and k. In this special system, the factor B_{AB} becomes zero, as can be seen from equation 3.199. Also it is shown in Chapter 5 that osmotic second virial coefficient B_{OS}, light-scattering second virial coefficient B_{LS}, and sedimentation equilibrium second virial coefficient B_{SE} simultaneously vanish when $B_{ik}^* = 0$ for all pairs of i and k. In this monograph, a solution which satisfies this condition for B_{ik}^* will be called *pseudoideal*. Some authors refer to it as ideal, but this name does not seem proper because B_{ik}^* generally does not vanish under thermodynamic ideality.

Mathematically, the condition of simultaneous vanishing of the second virial coefficients B_{OS}, B_{LS}, and B_{SE} (note that $B_{AB} = B_{LS}$, as pointed out in Section IV.B) does not ensure pseudoideality of a given solution, except for the case of binary solutions. However, since the values of individual B_{ik}^* are experimentally not measurable, it is assumed, in practice, that this condition, or even the vanishing of any one of these virial coefficients, is sufficient for a solution to be pseudoideal.

The pseudoideal solution thus defined is not a mere hypothetical existence. In fact, it has been shown experimentally for a great many synthetic polymers that the condition of pseudoideality can be realized if a combination of solvent and temperature is suitably chosen for a given polymer species. In polymer chemistry, such specific solvent and associated temperature are called *theta solvent* and *theta temperature*, or simply *theta conditions*.[46] For example, cyclohexane and 35°C are the best known examples of such conditions for polystyrene. Elias et al.[68] have tabulated the theta conditions for a great variety of polymeric substances. The state of polymer molecules under theta conditions has a very basic importance in the theory of dilute polymer solutions. For a comprehension of this and related subjects the reader should consult the classic monograph of Flory[46] or more recently published books by Tanford[32] and Yamakawa.[33]

The availability of theta conditions for a given macromolecular substance will facilitate determination of M_w by the Archibald method, because a plot of $1/M_{app}$ versus c_0 then has zero initial slope so that its extrapolation to infinite dilution may be made with a higher degree of certainty. It will be seen in Chapter 5 that the same argument applies for determinations of M_n by osmotic pressure measurements and of M_w by light-scattering or sedimentation equilibrium measurements. However, at present, it is in no way disadvantageous for ultracentrifugal studies that given solutions are not thermodynamically pseudoideal.

The pseudoideality condition is not generally available for solutions of macromolecules of biological interest. The second virial coefficients of these solutions are positive, in general, being of the order of 10^{-4} or larger in units of ml-mole/g^2. According to the theory,[32] this feature may be attributed chiefly to a compact and rigid structure of the solute molecule. It must be noted, however, that virtually zero or even negative second virial coefficients are sometimes obtained for native proteins in aqueous buffers. As will be expounded in Chapter 6, such results are usually interpreted as due to self-association of protein molecules.

APPENDIX A

Equation 3.46 may be rewritten

$$(r_M')^2 = \frac{r_1^2 n_c^{\ 0}}{\sum\limits_{i=1}^{q} (n_i)_{r=r_p}} \tag{A.1}$$

where $(n_i)_{r=r_p}$ is the excess refractive index of solute i in the plateau region. For a freely sedimenting boundary of an ultracentrifugally ideal solution equation 2.99 may be applied to $(c_i)_{r=r_p}$, the concentration of solute i in the plateau region. Since $(n_i)_{r=r_p} = R_i (c_i)_{r=r_p}$, we obtain

$$(n_i)_{r=r_p} = R_i c_i^0 \exp(-2s_i \omega^2 t) = n_i^0 \exp(-2s_i \omega^2 t) \qquad (A.2)$$

Substitution of this into equation A.1, followed by expansion of the exponential terms, yields

$$(r_M')^2 = \frac{r_1^2}{1 - 2 \sum_{i=1}^{q} \nu_i s_i \omega^2 t + \cdots} = \frac{r_1^2}{1 - 2\bar{s}\omega^2 t + \cdots} \qquad (A.3)$$

Thus, correct to the order of $\bar{s}\omega^2 t$,

$$r_M' = r_1(1 + \bar{s}\omega^2 t) \qquad (A.4)$$

Equation 3.109 may be written

$$(r_M')^2 = (r_m)^2 + \sigma_2 \qquad (A.5)$$

If the distribution of s is nearly symmetric and the average diffusion coefficient $\bar{D}$ is not too large, it is seen from equation 3.110 that σ_2 is very small. Then the second term on the right-hand side of equation A.5 may be neglected, and use of equation A.4 gives

$$r_m = r_1(1 + \bar{s}\omega^2 t) \qquad (A.6)$$

which is the desired result.

APPENDIX B

THE POISSON DISTRIBUTION

This is a one-parameter distribution function defined by

$$g(M) = \frac{\nu}{\nu+1} \left(\frac{P}{M_0} \right) \frac{e^{-\nu} \nu^{P-1}}{P!} \qquad \left(P = \frac{M}{M_0} \right) \qquad (B.1)$$

where M_0 is the molar weight of a monomer unit, and ν is an adjustable (positive) parameter. Successive average molecular weights are expressed as follows:

$$M_n = M_0 \nu, \qquad M_w = M_0(\nu+1), \qquad M_z = \frac{M_0(\nu^2 + 3\nu + 1)}{\nu+1}, \qquad \cdots \qquad (B.2)$$

Hence

$$\frac{M_w}{M_n} = 1 + \frac{1}{\nu}, \qquad \frac{M_z}{M_w} = 1 + \frac{\nu}{(\nu+1)^2}, \qquad \cdots \qquad (B.3)$$

It is seen from these relations that ν is equal to the number-average degree of polymerization, P_n, and that the ratio of M_w to M_n differs from unity only by P_n^{-1}. Thus the Poisson distribution represents a very narrow distribution of M. "Living" polymers are often characterized approximately by this distribution function. Figure 3.28 shows the $g(M)$ curves calculated from equation B.1 for three different ν.

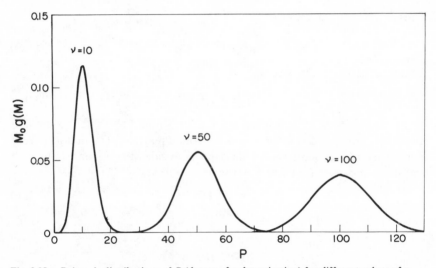

Fig. 3.28. Poisson's distributions of P (degree of polymerization) for different values of ν.

THE SCHULZ-ZIMM DISTRIBUTION

This is a two-parameter distribution function* defined by

$$g(M) = \frac{\alpha^{h+1}}{\Gamma(h+1)} \frac{P^h}{M_0} \exp(-\alpha P) \qquad (B.4)$$

where h and α are adjustable (positive) parameters and Γ denotes the

*Some authors refer to this as the exponential distribution.

gamma function. Successive average molecular weights are expressed as follows:

$$M_n = M_0\left(\frac{h}{\alpha}\right), \quad M_w = \frac{M_0(h+1)}{\alpha},$$

$$M_z = \frac{M_0(h+2)}{\alpha}, \quad M_{z+1} = \frac{M_0(h+3)}{\alpha}, \quad \cdots \quad \text{(B.5)}$$

from which

$$\frac{M_w}{M_n} = \frac{h+1}{h}, \quad \frac{M_z}{M_w} = \frac{h+2}{h+1}, \quad \frac{M_{z+2}}{M_{z+1}} = \frac{h+3}{h+2}, \quad \cdots \quad \text{(B.6)}$$

Thus the ratios of any pairs of average molecular weights are uniquely determined by h. In particular, when $h = 1$, $M_n : M_w : M_z : \cdots = 1 : 2 : 3 : \ldots$. The distribution for this special h is called "most probable." As h is increased, with the ratio h/α fixed constant, the ratios of all the average molecular weights tend to unity, which implies that the molecular weight distribution becomes narrower. Deviation of h^{-1} from zero thus can be taken as a measure of the spread of distribution. The general features of the $g(M)$ curve given by equation B.4 may be seen from Figure 3.29. Polymer samples prepared by radical polymerization or by condensation polymerization often have molecular weight distributions fitted approximately by equation B.4.

THE LOGARITHMIC NORMAL DISTRIBUTION

This is another two-parameter distribution function defined by

$$g(M) = \frac{\exp(-\beta^2/4)}{M_0 \beta P_* \sqrt{\pi}} \exp\left\{-\frac{[\ln(P/P_*)]^2}{\beta^2}\right\} \quad \text{(B.7)}$$

Here P_* and β are adjustable (positive) parameters. The name "logarithmic normal" stems from the fact that when plotted against $\ln P$, the $g(M)$ given by equation B.7 assumes a Gaussian form. As is the case with the Poisson distribution and the Schulz-Zimm distribution, equation B.7 gives a curve having one maximum, which occurs at $P = P_*$. Average molecular weights of a sample with this distribution are given by

$$M_n = M_0 P_* \exp\left(\frac{\beta^2}{4}\right), \quad M_w = M_0 P_* \exp\left(\frac{3\beta^2}{4}\right)$$

$$M_z = M_0 P_* \exp\left(\frac{5\beta^2}{4}\right), \quad M_{z+1} = M_0 P_* \exp\left(\frac{7\beta^2}{4}\right), \quad \cdots \quad \text{(B.8)}$$

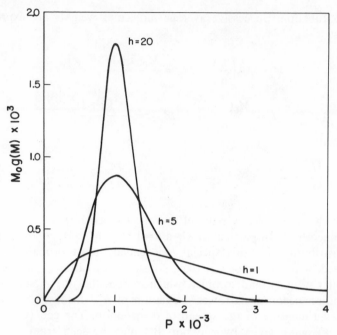

Fig. 3.29. Schulz-Zimm's distributions of P for $h/\alpha = 1$ and different values of h.

from which

$$\frac{M_w}{M_n} = \exp\left(\frac{\beta^2}{2}\right), \qquad \frac{M_z}{M_w} = \exp\left(\frac{\beta^2}{2}\right), \qquad \frac{M_{z+1}}{M_z} = \exp\left(\frac{\beta^2}{2}\right), \qquad \cdots$$

$$(B.9)$$

Thus the ratios of successive average molecular weights are constant. When β tends to zero, this constant approaches unity, implying that deviation of β from zero is a measure of the spread of logarithmic normal distribution. Sometimes, equation B.7 is referred to as the Lansing-Kraemer distribution function.[71] Figure 3.30 shows its shapes for different values of β.

THE TUNG DISTRIBUTION

This is a purely empirical, two-parameter distribution function defined as[72]

$$g(M) = \frac{mn}{M_0} P^{n-1} \exp(-mP^n) \qquad (B.10)$$

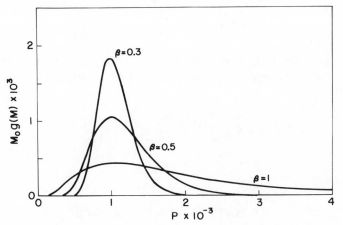

Fig. 3.30. Logarithmic normal distributions of P for $P_* = 1000$ and various values of β.

where m and n are adjustable (positive) parameters. Comparison with equation B.4 shows that this distribution is an extension of the Schulz-Zimm distribution. Average molecular weights calculated from equation B.10 are as follows:

$$M_n = M_0 m^{-1/n} \left[\Gamma\left(1 - \frac{1}{n}\right) \right]^{-1}, \qquad M_w = M_0 m^{-1/n} \Gamma\left(1 + \frac{1}{n}\right)$$

$$M_z = \frac{M_0 m^{-1/n} \Gamma(1 + 2/n)}{\Gamma(1 + 1/n)}, \qquad \cdots \tag{B.11}$$

The integral Tung distribution is

$$G(M) = \frac{1 - \exp(-mP^n)}{M_0} \tag{B.12}$$

This indicates that if $\ln\{\ln[1/(1 - M_0 G(M)]\}$ is plotted against $\ln P$, a straight line is obtained, and the parameters n and m can be determined from its slope and ordinate intercept.

APPENDIX C

Let us define the quantities A and B by

$$A = \left(\sum_{i=1}^{q} g_i M_i^{m+1} \right) \left(\sum_{j=1}^{q} g_j M_j^{m-1} \right) \tag{C.1}$$

$$B = \left(\sum_{i=1}^{q} g_i M_i^{m} \right)^2 \tag{C.2}$$

where, as in the text, g_i and M_i are the weight fraction and molecular weight of the ith fraction, and m is an arbitrary number. It can be shown easily that

$$A - B = \sum_{i=1}^{q-1} \sum_{j>i}^{q} g_i g_j (M_i M_j)^{n} f\left(\frac{M_i}{M_j} \right) \tag{C.3}$$

where

$$f(x) = \left(\sqrt{x} - \frac{1}{\sqrt{x}} \right)^2 \qquad (x>0) \tag{C.4}$$

The function $f(x)$ for $x>0$ vanishes at $x=1$ and is positive otherwise. Hence, $A - B > 0$ for any sample heterogeneous in molecular weight. For monodisperse samples we have $g_i g_j = 0$ if $i \neq j$, so that $A - B = 0$. Thus,

$$\frac{\sum_{i=1}^{q} g_i M_i^{m+1}}{\sum_{i=1}^{q} g_i M_i^{m}} \geqslant \frac{\sum_{i=1}^{q} g_i M_i^{m}}{\sum_{j=1}^{q} g_j M_j^{m-1}} \tag{C.5}$$

where the equality sign holds only for monodisperse samples. Inequality C.5 leads to the desired result: $M_n \leqslant M_w \leqslant M_z \leqslant M_{z+1} \leqslant \cdots$. For example, if m is set equal to zero, we obtain $M_n \leqslant M_w$. For $m=2$ we find $M_z \leqslant M_{z+1}$.

APPENDIX D

If the expression

$$\mu_i = (\mu_i^{\infty})_x + \frac{RT}{M_i} \ln (f_i^{\infty} x_i) \tag{D.1}$$

where x_i is the mole fraction of component i, is substituted into the identity

$$M_i \left(\frac{\partial \mu_i}{\partial n_j} \right)_{T,P,c_{k \neq j}} = M_j \left(\frac{\partial \mu_j}{\partial n_i} \right)_{T,P,c_{k \neq i}} \tag{D.2}$$

where n_i is the number of moles of component i, we obtain

$$\left(\frac{\partial \ln f_i^\infty}{\partial n_j} \right)_{T,P,c_{k\neq j}} = \left(\frac{\partial \ln f_j^\infty}{\partial n_i} \right)_{T,P,c_{k\neq i}} \tag{D.3}$$

On the other hand, if use is made of the relation

$$c_i = \frac{M_i n_i}{V} = \frac{M_i n_i}{\sum\limits_{j=0}^{q} \bar{V}_j n_j} \tag{D.4}$$

where $\bar{V}_j$ is the partial molar volume of component j, equation 3.221 may be written

$$\ln f_i^\infty = \frac{M_i}{\bar{V}_0 n_0} \sum_{k=1}^{q} A_{ik} M_k n_k + \text{higher terms in } n_k \qquad (k \neq 0) \tag{D.5}$$

The subscript 0 refers to the solvent component. Introducing equation D.5 into equation D.3 and letting the concentrations of all solute components approach zero, we arrive at the desired relation $A_{ik} = A_{ki}$, provided $i \neq 0$ and $k \neq 0$.

REFERENCES

1. T. Svedberg, *Kolloid-Z.* **85**, 119 (1938).
2. For a comparison of the optical systems used in the current ultracentrifuges, see an excellent table prepared by C. H. Chervenka, *A Manual of Methods for the Analytical Ultracentrifuge*, Spinco Division of Beckman Instruments, Stanford Industrial Park, Palo Alto, California, 1970, p. 4.
3. H. K. Schachman, *Biochemistry* **2**, 887 (1963).
4. S. Hanlon, K. Lamers, G. Lauterbach, R. Johnson, and H. K. Schachman, *Arch. Biochem. Biophys.* **99**, 157 (1962); H. K. Schachman, L. Gropper, S. Hanlon, and F. Putney, *ibid.* **99**, 175 (1962); K. Lamers, F. Putney, I. Z. Steinberg, and H. K. Schachman, *ibid.* **103**, 379 (1963).
5. H. K. Schachman and S. J. Edelstein, *Biochemistry* **5**, 2681 (1966).
6. I. Z. Steinberg and H. K. Schachman, *Biochemistry* **5**, 3728 (1966).
7. R. Trautman, V. N. Schumaker, W. F. Harrington, and H. K. Schachman, *J. Chem. Phys.* **22**, 555 (1954).
8. A. S. McFarlane, *Biochem. J.* **29**, 407, 660 (1935).
9. K. O. Pedersen, *Nature* **138**, 363 (1936); see also Ref. 19, p. 408 *et seq.*
10. J. P. Johnston and A. G. Ogston, *Trans. Faraday Soc.* **42**, 789 (1946).
11. H. K. Schachman, *Ultracentrifugation in Biochemistry*, Academic Press, New York, 1959, p. 108 *et seq.*

12. R. Trautman and V. N. Schumaker, *J. Chem. Phys.* **22**, 551 (1954).
13. L. G. Longsworth, *J. Am. Chem. Soc.* **67**, 1109 (1945).
14. V. P. Dole, *J. Am. Chem. Soc.* **67**, 1119 (1945).
15. W. F. Harrington and H. K. Schachman, *J. Am. Chem. Soc.* **75**, 3533 (1953).
16. A. Soda, T. Fujimoto, and M. Nagasawa, *J. Phys. Chem.* **71**, 4274 (1967).
17. J. Vinograd and R. Bruner, *Biopolymers* **4**, 131 (1966).
18. J. Vinograd, R. Bruner, R. Kent, and J. Weigle, *Proc. Natl. Acad. Sci., U. S.* **49**, 902 (1963).
19. T. Svedberg and K. O. Pedersen, *The Ultracentrifuge*, Oxford University Press, London and New York (Johnson Reprint Corporation, New York), 1940, p. 325 *et seq.*
20. H. Rinde, "The Distribution of the Sizes of Particles in Gold Sols Prepared According to the Nuclear Method," Thesis, Uppsala, 1928.
21. R. Signer and H. Gross, *Helv. Chim. Acta* **17**, 726 (1934).
22. W. B. Bridgman, *J. Am. Chem. Soc.* **64**, 2349 (1942).
23. R. L. Baldwin, *J. Phys. Chem.* **58**, 1081 (1954).
24. R. L. Baldwin and J. W. Williams, *J. Am. Chem. Soc.* **72**, 4325 (1950).
25. J. W. Williams, R. L. Baldwin, W. M. Saunders, and P. G. Squire, *J. Am. Chem. Soc.* **74**, 1542 (1952).
26. L. J. Gosting, *J. Am. Chem. Soc.* **74**, 1548 (1952).
27. H. Fujita, *Biopolymers* **7**, 59 (1969).
28. D. G. Sharp, M. H. Hebb, A. R. Taylor, and J. W. Beard, *J. Biol. Chem.* **142**, 217 (1942).
29. R. A. Alberty, *J. Am. Chem. Soc.* **70**, 1675 (1948).
30. R. L. Baldwin, *J. Phys. Chem.* **63**, 1570 (1959).
31. A. F. V. Eriksson, *Acta Chem. Scand.* **10**, 360 (1956).
32. C. Tanford, *Physical Chemistry of Macromolecules*, J. Wiley, New York, 1961.
33. H. Yamakawa, *Modern Theory of Polymer Solutions*, Harper & Row, New York, 1971.
34. R. L. Baldwin, *J. Am. Chem. Soc.* **76**, 402 (1954).
35. R. L. Baldwin, *Biochem. J.* **65**, 490 (1957).
36. Ref. 11, p. 145 *et seq.*
37. J. W. Williams, W. M. Saunders, and J. S. Cicirelli, *J. Phys. Chem.* **58**, 774 (1954); see also J. W. Williams and W. M. Saunders, *ibid.* **58**, 854 (1954).
38. R. L. Baldwin, L. J. Gosting, J. W. Williams, and R. A. Alberty, *Discuss. Faraday Soc.* **20**, 13 (1955).
39. N. Gralén and G. Lagermalm, *J. Phys. Chem.* **56**, 514 (1952).
40. A. Yamamoto, I. Noda, and M. Nagasawa, *Polymer J.* **1**, 304 (1970).
41. T. Kotaka and N. Donkai, *J. Polymer Sci.* *A-2* **6**, 1457 (1968).
42. I. H. Billick, *J. Polymer Sci.* **62**, 167 (1962).
43. M. Wales and S. Rehfeld, *J. Polymer Sci.* **62**, 179 (1962).
44. J. E. Blair and J. W. Williams, *J. Phys. Chem.* **68**, 161 (1964).
45. M. Wales, in *Characterization of Macromolecular Structure*, Natl. Acad. Sci., Washington, D. C., 1968, p. 343 *et seq.*
46. P. J. Flory, *Principles of Polymer Chemistry*, Cornell University Press, Ithaca, New York, 1953.

47. H. W. McCormick, *J. Polymer Sci.* **36**, 341 (1959).
48. W. J. Archibald, *J. Phys. Colloid Chem.* **51**, 1204 (1947).
49. G. Kegeles, S. M. Klainer, and W. J. Salem, *J. Phys. Chem.* **61**, 1286 (1957).
50. H. Fujita, H. Inagaki, T. Kotaka, and H. Utiyama, *J. Phys. Chem.* **66**, 4 (1962).
51. J. G. Kirkwood and R. J. Goldberg, *J. Chem. Phys.* **18**, 54 (1950); W. H. Stockmayer, *ibid.* **18**, 58 (1950).
52. T. Kotaka, N. Donkai, H. Ohnuma, and H. Inagaki, *J. Polymer Sci.* *A*-2 **6**, 1803 (1968).
53. R. Trautman, *J. Phys. Chem.* **60**, 1211 (1956).
54. R. Trautman and C. F. Crampton, *J. Am. Chem Soc.* **81**, 4036 (1959).
55. H. Inagaki, S. Kawai, and A. Nakazawa, *J. Polymer Sci.* **A1**, 3303 (1963).
56. Y. Toyoshima and H. Fujita, *J. Phys. Chem.* **68**, 1378 (1964).
57. S. M. Klainer and G. Kegeles, *J. Phys. Chem.* **59**, 952 (1955).
58. A. Ginsburg, P. Appel, and H. Schachman, *Arch. Biochem. Biophys.* **65**, 545 (1956).
59. D. B. Smith, G. C. Wood, and P. A. Charlwood, *Can. J. Chem.* **34**, 364 (1956).
60. N. E. Weston and F. W. Billmayer, *J. Phys. Chem.* **67**, 2728 (1963).
61. F. E. LaBar, *Biochemistry* **5**, 2362 (1966); see also F. E. LaBar, *ibid.* **5**, 2368 (1966).
62. A. Ehrenberg, *Acta Chem. Scand.* **11**, 1257 (1957).
63. J. M. Peterson and R. M. Mazo, *J. Phys. Chem.* **65**, 566 (1961).
64. V. H. Paetkau, *Biochemistry* **6**, 2767 (1967).
65. G. H. Weiss and D. A. Yphantis, *J. Chem. Phys.* **42**, 2117 (1965).
66. D. A. Yphantis, *J. Phys. Chem.* **63**, 1742 (1959).
67. For example, J. G. Kirkwood and I. Oppenheim, *Chemical Thermodynamics*, McGraw-Hill, New York, 1961.
68. H.-G. Elias, G. Adank, Hj. Dietschy, O. Etter, U. Gruber, and F. W. Ibrahim, in *Polymer Handbook* (J. Brandrup and E. H. Immergut, Eds.), Interscience, New York, 1966, IV-163.
69. G. V. Schultz, *Z. Phys. Chem.* **B43**, 25 (1939).
70. B. H. Zimm, *J. Chem. Phys.* **16**, 1099 (1948).
71. W. P. Lansing and E. O. Kraemer, *J. Am. Chem. Soc.* **57**, 1369 (1935).
72. L. H. Tung, *J. Polymer Sci.* **20**, 495 (1956).

SEDIMENTATION TRANSPORT IN CHEMICALLY REACTING SYSTEMS

I. INTRODUCTION

A. THE SCOPE OF THIS CHAPTER

Up to this point we have proceeded with the assumption that any component in a given system does not react chemically with itself or with coexisting components. In what follows, we remove this assumption, and formulate the sedimentation transport of solutes which interact chemically. Such a formulation not only is theoretically interesting but also has considerable biochemical significance, because many substances of biological interest, especially proteins, are known to undergo self-association, isomerization, bimolecular complex formation, and so on, under appropriate solvent conditions. Thus, proteins such as insulin, adrenocorticopropin, chymotrypsin, lysozyme, γ-globulin, and β-lactoglobulin self-associate to form one or more oligomers. The specific combination of proteolytic enzymes with their substrates and that of antigens with antibodies are examples of complex formation. In view of the nature of this book, we do not enter into the details of this subject. The reader is referred to an excellent review written by Reithel.[1] Several physicochemical techniques are available for studying chemical interactions in dilute solutions. These are sedimentation velocity and equilibrium, diffusion, electrophoresis, chromatography, light scattering, osmometry, depolarization of fluorescence, and so forth. No single approach is sufficient to draw a definite conclusion about the nature of reactions involved in a given system. The review article by Nichol et al.[2] is one of the most pertinent sources of information about these methods.

In Section II, we derive continuity equations for chemically reacting systems. The equations are written for reacting solutes and also for newly defined entities called constituents. Sections III, IV, and V are devoted to integration of the continuity equations for three typical reactions: self-association, isomerization, and biomolecular complex formation. In these efforts, account of the mathematical processes is limited to a minimum, and emphasis is placed on clarifying how much the shape of the sedimentation boundary curve is affected by chemical interactions between sedimenting solutes.

B. EFFECTS OF REACTION RATES

For a given type of reaction, the extent to which the sedimentation boundary is affected by chemical interactions depends on the relative rates of forward and backward reactions and also on their magnitudes relative to the rate of sedimentation. For example, let us consider reversible association-dissociation between a monomer P_1 and its dimer P_2, which is schematically represented by

$$2P_1 \underset{k_b}{\overset{k_f}{\rightleftarrows}} P_2 \tag{4.1}$$

with k_f and k_b being the rate constants for association of P_1 and dissociation of P_2, respectively.

If $k_f \ll k_b$ (or $k_f \gg k_b$), the system will behave ultracentrifugally as if it contained only monomer P_1 (or dimer P_2) as a solute, because dimers (or monomers), when produced, immediately dissociate (or associate) to P_1 (or P_2). If k_f and k_b are much smaller than the parameter characterizing the rate of sedimentation, no interconversion between P_1 and P_2 will actually take place during the period of time of a sedimentation experiment. Thus, the system will behave ultracentrifugally as if it contained two nonreacting solutes, and the sedimentation boundaries may be analyzed by the methods described in Chapter 3.

On the other hand, if k_f and k_b are sufficiently large (but their ratio remains finite), that is, both association and dissociation occur much more rapidly than sedimentation transports of the two solute species, one may consider that chemical equilibrium is attained instantaneously between P_1 and P_2 at any position in the system. Thus, the two solutes may no longer be regarded as independent thermodynamic components, and they will behave ultracentrifugally as a single substance. Is the sedimentation boundary in this extreme case essentially similar to that of a binary solution consisting of a nonreacting solute and a solvent, or does it exhibit features quite different from those expected for the latter system? These questions led Gilbert and collaborators to a series of theoretical investigations which have contributed very importantly to the formation of our current concepts on sedimentation and electrophoresis of chemically reacting solutes.

Finally, when the rates of association and dissociation are comparable to that of sedimentation, the situation naturally becomes intermediate between the above-mentioned extreme cases of k_f and k_b. In this case, P_1 and P_2 are converted to P_2 and P_1, respectively, at certain finite rates which are not large enough for the two solutes to establish chemical equilibrium in a short interval of time. Such rate processes make mathematical treatment of the problem very difficult. Known solutions, either analytical or numerical, to the continuity equations for systems involving reactions of finite rates are still very limited in number and in scope. The description of such systems, therefore, cannot help becoming quite fragmentary, as will be seen below.

II. BASIC EQUATIONS

A. CONTINUITY EQUATIONS

When solute components interact chemically with each other, the continuity equations derived in Chapter 1 for nonreacting solutes must be corrected for the production and disappearance of individual solutes. In order to do this, we assume, for simplicity of mathematical treatments and as a reasonable physical approximation, that the partial specific volume of any component or any reacting species in the system is independent of concentration and pressure.

Now, suppose that when we dissolve q (nonelectrolyte) solute components in a solvent (component 0), p additional chemical species are produced by chemical reactions. Then we have $q + p$ solutes in component 0. We label these solutes as $1, 2, \ldots, n$ $(= q + p)$, and denote the c-concentration of solute k at a radial position r and time t by $c_k(r, t)$.

Referring to a cylindrical slice of volume in the ultracentrifuge cell shown in Fig. 1.1, the rate of change with time of the amount of solute k in this slice must be equal to the sum of the following two contributions:

1. The amount (in grams) of solute k passing, in the centrifugal direction, through the surface at r in unit time minus the amount (in grams) passing, in the same direction, through the surface at $r + \delta r$ in the same interval of time.

2. The amount (in grams) of solute k produced in this slice by chemical reactions in unit time at the time being considered. This contribution is absent in nonreacting systems.

We denote the production of solute k *per unit time per unit volume of solution* by Q_k, and go to the limit $\delta r \to 0$. Then the following equation may be derived:

$$\frac{\partial c_k}{\partial t} = - \frac{\partial}{r \, \partial r} [r (J_k)_c] + Q_k \qquad (k = 1, 2, \ldots, n) \qquad (4.2)$$

where $(J_k)_c$ (in g/sec-cm^2) is the flow of solute k relative to the cell. Equation 4.2 is the desired continuity equation for solute k when chemical reactions occur.

For $(J_k)_c$ we may substitute equation 1.102, with q being replaced by n. For Q_k, which may be a function of $c_1, c_2, \ldots, c_n$, detailed knowledge about the chemical reactions is required. Among the solutes originally dissolved in the solvent there may be some which do not take part in chemical reaction. It is evident that for such solutes the terms Q_k are identically zero. In the present discussion, we shall assume, for simplicity of treatments, that no such solute exists in the solution, that is, all solutes are either reactants or products, unless otherwise stated.*

Equation 4.2 is summed over all solutes to give

$$\frac{\partial c}{\partial t} = -\frac{\partial}{r\,\partial r}(rJ_c) + \sum_{k=1}^{n} Q_k \tag{4.3}$$

where c is the total solute concentration at the position and time considered, that is,

$$c = c_1 + c_2 + \cdots + c_n \tag{4.4}$$

and J_c is the total flow of solutes relative to the cell at the same position and time, that is,

$$J_c = (J_1)_c + (J_2)_c + \cdots + (J_n)_c \tag{4.5}$$

Since the sum term on the right-hand side of equation 4.3 vanishes by virtue of the law of conservation of mass, equation 4.3 reduces to

$$\frac{\partial c}{\partial t} = -\frac{\partial}{r\,\partial r}(rJ_c) \tag{4.6}$$

which indicates that even when individual solutes react chemically, the total solute concentration and the flow of all solutes relative to the c satisfy the continuity equation of the same form as that for a nonreactiı solute component.

Substitution for $(J_k)_c$ in equation 4.5 from equation 1.102 and rearrangement yield

$$J_c = \bar{s}c\omega^2 r - \bar{D}\frac{\partial c}{\partial r} \tag{4.7}$$

where $\bar{s}$ and $\bar{D}$ are defined by

$$\bar{s} = \frac{\sum_{k=1}^{n}(s_k)_v c_k}{\sum_{k=1}^{n} c_k} \tag{4.8}$$

*Furthermore, we assume that all reactions considered below are reversible.

and

$$\overline{D} = \frac{\sum_{k=1}^{n}\sum_{j=1}^{n}(D_{kj})_{\text{v}}(\partial c_j / \partial r)}{\sum_{k=1}^{n}(\partial c_k / \partial r)} \tag{4.9}$$

It can be recognized that the quantity $\bar{s}$ represents the *weight-average* of sedimentation coefficients (referred to the center of volume) of all solutes present in a solution in which the concentration of solute i is c_i ($i = 1, 2, \ldots, n$). For the quantity $\overline{D}$ no such simple physical interpretation can be given. Introducing equation 4.7 into equation 4.6, we obtain

$$\frac{\partial c}{\partial t} = \frac{\partial}{r\,\partial r}\left(r\overline{D}\,\frac{\partial c}{\partial r} - r^2\omega^2\bar{s}c\right) \tag{4.10}$$

This is formally the same as the Lamm equation for a nonreacting solute in binary solutions. However, it should be noted that both $\bar{s}$ and $\overline{D}$ are generally neither constant nor functions of c only.

B. CONSTITUENT QUANTITIES

In complex-forming systems, any solute species may be regarded as composed of a finite number of constituents which do not undergo any chemical change during the complex formations. For example, in the case in which a macromolecule P of molecular weight M_P forms a series of complexes, $PA, PAA, \ldots, PAA \cdots A$, with a smaller molecule A of molecular weight M_A, we can consider P and A as constituents of the complexes.

Let us denote f_α^k the weight fraction of constituent α in complex k and by c_k^α the number of grams of this constituent per unit volume of the solution in which the c-scale concentration of complex k is c_k. Then we have

$$c_k^\alpha = f_\alpha^k c_k \tag{4.11}$$

We define a quantity $(J_k^\alpha)_c$ by

$$(J_k^\alpha)_c = f_\alpha^k (J_k)_c \tag{4.12}$$

which represents the number of grams of constituent α transported by complex k per unit time through unit cross section fixed to the ultracentrifuge cell. Therefore the quantity $(J^\alpha)_c$ defined by

$$(J^\alpha)_c = \sum_{k=1}^{n} (J_k^\alpha)_c \tag{4.13}$$

is the total mass in grams of constituent α passing in unit time through the cross section considered, and may be termed the flow of constituent α relative to the cell.

In the above-mentioned example, let us assume that m is the maximum number of molecules of A that are bound on P and let us label the total solutes as follows:

Solute	P	A	PA	PAA	$\cdots$	PAA$\cdots$A
i	-1	0	1	2	$\cdots$	m

Then the weight fraction of constituent A in solute i, that is, $f_A{}^i$, is given by

$$f_A{}^i = \begin{cases} 0 & (i = -1) \\ 1 & (i = 0) \\ \dfrac{iM_A}{M_P + iM_A} & (i = 1, 2, \ldots, m) \end{cases} \tag{4.14}$$

The weight fraction of constituent P in solute i, that is, $f_P{}^i$, is represented by

$$f_P{}^i = \begin{cases} 1 & (i = -1) \\ 0 & (i = 0) \\ \dfrac{M_P}{M_P + iM_A} & (i = 1, 2, \ldots, m) \end{cases} \tag{4.15}$$

Now, substitution of equation 1.102 for $(J_k)_c$ in equation 4.12, with the understanding that q in equation 1.102 should be replaced here by n, yields

$$(J_k{}^\alpha)_c = s_k c_k{}^\alpha \omega^2 r - \overline{D}_k \frac{\partial c_k{}^\alpha}{\partial r} \tag{4.16}$$

where $s_k = (s_k)_V$, and $\overline{D}_k$ stands for a certain average diffusion coefficient defined by

$$\overline{D}_k = \sum_{j=1}^{n} (D_{kj})_V \left(\frac{\partial c_j}{\partial r} \right) \bigg/ \frac{\partial c_k{}^\alpha}{\partial r} \tag{4.17}$$

Introducing equation 4.16 into equation 4.13, we obtain

$$(J^\alpha)_c = s^\alpha c^\alpha \omega^2 r - D^\alpha \frac{\partial c^\alpha}{\partial r} \tag{4.18}$$

with

$$c^{\alpha} = \sum_{k=1}^{n} c_k{}^{\alpha} \tag{4.19}$$

$$s^{\alpha} = \frac{\sum_{k=1}^{n} s_k c_k{}^{\alpha}}{c^{\alpha}} \tag{4.20}$$

$$D^{\alpha} = \sum_{k=1}^{n} \bar{D}_k \left(\frac{\partial c_k{}^{\alpha}}{\partial r} \right) \Big/ \frac{\partial c^{\alpha}}{\partial r} \tag{4.21}$$

The quantities c^{α}, s^{α}, and D^{α} are termed the *constituent concentration*, the *constituent sedimentation coefficient*, and the *constituent diffusion coefficient* of constituent α.

Multiplication on both sides of equation 4.2 by $f_{\alpha}{}^k$ and summation over all solutes then gives

$$\frac{\partial c^{\alpha}}{\partial t} = -\frac{\partial}{r \partial r} [r(J^{\alpha})_c] + \sum_{k=1}^{n} f_{\alpha}{}^k Q_k \tag{4.22}$$

where equations 4.12, 4.13, and 4.19 have been used. It should be noted that the sum term on the right-hand side of this equation does not vanish, except in special cases.

Concepts of constituent quantities, introduced by Tiselius[3] some 40 years ago, are useful only when the constituent concentrations or their gradients can be determined experimentally. Neither the Rayleigh interference method nor the schlieren method is of use for this purpose, since these are only capable of measuring the excess refractive index of the total solute or its gradient.

III. SELF-ASSOCIATION

A. CONTINUITY EQUATIONS FOR MONOMER-j-MER ASSOCIATION

The most typical of the chemical reactions which occur in protein solutions is self-association. In this reaction, a monomeric protein P_1 associates to form dimer P_2, trimer P_3,..., and j-mer P_j. Reithel[1] classified a number of proteins into three types in accordance with their association behavior, as shown in Table 4.1. In some cases, as exemplified by hemoglobin and β-lactoglobulin, a protein molecule dissociates into a small number of subunits which consist of either a single polypeptide chain or of

TABLE 4.1   Association Behavior of Proteins[1]

A. Proteins whose structural subunits are known to associate via disulfide links to some degree

Plasma albumin	Cytochrome c_1
α-Chymotrypsin	Insulin
γ-Globulins	Lactic dehydrogenase
Macroglobulins	β-Mercaptopyruvate transulfurase
Papain	Cold-insoluble fraction soybean protein
Thyroglobulin	
Urease	Glutenin
K-Casein	Thetin homocysteine methylpherase

B. Proteins known to associate via disulfide links

Plasma albumin	Papain
Egg albumin	Thetin homocysteine methylpherase
Insulin	Glucose 6-phosphate dehydrogenase
	β-Galactosidase

C. Proteins known to associate in absence of disulfide links

Plasma albumin	β-Lactoglobulin
Flagellins	Lysozyme
Glutamic acid dehydrogenase	Phosphorylase
Cytochrome c	Ribonuclease

two or more polypeptide chains bound together. If the constituent chains of these subunits are almost identical, one may regard the system as a self-associating one, taking the smallest subunit as a monomer species.

Evidently, theoretical treatments of transport processes of any chemically reacting system become more complex as the number of reacting species present in the solution is increased. Therefore, we begin our discussion with the simplest one, in which there exist only a monomer and its single oligomer species, say j-mer.

The monomer and the j-mer are labeled solutes 1 and j. Their continuity equations are derived by substitution of equation 1.102 into equation 4.2, giving

$$\frac{\partial c_1}{\partial t} = \frac{\partial}{r\,\partial r}\left(rD_{11}\frac{\partial c_1}{\partial r} + rD_{1j}\frac{\partial c_j}{\partial r} - r^2\omega^2 s_1 c_1\right) + Q_1 \qquad (4.23)$$

$$\frac{\partial c_j}{\partial t} = \frac{\partial}{r\,\partial r}\left(rD_{j1}\frac{\partial c_1}{\partial r} + rD_{jj}\frac{\partial c_j}{\partial r} - r^2\omega^2 s_j c_j\right) + Q_j \qquad (4.24)$$

Here the subscript V previously affixed to s and D has been omitted for simplicity. (This convention is used throughout the present chapter.) It is usually customary to represent Q_1 and Q_j by the expressions

$$Q_1 = -k_{1j}c_1^{\ j} + k_{j1}c_j \tag{4.25}$$

$$Q_j = k_{1j}c_1^{\ j} - k_{j1}c_j \tag{4.26}$$

where k_{1j} is the rate constant for association $jP_1 \rightarrow P_j$ and k_{j1} is that for dissociation $P_j \rightarrow jP_1$. These expressions satisfy the requirement of the conservation of mass, $Q_1 + Q_j = 0$. Unless otherwise stated, all the coefficients in the equations above, s_1, s_j, D_{11}, D_{1j}, D_{j1}, D_{jj}, k_{1j}, and k_{j1}, must be treated as functions of c_1 and c_j.

There is no possibility that one obtains general solutions to the set of differential equations for c_1 and c_j given above. Even if all coefficients are constant, the difficulty is not relaxed, as long as one seeks analytic solutions. However, numerical solutions may be worked out if use is made of high-speed computers. For example, Belford and Belford[4] have obtained such solutions for the case of monomer-dimer association.

B.   MONOMER-j-MER SYSTEMS WITH INFINITE RATE CONSTANTS AND NEGLIGIBLE DIFFUSION

If the rate constants k_{1j} and k_{j1} are made indefinitely large with their ratio kept constant, it follows from equation 4.25 or 4.26 that

$$c_j = K_j c_1^{\ j} \tag{4.27}$$

where

$$K_j = \lim_{\substack{k_{1j} \rightarrow \infty \\ k_{j1} \rightarrow \infty}} \frac{k_{1j}}{k_{j1}} \tag{4.28}$$

This is because without relation 4.27 one is led to the physically unallowable divergence of Q_1 and Q_j to infinity. However, equation 4.27 does not mean that Q_1 and Q_j vanish. Instead, the values of these quantities become indeterminate at the limit of indefinitely large k_{1j} and k_{j1}. One can only say that the limiting values are equal in magnitude and are opposite in sign, because the condition $Q_1 + Q_j = 0$ must be satisfied however large the rate constants may be.

Equation 4.27 has the familiar form of equilibrium relation for monomer–j-mer association, but the "equilibrium constant" K_j may be a function of c_1 and c_j. As is shown in Chapter 6, Section I.A, K_j becomes constant only when the activity coefficients on the c-concentration scale of monomer and j-mer, y_1 and y_j, are represented by

$$\ln y_1 = M_1 Bc, \qquad \ln y_j = M_j Bc$$

Here M_1 and M_j are the molecular weights of monomer and j-mer (hence $M_j = j M_1$), B is a nonideality parameter, and c is the total solute concentration, that is,

$$c = c_1 + c_j \tag{4.29}$$

Regardless of whether K_j is a function of c_1 and c_j or a constant, equation 4.27 implies that when association and dissociation are very fast, chemical equilibrium is attained between monomer and j-mer at any instant and at any position in the solution, and that there exists a unique relation between the local concentrations of the two species. From this latter fact and equation 4.29 it follows that c_1 and c_j can be expressed in terms of c only. Under this condition, therefore, the sedimentation coefficients and K_j can be treated as functions of c.

Now, since Q_1 and Q_j are left indeterminate for indefinitely large k_{1j} and k_{j1}, no information can be drawn from each of equations 4.23 and 4.24. However, the addition of these two equations eliminates such unknown terms, yielding

$$\frac{\partial c}{\partial t} = \frac{\partial}{r\,\partial r}\left[\overline{D}(c)r\frac{\partial c}{\partial r} - r^2\omega^2\bar{s}(c)c\right] \tag{4.30}$$

where

$$\bar{s}(c) = \frac{s_1 c_1 + s_j c_j}{c_1 + c_j} \tag{4.31}$$

$$\overline{D}(c) = \frac{(D_{11} + D_{j1})(\partial c_1/\partial r) + (D_{1j} + D_{jj})(\partial c_j/\partial r)}{(\partial c_1/\partial r) + (\partial c_j/\partial r)} \tag{4.32}$$

Equation 4.30 is a special case of the general equation 4.6. When combined with equations 4.27 and 4.29, this differential equation allows c_1 and c_j to be determined as functions of r and t. However, it does not appear that this determination can be carried out analytically, except for a special case in which $\overline{D}$ is zero, that is, effects of diffusion are negligible.

This special case was investigated for the first time by Gilbert[5] in his celebrated paper published in 1955. His treatment was developed with the transport equations simplified by the rectangular-cell approximation, but it can be shown that the corresponding solution, free from such an approximation, is obtainable if a somewhat more sophisticated method is used. The solution thus obtained reads

$$\theta_1 = 0 \qquad (0 < \xi < \tau) \tag{4.33}$$

$$\theta_1 = \frac{1 - \exp\left[-((j-1)/j)(\alpha_j/(\alpha_j - \beta_j))(\xi - \tau)\right]}{\alpha_j \left\{\exp\left[((j-1)/(\alpha_j - \beta_j))(\alpha_j \tau - \beta_j \xi)\right] - 1\right\}} \tag{4.34}$$

$$[\tau < \xi < \xi^*(\tau)]$$

$$\theta_1^{-1}\left(\frac{1 + \alpha_j \theta_1}{1 + \alpha_j}\right)^{[1 - j(\beta_j/\alpha_j)]} = e^{(j-1)\tau} \tag{4.35}$$

$$\left[\xi^*(\tau) < \xi < \ln\left(r_2/r_1\right)^2\right]$$

where

$$\theta_1 = \left(\frac{c_1}{c_1^0}\right)^{j-1}, \qquad \tau = 2s_1\omega^2 t, \qquad \xi = \ln\left(\frac{r}{r_1}\right)^2$$

$$\alpha_j = \left(\frac{c_j^0}{c_1^0}\right)\left(\frac{s_j}{s_1}\right), \qquad \beta_j = \frac{c_j^0}{c_1^0} \tag{4.36}$$

and

$$\xi^*(\tau) = -\left(\frac{1}{j-1}\right)\ln\theta_1 + \ln\left(\frac{1 + \alpha_j}{1 + \alpha_j\beta_j}\right) \tag{4.37}$$

with θ_1 being given as a function of τ through equation 4.35. The c_1^0 and c_j^0 denote the concentrations of monomer and j-mer in the initial uniform solution. These formulas are valid for constant sedimentation coefficients and equilibrium constants.

From equations 4.33 through 4.35 it follows that as for the distribution of monomer species, the solution column is separated into three distinct regions; region I: $0 < \xi < \tau$; region II: $\tau < \xi < \xi^*(\tau)$; region III: $\xi^*(\tau) < \xi < \ln(r_2/r_1)^2$. In region I, no monomer exists; in region II, the monomer concentration increases with radial distance; and in region III, it is uniform. Throughout these regions there occurs no discontinuity in monomer concentration.

The expression for the total solute concentration c is obtained by inserting the results obtained above for θ_1 in the equation

$$c = c_1{}^0(\theta_1)^{1/(j-1)}(1 + \beta_j\theta_1) \tag{4.38}$$

which is derived by substitution of equation 4.27 into equation 4.29, followed by expression of c_1 in terms of θ_1. The resulting expression shows that the distribution of c in the solution column has a similar feature to that of c_1, that is, c is zero in region I, changes with distance in region II, and is uniform in region III. These regions, therefore, may be referred to as the solvent region, the sedimentation boundary between solvent and solution, and the plateau region, respectively. It is to be noted that *the sedimentation boundary here has a finite width, in spite of the absence of diffusion.*

It can be shown from equations 4.27, 4.29, and 4.31 that

$$\frac{d\bar{s}(c)}{dc} = \frac{K_j(j-1)(s_j - s_1)(c_1)^{j-2}}{\left[1 + K_j(c_1)^{j-1}\right]^2\left[1 + K_j j(c_1)^{j-1}\right]} \tag{4.39}$$

Hence, if $s_j > s_1$, as should be the case physically, and if these sedimentation coefficients are independent of concentration, $d\bar{s}/dc > 0$, so that $\bar{s}$ is an *increasing* function of c.

1. Features of Gradient Curves

The expression for the gradient curve, derived by differentiating equation 4.38 with respect to r, has a very complicated form. For early stages of sedimentation,* that is, for values of $\tau \ll 1$, it is simplified as follows:

$$\frac{\partial c}{\partial r} = 0 \qquad \text{(region I: } 0 < \xi < \tau) \tag{4.40}$$

$$\frac{\partial c}{\partial r} = \frac{2c_1{}^0(\alpha_j - \beta_j)^2}{\tau r_1(j-1)}(j)^{-1/(j-1)}\frac{\left[(\xi/\tau) - 1\right]^{(2-j)(j-1)}}{\left[\alpha_j - \beta_j(\xi/\tau)\right]^{(2j-1)/(j-1)}} \tag{4.41}$$

$$\left[\text{region II: } \tau < \xi < \xi^*(\tau)\right]$$

$$\frac{\partial c}{\partial r} = 0 \qquad \left[\text{region III: } \xi^*(\tau) < \xi < \ln(r_2/r_1)^2\right] \tag{4.42}$$

*Restricting the consideration to $\tau \ll 1$ is equivalent to the rectangular-cell approximation.

where

$$\xi^*(\tau) = \left(\frac{1 + \alpha_j j}{1 + \beta_j j}\right)\tau \tag{4.43}$$

For the monomer-dimer case equation 4.41 reduces to

$$\frac{\partial c}{\partial r} = c_1{}^0(\alpha_2 - \beta_2)^2(\tau r_1)^{-1}\left(\alpha_2 - \beta_2\frac{\xi}{\tau}\right)^{-3} \tag{4.44}$$

where

$$\tau < \xi < \left(\frac{1 + 2\alpha_2}{1 + 2\beta_2}\right)\tau \tag{4.45}$$

Figure 4.1 shows a sketch of the concentration gradient curve represented by equation 4.44. It is seen that the gradient increases with distance and attains a maximum at the upper edge of the boundary region.

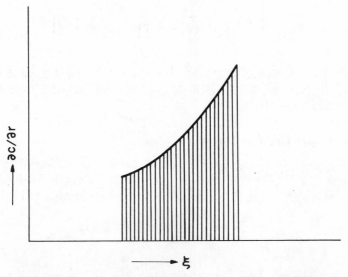

Fig. 4.1. Sedimentation boundary in a diffusion-free monomer-dimer equilibrium system.

For $j > 2$ the right-hand side of equation 4.41 has a minimum at $\xi = \xi_{\mathrm{m}}$, where

$$\xi_{\mathrm{m}} = \frac{2j - 1}{3(j - 1)}\left[1 + \frac{j - 2}{2j - 1}\left(\frac{\alpha_j}{\beta_j}\right)\right]\tau \tag{4.46}$$

Subject to the condition $s_1 < s_j$, this minimum appears inside the boundary region (i.e., region II) if

$$\beta_j \left(= \frac{c_j^0}{c_1^0} \right) > \frac{j-2}{j(2j-1)} \tag{4.47}$$

In this case, the gradient curve exhibits two maxima, one at the upper edge and the other at the lower edge of the boundary region. Thus, it has the general shape shown in Fig. 4.2a. When condition 4.47 is not obeyed, the gradient curve has the feature shown in Fig. 4.2b. The gradient is infinite at the lower edge of the boundary region and decreases monotonically with distance. Regardless of whether or not it exhibits two maxima, the gradient curve for the system with $j > 2$ is sharply contrasted to that for the system

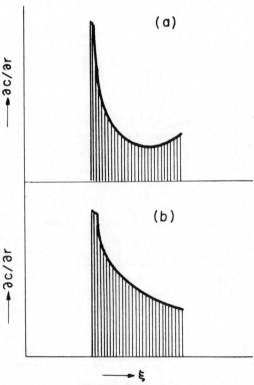

Fig. 4.2. Sedimentation boundaries in diffusion-free monomer-trimer equilibrium systems which correspond (a) to the case $c_3^0/c_1^0 > \frac{1}{15}$ and (b) to the case $c_3^0/c_1^0 < \frac{1}{15}$. The boundary shape changes from type (a) to type (b) as the initial total concentration decreases.

with $j = 2$. It should be noted, however, that this characteristic result has been derived under the neglect of diffusion.

Since one has the relation $c_j^0 = K_j(c_1^0)^j$, the inequality 4.47 may be written

$$K_j(c_1^0)^{j-1} > \frac{j-2}{j(2j-1)} \tag{4.48}$$

On the other hand, the total concentration in the initial solution, c_0, is expressed in terms of c_1^0 as

$$c_0 = c_1^0\left[1 + K_j(c_1^0)^{j-1}\right] \tag{4.49}$$

which shows that c_1^0 decreases as c_0 is lowered. Thus the inequality 4.48 ceases to be valid for values of c_0 smaller than the value which corresponds to

$$c_1^0 = \left[\frac{j-2}{K_j j(2j-1)}\right]^{1/(j-1)} \tag{4.50}$$

Therefore, when a given system contains a monomer and an oligomer higher than the dimer, the gradient curve eventually changes its shape from the two-peak type to the single-peak type as the initial concentration is diluted below a certain limit.

2. Sedimentation Rates

The theory just discussed for equilibrium monomer–j-mer systems gives the following predictions for the rate of movement of the peaks that appear at the upper and lower edges of the boundary region.

1. The peak at the lower edge, that is, the slower peak, yields a sedimentation coefficient s^s which agrees with s_1, for it moves with time in accordance with the relation $\xi = \tau$, which is written in terms of original variables

$$\ln\left(\frac{r^s}{r_1}\right) = s_1\omega^2 t \tag{4.51}$$

where r^s is the radial position of the slower peak.

2. The peak at the upper edge, the faster peak, gives a sedimentation

coefficient s^f which depends on c_1^0 as

$$s^f = \frac{s_1 + jK_js_j(c_1^0)^{j-1}}{1 + jK_j(c_1^0)^{j-1}} \tag{4.52}$$

This prediction follows from equation 4.43 which relates the position of the faster peak r^f to time t by

$$\ln\left(\frac{r^f}{r_1}\right) = \left(\frac{1+j\alpha_j}{1+j\beta_j}\right)s_1\omega^2 t \tag{4.53}$$

Equation 4.52 is combined with equation 4.49 to give s^f as a function of c_0. If, as should be the case in general, $s_j > s_1$, this function gives a curve of s^f which increases monotonically with the increase in c_0 and approaches s_j. It should be noted that for $j > 2$, the peak at the upper edge of the boundary region disappears for c_0 lower than the value corresponding to the c_1^0 given by equation 4.50. Thus, for $j > 2$, the dependence of s^f on c_0 calculated from equations 4.52 and 4.49 may not be extended below a certain c_0 value. This fact is allowed for in Fig. 4.3, which illustrates the s^f

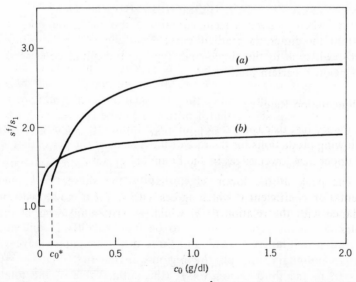

Fig. 4.3. Changes in the sedimentation coefficient s^f associated with the leading edge of the boundary region with the initial concentration c_0 of the solution for diffusion-free monomer–j-mer equilibrium systems. (a) Monomer–trimer system with $s_3/s_1 = 3$ and $K_3 = 5$ (dl²/g²). (b) Monomer–dimer system with $s_2/s_1 = 2$ and $K_2 = 5$ (dl/g). Note that the curve (a) may not be extended below $c_0 = c_0^* = 0.087$ (g/dl).

versus c_0 relations for a monomer-trimer system with $K_3 = 10$ and $s_3/s_1 = 3$ and for a monomer-dimer system with $K_2 = 10$ and $s_2/s_1 = 2$.

The theoretical treatment given above assumes that s_1 and s_j are independent of solute concentration. This assumption may not be realistic for systems with macromolecular solutes. An extension of the theory to concentration-dependent systems is discussed in Section III.D.1.

Another shortcoming of this theory is that it has ignored diffusion. This effect is considered in Section II.D.2, but for the present we only anticipate that diffusion will tend to diminish the extreme asymmetry of gradient distributions as illustrated in Figs. 4.1 and 4.2 and that, in some cases, it may lead to an essentially symmetric, single-peaked boundary curve even if self-association of solutes is taking place. In such cases, it becomes almost hopeless to deduce from sedimentation transport experiments any meaningful information about the chemical reactions occurring in the solution.

In sum, these considerations suffice to indicate how hazardous it is to attempt inadvertently an analysis of sedimentation boundary curves by the traditional procedures (as explained in Chapter 3) when chemical interactions between solutes may be suspected.

C. GENERAL SELF-ASSOCIATING SYSTEMS WITH INFINITE RATE CONSTANTS AND NEGLIGIBLE DIFFUSION

In actual cases, except under special conditions, it is quite likely that if j is larger than 2, the system contains one or more intermediate oligomers coexisting with the monomer and j-mer species. Continuity equations for such general self-associating systems are more complex and difficult to treat mathematically than the continuity equation for the monomer–j-mer system. Gilbert[6] has shown that if diffusion can be ignored and if the rate constants for all reactions are so large that equilibrium may be regarded as being attained instantaneously between any pair of solutes, a closed expression is obtainable from which the sedimentation boundary of a self-associating system containing an arbitrary number of oligomers or aggregates can be calculated. This theory of Gilbert is again based on the rectangular-cell approximation, and all sedimentation coefficients and equilibrium constants are assumed to be independent of concentration. The solution column is divided into three distinct regions: solvent, sedimentation boundary, and plateau regions. In the first region, $c = 0$ and $\partial c / \partial r = 0$; in the third region, $c = c_0$ (the initial value of the total concentration c) and $\partial c / \partial r = 0$; and in the second region, $c(r, t)$ is determined from the relation

$$\frac{r - r_1}{t} = \frac{Q}{P} \tag{4.54}$$

and the concentration gradient is represented by

$$\frac{\partial c}{\partial r} = \frac{P^2}{tQ[(Q'/Q)-(P'/P)]} \tag{4.55}$$

The lower edge r_* and upper edge r^* of the boundary region are given by

$$r_* = r_1 + u_1 t \tag{4.56}$$

$$r^* = r_1 + t\left(\frac{Q}{P}\right)_{c_1 = c_1^0} \tag{4.57}$$

In these equations, the various symbols have the following significance:

$$P = \sum_{k=1}^{\infty} kK_k c_1^{k-1} \tag{4.58}$$

$$Q = \sum_{k=1}^{\infty} ku_k K_k c_1^{k-1} \tag{4.59}$$

$$P' = \frac{dP}{dc_1} \tag{4.60}$$

$$Q' = \frac{dQ}{dc_1} \tag{4.61}$$

$$u_k = s_k \omega^2 r_1 \tag{4.62}*$$

and

$$c = \sum_{k=1}^{\infty} c_k = \sum_{k=1}^{\infty} K_k c_1^{k} \tag{4.63}$$

It should be noted that P, Q, P', and Q' are implicitly dependent only on c, because, as can be seen from equation 4.63, c_1 is a function only of c.

Concentration gradient curves at various fixed values of time t can be obtained by eliminating c_1 from equations 4.54 and 4.55. They should superimpose on a single curve if $t(\partial c/\partial r)$ is plotted against a reduced cell coordinate $(r - r_1)/t$. In general, the elimination of c_1 cannot be done analytically.

---

*The quantity u_k defined by this equation is hereafter called the mobility of solute species k. It always appears as a transport coefficient corresponding to s_k when the ultracentrifuge equations for a sector-shaped cell are simplified by the rectangular-cell approximation.

1. Special Cases

1. Monomer–j-mer Association

In this case, equations 4.54 and 4.55 reduce to

$$\frac{x}{t} = \frac{u_1 + j u_j K_j \vartheta}{1 + j K_j \vartheta} \tag{4.64}$$

and

$$t \frac{\partial c}{\partial r} = \frac{(1 + j K_j \vartheta)^2}{j(j-1) K_j \vartheta^{(j-2)/(j-1)} [u_j - (x/t)]} \tag{4.65}$$

where

$$x = r - r_1 \tag{4.66}$$

$$\vartheta = (c_1)^{j-1} \tag{4.67}$$

The variable ϑ can be eliminated easily from equations 4.64 and 4.65 to give

$$t \frac{\partial c}{\partial r} = \frac{(u_j - u_1)^2}{(j-1)} (K_j j)^{-1/(j-1)} \frac{[(x/t) - u_1]^{(2-j)/(j-1)}}{[u_j - (x/t)]^{(2j-1)/(j-1)}} \tag{4.68}$$

which is in perfect agreement with an expression obtained for this special case by Gilbert in his earlier paper.[5] This is also derived from equation 4.41 by replacing ξ with its first approximation $2(r - r_1)/r_1$.

2. Monomer-Dimer-Trimer Association

In relation to their study on α-chymotrypsin, this special case was treated by Rao and Kegeles[7] prior to the appearance of Gilbert's general theory. Their expression for concentration gradients is

$$t \frac{\partial c}{\partial r} = \frac{(K_2^3/K_3^2) F(\delta)}{18 \delta^2 (1-\delta)^4 (u_3 - u_1)} \tag{4.69}$$

where

$$F(\delta) = \frac{[D(\delta) - \beta + \delta]^2 [D(\delta) + \beta + \delta(1 - 2\beta)]^2}{D(\delta)} \tag{4.70}$$

with

$$\beta = \frac{u_2 - u_1}{u_3 - u_1} \tag{4.71}$$

$$\delta = \frac{(x/t) - u_1}{u_3 - u_1} \tag{4.72}$$

and

$$D(\delta) = \left[(\beta - \delta)^2 + \left(\frac{K_3}{K_2^2} \right) \delta(1 - \delta) \right]^{1/2} \tag{4.73}$$

As should be expected, these results can be derived readily from equations 4.54 and 4.55.

Rao and Kegeles[7] computed the concentration gradient curve for an initial concentration of 18.9 g/l and $(u_3 - u_1)t = 0.9195$ cm, using $u_2/u_1 = (2)^{2/3}$, $u_3/u_1 = (3)^{2/3}$, $K_2 = 1/11.1$ (l/g), and $K_3 = 1/50.0$ (l^2/g^2). This choice of the mobility ratios is based on the assumption[8] that the molecules of monomer, dimer, and trimer are compact in structure and spherical in shape. Rao and Kegeles also calculated for comparison the case of monomer-trimer association, with $K_2 = 0$ and all other parameters unchanged. Later, Bethune and Kegeles[9] corrected the earlier results for small numerical errors. Here our calculated curves are shown in Figs. 4.4a and 4.4b. It is seen that the curve for the monomer-dimer-trimer association (Fig. 4.4b) is not very much different in character from the curve for the monomer-dimer association shown in Fig. 4.1, and that it does not exhibit even the slightest evidence of the existence of three different solute species in the system.

The extreme asymmetry of the curves in these figures is incompatible with the nearly symmetric, single-peaked boundary curve observed by Rao and Kegeles[7] for an α-chymotrypsin preparation in pH 6.2 phosphate buffer of ionic strength 0.1 at $c_0 = 18.9$ g/l and $\omega = 59{,}780$ rev/min. In this connection, we remark that from Archibald-type ultracentrifugal experiments Kegeles and Rao[10] concluded that more than two reacting species were to be contained in this protein system. Bethune and Kegeles[9] attempted to include the diffusion terms in the computation, assuming appropriate values for the diffusion coefficients. However, the calculated boundary curves still exhibited an appreciable asymmetry. Bethune and Kegeles state that if the (normal) concentration dependence of the sedimentation coefficients of individual solute species were included in the computation, it would speed up the monomer species in the dilute region

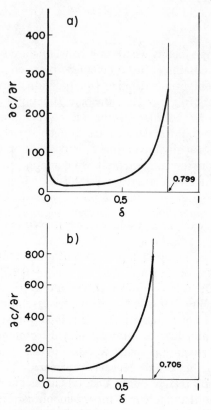

Fig. 4.4. Sedimentation boundaries for (*a*) monomer-trimer and (*b*) monomer-dimer-trimer equilibrium systems in the absence of diffusion. See text for the numerical values of the parameters used in the calculations; except for K_2 [0 for (*a*) and $1/11.1$ (l/g) for (*b*)], all other conditions are the same for the two systems.

and retard the dimer and trimer species in the region of high concentration to give a more symmetric gradient curve.

For an α-chymotrypsin preparation in pH 7.90 phosphate buffer of ionic strength 0.01, Massey et al.[11] observed bimodal gradient curves, in contrast to the above-mentioned observation by Rao and Kegeles[7] at a lower pH and at a higher ionic strength. The difference is presumably attributed to charge effects and specific ion binding. Earlier, Gilbert[5] pointed out that the essential features of the data of Massey et al. could be explained by a model in which a monomer and its hexamer are at chemical equilibrium. Later, he[6] attempted a detailed reanalysis of the same data with success, this time assuming the presence of all intermediate aggregates but still neglecting species higher than the hexamer.

D. CONCENTRATION DEPENDENCE OF SEDIMENTATION COEFFICIENTS

In self-associating systems where rate constants are so large that reequilibrium is established almost instantaneously between any pair of reacting species, the sedimentation coefficients of individual species, even though otherwise dependent on $c_1, c_2, \ldots$, become functions of the total solute concentration c only because of the reason mentioned earlier in this chapter. Hence, the $\bar{s}$ and $\bar{D}$ defined, respectively, by equations 4.8 and 4.9 become functions of a single variable c, and equation 4.10 coincides with the Lamm equation for nonreacting two-component systems. Thus, the following important theorem can be established: *Any self-associating system in which equilibrium is attained instantaneously among all solutes is ultracentrifugally equivalent to a binary solution of a concentration-dependent, nonreacting solute.*

From this it follows that if s is replaced by $\bar{s}$, equation 2.205 applies for chemically reacting systems of this type, that is,

$$\bar{s}(c_p) = \frac{1}{\omega^2} \frac{d \ln r_M}{dt} \tag{4.74}$$

where c_p is the total solute concentration in the plateau region, and r_M is the equivalent boundary position defined by equation 2.206. The reader is referred to Chapter 2 for the actual procedure of determining $\bar{s}$ as a function of c by use of equation 4.74. In this connection, we may note two points. One is that equation 4.74 is applicable whatever the shapes of observed boundary curves may be. The other point is that this equation is of no use unless the specific refractive index increments of all solute species are equal, that is,

$$R_1 = R_2 = R_3 = \cdots \tag{4.75}$$

because otherwise the quantity r_M cannot be evaluated from observed schlieren patterns.

From equations 4.8 and 4.63 it follows that $\bar{s}(c)$ is a monotonically increasing function of c when the sedimentation coefficients of all individual solutes are constant and have the magnitudes such that $s_1 < s_2 < s_3 \ldots$. However, in many instances, observed values of $\bar{s}$ for self-associating proteins follow a curve which initially rises steeply, passes through a broad maximum, and then declines gradually in a linear or nearly linear fashion. Examples are shown in Fig. 4.5.[12] It is usual to account for such experimental results by postulating two opposing effects, the association effect and a normal decrease of the sedimentation coefficient of each solute with

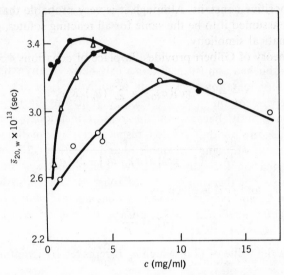

Fig. 4.5. Plots of $\bar{s}$ versus c for α-chymotrypsin at 25°C.[12] The actually measured values of $\bar{s}$ are reduced to the state "in water at 20°C" by the conventional procedure (see Chapter 2, Section I.C.1) and plotted as shown. ●, at pH 3.86 and ionic strength 0.2; △, at pH 4.99 and ionic strength 0.2; ▲ with another sample; ○, at pH 6.20 and ionic strength 0.2; ◐, from an experiment at 10°C.

the increase in total concentration c. There will be cases in which the two effects happen to cancel each other, giving rise to a plot of $\bar{s}$ versus c which is almost horizontal over a fairly wide range of concentration. Anyway, solute associations may be suspected if one obtains a sedimentation coefficient versus concentration curve that has a *positive initial slope*.

These considerations suggest the importance of extending the measurements of $\bar{s}$ down to very low concentrations in cases where the solute is suspected to undergo aggregation, even if the observed schlieren patterns are single-peaked and typically symmetric. In fact, as has been summarized in tabular form by Nichol et al.,[2] there have been reported a good many protein systems in which boundary curves were essentially of this type but $\bar{s}$ increased with c at low concentrations.

1. Extension of Gilbert's Theory to Concentration-Dependent Systems

In view of the importance of taking the normal concentration dependence of s of each solute into account, Gilbert[13,14] attempted an extension of his earlier theory,[6] assuming for u_k a linear c-dependence of the form

$$u_k = (u_k)_0(1 - gc) \tag{4.76}$$

Here g is a positive constant. Although it is very probable that g depends on k, Gilbert assumed it to be the same for all reacting solutes, presumably for a mathematical simplicity.

The new theory of Gilbert provides, in place of equations 4.54 and 4.55,

$$\frac{x}{t} = (1 - gc)(\sigma)_0 - g \sum_{k=1}^{\infty} (u_k)_0 K_k c_1^{\ k} \tag{4.77}$$

and

$$t \frac{\partial c}{\partial x} = \frac{1}{(1 - gc)[d(\sigma)_0/dc] - 2g(\sigma)_0} \tag{4.78}$$

where $x = r - r_1$, and $(\sigma)_0$ is given by

$$(\sigma)_0 = \frac{Q_0}{P} \tag{4.79}$$

with Q_0 being the value of Q in which u_k is replaced by its infinite-dilution value $(u_k)_0$.

From equation 4.78 it follows that the concentration gradient becomes infinite at a concentration given by

$$c = \frac{1}{g} - 2(\sigma)_0 \frac{dc}{d(\sigma)_0} \tag{4.80}$$

At any concentration higher than this, the boundary must have an infinitely sharp leading edge, that is, the gradient curve must terminate on the solution side with a hypersharp spike. If the concentration is increased, this spike moves back over the curve until all of the sedimentation boundary has been swallowed up, leaving only a single-step discontinuous boundary between solvent and solution.

Protein preparations are often contaminated with impurities which are difficult to remove by ordinary methods of purification. Gilbert[14] treated the case in which a given preparation is a mixture of two substances, one of which can undergo self-association, whereas the other, present as an impurity at a constant fraction of the whole, cannot. If it is assumed that neither chemical nor physical interaction exists between the associating substance A and the impurity B, the gradient curve still can be calculated from equations 4.77 and 4.78 so modified that the term gc is replaced by $g(c + \theta c_0)$. Here θ denotes the number of grams of B per gram of A present in the initial solution. Thus the initial concentration of the total solute is represented by $(1 + \theta)c_0$.

Gilbert and Gilbert,[15] and then Gilbert,[14] applied the new theory to analyze the sedimentation coefficient data obtained by Timasheff and Townend[16] for a preparation of β-lactoglobulin A at pH 4.65, ionic strength 0.1, and 2°C. Their analysis is described below in some detail, because it provides a concrete example of what one actually has to do in carrying through this kind of data evaluation.

The light-scattering data of Townend and Timasheff[16,17] indicate that, in the range of pH 3.7 to 5.2 at low temperatures and at ionic strength 0.1, the dimer of β-lactoglobulin A is stable enough for its dissociation into monomers to be discounted and it reversibly aggregates to tetramers, hexamers, and octamers. Hence, for the theoretical treatment of this protein system, the dimer may be taken as the basic associating unit, that is, as an effective monomer, and the three higher species formed may be termed dimer, trimer, and tetramer, respectively. Timasheff and Townend[16] set forth a model in which the tetramer is assumed to be roughly a cubic aggregate of four monomers held together in parallel by the equivalent of four bonds of equal strength. In building up the aggregate, two monomers are bound together by one such bond, a further monomer is then attached through a similar bond, and then the aggregate is completed by the attachment of a fourth monomer. It is to be noted that not one but two bonds are made as the fourth monomer completes the square array. Thus one may write the following forms for the equilibrium constants K_2, K_3, and K_4:

$$K_2 = 2k, \qquad K_3 = 3k^2, \qquad K_4 = 4\gamma k^3 \qquad (4.81)$$

where γ is a numerical factor related to the extent to which the extra bond introduced in completing the tetramer stabilizes the system. A larger value of γ thus accentuates the formation of the tetramer. The constant k is related to the molar equilibrium constant K for the reaction between monomer and dimer by

$$k = \frac{1000K}{M_1} \qquad (4.82)$$

where M_1 is the molecular weight of the monomer—this time, two molecules of β-lactoglobulin A.

Now, Gilbert[14] assigned the following numerical values to parameters associated with the Timasheff-Townend association scheme:

$$M_1 = 36,000$$

$$(s_1)_0 = 2.87 \quad \text{(in svedbergs)}[18]$$

$$(s_j)_0 = 2.87(j)^{2/3}(1.044/f_j) \quad \text{(in svedbergs)} \qquad (j \geqslant 2),$$

$$\text{with} \quad f_2 = 1.042, \quad f_3 = 1.024, \quad \text{and} \quad f_4 = 1.024$$

$$\gamma k^3 = 5 \times 10^{11} \; (\text{l}^3/\text{mole})$$

$$\gamma = 100$$

He also assumed that

$$g = 0.058 \ (\mathrm{dl/g})$$

$\theta = \frac{1}{9}$ (i.e., 10% of impurity is present which cannot aggregate)

and that the impurity had the same sedimentation characteristics as the monomer species. The number 1.044 stands for the friction ratio of the protein dimer considered as a prolate ellipsoid of axial ratio $2:1$. The quantity f_j represents the friction ratio of the j-mer, and $(j)^{2/3}$ takes into account the effect of particle weight. The order of magnitude of γ was predetermined by the light-scattering data which required tetramer to be the only important oligomer. The results were found to be insensitive to the absolute value of γ if it is as large as chosen here.

When the parameters were given as above, it was found that the calculated gradient curve shows two peaks for all concentrations above $0.137 \ \mathrm{g/l}$ of total protein. The sedimentation coefficients derived from the leading and trailing peaks are shown by solid lines in Fig. 4.6, where the open circles are the experimental values of Timasheff and Townend.[16] The agreement between theory and experiment is satisfactory. The dashed lines have been calculated with g and θ taken to be zero. It is seen that the inclusion of concentration dependence of the sedimentation coefficients of

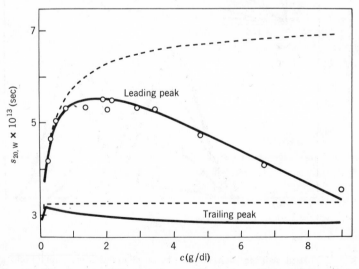

Fig. 4.6. Sedimentation coefficients (reduced to the state "in water at 20°C") associated with the leading and trailing peaks of the calculated boundaries for β-lactoglobulin A containing 10% nonaggregating protein. Solid line, theory with concentration-dependent s_i; dashed line, theory with constant s_i and neglect of nonaggregating impurity; circles, experimental data of Timasheff and Townend[16] at pH 4.65 and ionic strength 0.1.

individual solute species into the computation brings about a substantial improvement in the fit of theory to experiment.

2. Correction for Diffusion

The Gilbert theory is concerned with a limiting case in which boundary spreading due to diffusion is negligible. Any attempt to correct it for the effect of diffusion encounters insuperable analytic difficulties, but the basic differential equations may be solved numerically if a high-speed computer is used.

Bethune and Kegeles[9, 19] have drawn an analogy between countercurrent distribution and boundary transport and have developed a computer procedure which allows numerical solution of the equations descriptive of the former in which solutes undergo rapidly reequilibrating chemical reactions. The results can be transcribed to the concentration gradient

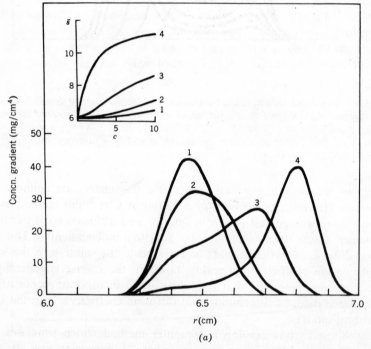

Fig. 4.7a. Calculated sedimentation boundaries of monomer-trimer equilibrium systems with different K_3 at 40 min after initiation of centrifugation. The K_3 values for curves 1 through 4 are 1×10^{-3}, 3×10^{-3}, 2×10^{-2}, and 1, respectively, in cc^2/mg^2. The numerical values used for other parameters are $\omega = 59,870$ rev/min, r_0 (position of initial sharp boundary between solution and solvent) $= 6.085$ cm, $s_1 = 6.0 \times 10^{-13}$ sec, $s_3 = (3)^{2/3}s_1$, $D_{11} = 8.0 \times 10^{-7}$ cm^2/sec, $D_{33} = D_{11}/(3)^{1/3}$, $D_{13} = D_{31} = 0$.

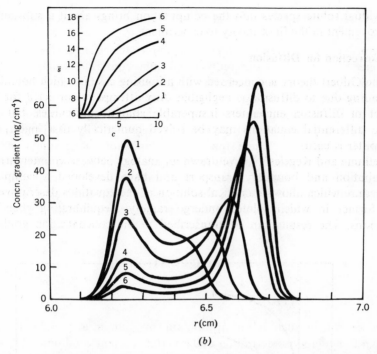

Fig. 4.7b. Calculated sedimentation boundaries of monomer-hexamer equilibrium systems with different K_6 at 20 min after initiation of centrifugation. The K_3 values for curves 1 through 6 are 3.78×10^{-6}, 1.6×10^{-5}, 1.2×10^{-4}, 3.88×10^{-3}, 3.78×10^{-2}, and 1.2, respectively, in cc^5/mg^5. Other parameters are chosen as in Fig. 4.7a except that $s_6 = (6)^{2/3} s_1$, $D_{66} = D_{11}/(6)^{1/3}$, $D_{16} = D_{61} = 0$.

curves in sedimentation experiments if the parameters are suitably reinterpreted. However, as has been pointed out by Cox,[20] this method has the serious disadvantage that the sedimentation and diffusion coefficients of a particular solute species cannot be adjusted independently. Thus, the lowest diffusion coefficient must be assigned to the solute with the lowest sedimentation coefficient. In reality, however, the reverse is generally the case. Cox also mentions that it is difficult to incorporate concentration dependence of the sedimentaion and diffusion coefficients into the countercurrent model.

Cox himself[20] has develop a computer method which simulates sedimentation and diffusion in systems where a monomer and its j-mer undergo rapidly reequilibrating association-dissociation reactions. No such unrealistic restriction as in the Bethune-Kegeles approach is involved in it. In Figs. 4.7a and 4.7b are reproduced his calculated gradient curves for monomer-trimer systems and monomer-hexamer systems. The numerical values used for the parameters are given in the legend to each figure. The inserts in the two figures show the $\bar{s}$ versus c (total solute concentration) relationships for the respective gradient curves which, in a given reaction scheme, correspond to different equilibrium constants. Comparison of Figs. 4.7a and 4.4a tells how greatly the shape of boundary curves in monomer-trimer associating systems is affected by diffusion as well as the magnitude of the equilibrium constant. In particular, it is to be noted that the spike at the lower edge of the boundary region in Fig. 4.4a is no longer visible in Fig. 4.7a, where the trailing limbs of the gradient curves approach the base line smoothly and continuously. This and other results presented by Cox clearly indicate that, though its theoretical value is undoubtedly great, the Gilbert theory for diffusion-free self-associating systems must be applied with great care in the interpretation of actual data.

IV. ISOMERIZATION

A. BASIC EQUATIONS

Some proteins such as serum albumin, ovalbumin, and γ-globulin are known to undergo isomerization under certain solvent conditions. Thus, in such a system, two isomeric forms of a protein, A and B, coexist in relative amounts dependent on the total protein concentration and the solvent conditions such as pH, ionic strength, temperature, and so forth. If the sedimentation and diffusion coefficients of A and B are equal, the solution behaves ultracentrifugally as if it contained only a single nonreacting solute. Therefore, in what follows, we consider cases in which at least one of these coefficients has different values for the two solutes.

We label A and B as solutes 1 and 2, and designate the rate constants for the conversion from A to B and that from B to A by k_1 and k_2, respectively. The specific rates of production by isomerization, Q_1 and Q_2, may then be represented by

$$Q_1 = -k_1c_1 + k_2c_2, \qquad Q_2 = k_1c_1 - k_2c_2 \tag{4.83}$$

The continuity equations for isomer systems are therefore written

$$\frac{\partial c_1}{\partial t} = \frac{\partial}{r\,\partial r}\left(\bar{D}_1 r\,\frac{\partial c_1}{\partial r} - s_1\omega^2 r^2 c_1\right) - k_1c_1 + k_2c_2 \tag{4.84}$$

$$\frac{\partial c_2}{\partial t} = \frac{\partial}{r\,\partial r}\left(\bar{D}_2 r\,\frac{\partial c_2}{\partial r} - s_2\omega^2 r^2 c_2\right) + k_1c_1 - k_2c_2 \tag{4.85}$$

By the same kind of argument as in Section III.B, it can be concluded that if k_1 and k_2 are infinitely large, c_1 and c_2 are related to each other by

$$c_2 = Kc_1 \qquad (4.86)$$

where

$$K = \lim_{\substack{k_1 \to \infty \\ k_2 \to \infty}} \frac{k_1}{k_2} \qquad (4.87)$$

Equation 4.86 is the equilibrium relation for isomerization, implying that when k_1 and k_2 are very large, chemical equilibrium is established almost instantaneously between the isomers. The concentrations c_1 and c_2 as functions of r and t under this condition are determined by equation 4.86 coupled with the differential equation for the total solute concentration $c = c_1 + c_2$,

$$\frac{\partial c}{\partial t} = \frac{\partial}{r\,\partial r}\left(\overline{D}r\,\frac{\partial c}{\partial r} - \bar{s}\omega^2 r^2 c\right) \qquad (4.88)$$

which is derived by addition of equations 4.84 and 4.85. Here $\bar{s}$ and $\overline{D}$ are given by

$$\bar{s} = \frac{s_1 + Ks_2}{1 + K} \qquad (4.89)$$

$$\overline{D} = \frac{\overline{D}_1 + K\overline{D}_2}{1 + K} \qquad (4.90)$$

In general, these quantities are functions of c_1 and c_2, but when equation 4.86 holds, they depend only on c. Thus, when k_1 and k_2 are indefinitely large, equation 4.88 becomes identical to the Lamm equation for a concentration-dependent single solute. In other words, the system behaves ultracentrifugally as if it were a binary solution. We have reached a similar conclusion for self-associating systems when the reaction rate constants are indefinitely large. However, we must note the difference that, in the case of self-association, $\bar{s}$ becomes an increasing function of c even if the sedimentation coefficients of the individual solute species are independent of concentration, whereas, in the case of isomerization, $\bar{s}$ is constant if s_1 and s_2 are independent of concentration. Thus, in the latter case with indefinitely large rate constants, the sedimentation boundary undergoes no spreading if diffusion is absent.

B. VAN HOLDE'S SOLUTION FOR THE CASE OF NO DIFFUSION

Solutions to the set of equations 4.84 and 4.85 with finite rate constants were obtained analytically for the case in which $D_1 = D_2 = 0$, s_1, s_2, k_1, and k_2 are constant, and the rectangular-cell approximation is valid.[21] They are as follows:

$$\frac{c_1}{c_1^{0}} = 1 - k_1 \exp\left[(k_1 - k_2)w\right]F(t,w) \tag{4.91}$$

$$\frac{c_2}{c_2^{0}} = 1 - k_1 \exp\left[(k_1 - k_2)w\right]F(t,w)$$

$$- \exp\left[(k_1 - k_2)w\right]\exp\left(-k_1 t\right)I_0\left(2[k_1 k_2 w(t-w)]^{1/2}\right) \tag{4.92}$$

where

$$F(t,w) = \int_{w}^{t} \exp\left(-k_1 \theta\right)I_0\left(2[k_1 k_2 w(\theta - w)]^{1/2}\right)d\theta \tag{4.93}$$

and

$$w = \frac{x - u_1 t}{u_2 - u_1} \tag{4.94}$$

Here $x = r - r_1$, and u_i $(i = 1, 2)$ is the mobility of isomer i defined as $s_i \omega^2 r_1$. The symbol I_0 denotes the modified Bessel function of the first kind and zeroth order. The quantity w, having a dimension of time, represents the distance from the "unperturbed" boundary position* of isomer 1 divided by the difference in mobility between the two isomeric forms. The whole analysis implicitly assumes u_2 to be larger than u_1.

From the formulas just given it follows that the concentrations of solutes 1 and 2 are zero behind the unperturbed boundary for solute 1 (the region where $w < 0$) and that they remain at the initial values c_1^{0} and c_2^{0} beyond the unperturbed boundary for solute 2 which is located at $w = t$ on the w-axis. As these boundary positions are approached from within the region

---

*The boundary position of solute species i would be $x = u_i t$ if the solute sedimented without being disturbed by coexisting solutes. Such a position is often called the unperturbed boundary position of solute species i.

$0 < w < t$, the concentrations c_1 and c_2 tend to the following values:

$$\left. \begin{array}{l} c_1 = c_1{}^0 \exp(-k_1 t) \\ c_2 = 0 \end{array} \right\} (w \to 0) \tag{4.95}$$

$$\left. \begin{array}{l} c_1 = c_1{}^0 \\ c_2 = c_2{}^0 [1 - \exp(-k_2 t)] \end{array} \right\} (w \to t) \tag{4.96}$$

Therefore, the total concentration c $(= c_1 + c_2)$ changes discontinuously by $c_1{}^0 \exp(-k_1 t)$ and $c_2{}^0 \exp(-k_2 t)$ when one crosses the unperturbed boundaries for 1 and 2, respectively, from left to right. Thus, the concentration gradient curve has infinitely sharp spikes at these two boundary positions.

Some calculated distributions of c and $\partial c / \partial x$ for typical sets of $k_1 t$ and $k_2 t$ are depicted in Figs. 4.8 and 4.9. It is seen that when there is a considerable difference between the magnitudes of k_1 and k_2, the gradient curve is essentially single-peaked and that, under some circumstances in which k_1 and k_2 have comparable magnitudes, there appear three maxima in the concentration gradient distribution. This latter feature is a good example of how our intuition often fails us in thinking about the transport of chemically interacting solutes. In fact, if we were ignorant of the chemical reaction involved, it would be quite likely that we assigned a third solute component to the intermediate maximum gradient.

C. STUDIES OF THE GENERAL CASES

The first attempt to integrate equations 4.84 and 4.85 with the diffusion terms retained was made by Cann et al.[22] in relation to formally the same problem in electrophoresis. The Fourier transform method was used to obtain an exact analytic solution, but the evaluation of the Fourier integrals was found to be so difficult that Cann et al. had to be satisfied with providing an asymptotic solution valid for the special case in which $k_1 = k_2$ and $\bar{D}_1 = \bar{D}_2$. Unfortunately, it was pointed out by Scholten[23] that this asymptotic solution had been considerably in error due to the neglect of terms of the same order as those retained.

Bak and Kauman[24] also investigated the set of equations 4.84 and 4.85, again in connection with electrophoresis but with a different initial condition from that chosen by Cann et al.[22] and also by Van Holde,[21] that is, the initial distributions of the isomers were taken to be of a delta function. Scholten and Mysels[25] corrected the Bak-Kauman theory for a number of mathematical errors.

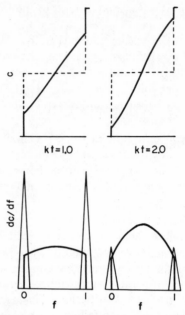

Fig. 4.8. Total concentration and concentration gradient as functions of a reduced cell coordinate $f=(x-u_1t)/(u_2-u_1)t$ in diffusion-free isomerization systems where $k_1t=1.0$, $k_2t=1.0$ and $k_1t=2.0$, $k_2t=2.0$; $f=0$ and 1 correspond to the "unperturbed" boundaries for solute species 1 and 2, respectively. The dashed line indicates the concentration profile for the case where reactions are very slow. In the dc/df curve, the gradients at the lower and upper edges of the boundary are actually infinite but are arbitrarily represented as triangles whose areas are proportional to the concentration jumps at these positions.

Scholten[23] computed the Fourier integrals involved in the solution of Cann et al. with the aid of a digital computer, and obtained graphs of concentration gradient distributions for about 40 combinations of parameters in the special case where $k_1 = k_2$ and $\overline{D}_1 = \overline{D}_2$, and for about 20 combinations of parameters in the general case in which the diffusion coefficients and the rate constants are different. Independently, Cann and Bailey[26] solved the problem numerically with the basic equations replaced by appropriate difference equations, though the calculations were restricted to cases in which $\overline{D}_1 = \overline{D}_2$.

The great variety of concentration gradient distributions presented by these authors, though not reproduced here for lack of space, is sufficient to discourage us from attempting to analyze sedimentation boundary curves when the rates of reaction are finite and the effects of diffusion are appreciable.

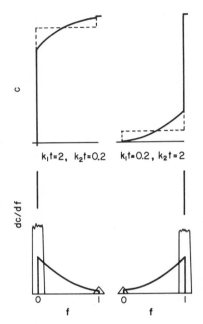

$k_1t=2, \ k_2t=0.2 \quad k_1t=0.2, \ k_2t=2$

Fig. 4.9. Total concentrations and concentration gradients in diffusion-free isomerization systems where $k_1t = 2.0$, $k_2t = 0.2$ and $k_1t = 2.0$.

V. COMPLEX FORMATION

A. INTRODUCTION

Presumably, self-association and isomerization are the simplest of a great variety of chemical reactions which occur in macromolecular solutions. As has been shown above, even the presence of such a relatively simple reaction makes it very difficult to calculate sedimentation boundary curves on the basis of the ultracentrifuge differential equations, and various assumptions and approximations, some of which are quite drastic or even unrealistic, have to be introduced in order to reduce the difficulty to manageable proportions. Obviously, each new reacting species included in the system adds to the degree of mathematical complication. Therefore, it is not difficult to imagine that one must confront an almost unsurmountable difficulty in theoretical treatments of the sedimentation transport of complex-forming systems, in which two or more kinds of substances react to form one or more new compounds. The simplest of such reactions is

$$A + B \rightleftarrows AB \tag{4.97}$$

Sedimentation and electrophoresis of a system involving this type of complex formation have been investigated theoretically by several authors, though the treatments were restricted to the reactions which occur so rapidly that reactants (A and B) and product (AB) establish chemical equilibrium instantaneously at all points in the solution. The most fundamentally important of such theories is presumably the one due to Gilbert and Jenkins.[27,28]

B.   THE THEORY OF GILBERT AND JENKINS

For a treatment of complex-forming systems it is convenient to use molar quantities rather than specific quantities. Let us denote the molar concentrations of species A, B, and AB by C_A, C_B, and C_C, their mobilities by u_A, u_B, and u_C, and their molar rates of production by R_A, R_B, and R_C. If the rectangular-cell approximation is used, all diffusion terms are neglected, and the mobilities are assumed to be independent of concentration, it is shown that the continuity equations for the three species are given by (with x defined as $r - r_1$)

$$\frac{\partial C_A}{\partial t} + u_A \frac{\partial C_A}{\partial x} = R_A \tag{4.98}$$

$$\frac{\partial C_B}{\partial t} + u_B \frac{\partial C_B}{\partial x} = R_B \tag{4.99}$$

$$\frac{\partial C_C}{\partial t} + u_C \frac{\partial C_C}{\partial x} = R_C \tag{4.100}$$

Since matter has to be conserved, it follows that R_A, R_B, and R_C are related by the equations

$$R_A = R_B = - R_C = - R \tag{4.101}$$

Actually, R may be represented by

$$R = K_f C_A C_B - K_b C_C \tag{4.102}$$

where K_f is the molar rate constant for the reaction $A + B \rightarrow AB$, and K_b is that for the reaction $AB \rightarrow A + B$.

In the limit where these reaction rates are indefinitely large but their ratio remains finite, we have the equilibrium relation

$$K = \frac{C_C}{C_A C_B} \tag{4.103}$$

where K is defined by

$$K = \lim_{\substack{K_f \to \infty \\ K_b \to \infty}} \frac{K_f}{K_b} \qquad (4.104)$$

In what follows, K is assumed to be independent of concentration, and, after Gilbert and Jenkins, its reciprocal is designated by k.

For indefinitely large reaction rates the quantity R defined by equation 4.102 becomes indeterminate, but the set of equations 4.98, 4.99, and 4.100 is now supplemented by a new condition 4.103, together with equations 4.101. We transform these differential equations by means of the substitutions

$$\xi = \frac{(x/t) - \frac{1}{2}(u_A + u_B)}{\frac{1}{2}(u_A - u_B)}, \qquad \eta = \frac{1}{t} \qquad (4.105)$$

where, since A and B are on an equal footing, we may assume $u_A > u_B$ without loss of generality. Eliminating R_A, R_B, and R_C from the transformed equations with the aid of equations 4.101, we obtain

$$(1 - \xi)\frac{\partial C_A}{\partial \xi} + (1 + \xi)\frac{\partial C_B}{\partial \xi} = \eta \frac{\partial}{\partial \eta}(C_A - C_B) \qquad (4.106)$$

$$(1 - \xi)\frac{\partial C_A}{\partial \xi} + (\lambda - \xi)\frac{\partial C_C}{\partial \xi} = \eta \frac{\partial}{\partial \eta}(C_A + C_B) \qquad (4.107)$$

where

$$\lambda = \frac{(u_C - u_A) + (u_C - u_B)}{u_A - u_B} \qquad (4.108)$$

These equations coupled with equation 4.103 should allow the concentrations C_A, C_B, and C_C to be determined as functions of ξ and η, but there appears no hope that the integration can be carried through analytically.

For vanishingly small values of η, that is, for indefinitely large t, equations 4.106 and 4.107 reduce to "asymptotic" forms

$$(1 - \xi)\frac{dC_A}{d\xi} + (1 + \xi)\frac{dC_B}{d\xi} = 0 \qquad (4.109)$$

$$(1 - \xi)\frac{dC_A}{d\xi} + (\lambda - \xi)\frac{dC_C}{d\xi} = 0 \qquad (4.110)$$

Gilbert and Jenkins[27,28] found that this set of ordinary differential equations coupled with equation 4.103 can be integrated in closed form, and examined the behavior of the "asymptotic" solutions of the problem so obtained for a variety of combinations of parameters involved. In what follows, we summarize their results for the case in which $u_B < u_A < u_C$, a situation supposed to be encountered usually in ultracentrifugal experiments. For other cases, important in relation to electrophoresis, the reader should consult Gilbert and Jenkins.[28]

1. Expressions for the Concentration Distributions of Individual Solute Species

In the case in which $u_B < u_A < u_C$, the sedimentation boundary appears either in the range $u_B t < x < x_0(t)$ or in the range $u_A t < x < x_0(t)$, depending on whether the parameter Φ defined below is positive or negative. The side centripetal to the lower edge of the boundary region ($u_B t$ or $u_A t$) is devoid of any solute species, while the side centrifugal to the upper edge of the boundary region (x_0) is the plateau region where the concentrations of A, B, and AB are uniform and maintain their initial values C_A^0, C_B^0, and C_C^0.

The boundary region is split into two parts at a position denoted here by $x'(t)$. The distributions of the three solute species in these two parts as well as the positions $x'(t)$ and $x_0(t)$ are governed by the magnitudes of u_A, u_B, u_C, C_A^0, C_B^0, and k through the parameters Φ and Φ_0 which are defined by the following set of relations:

$$C_A^0 = \frac{k}{\lambda - 1}\left(1 - \frac{\Phi - \Phi_0}{\sinh \Phi_0}\right)\cosh^2 \frac{\Phi_0}{2} \tag{4.111}$$

$$C_B^0 = \frac{k}{\lambda + 1}\left(1 + \frac{\Phi - \Phi_0}{\sinh \Phi_0}\right)\sinh^2 \frac{\Phi_0}{2} \tag{4.112}$$

First, $x_0(t)$ is determined from the relation

$$\tanh \frac{\Phi_0}{2} = \left(\frac{\lambda + 1}{\lambda - 1}\right)^{1/2}\left(\frac{x_0(t) - u_A t}{x_0(t) - u_B t}\right)^{1/2} \tag{4.113}$$

which indicates that $x_0(t)/t$ should be constant, that is, $x_0(t)$ should increase linearly with time, with zero initial value.

If the value of Φ determined from equations 4.111 and 4.112 is positive, $x'(t)$ is obtained from

$$\tanh \frac{\phi'}{2} = \left(\frac{\lambda+1}{\lambda-1}\right)^{1/2}\left(\frac{x'(t)-u_A t}{x'(t)-u_B t}\right)^{1/2} \tag{4.114}$$

where ϕ' denotes the root of the equation

$$\sinh \phi' + \phi' = \Phi \tag{4.115}$$

If Φ is negative, $x'(t)$ is determined by substituting the root of the equation

$$\sinh \phi' - \phi' = |\Phi| \tag{4.116}$$

into equation 4.114.

Regardless of whether Φ is positive or negative, the concentration distributions for x between $x'(t)$ and $x_0(t)$ are given by

$$C_A = \frac{k}{\lambda-1}\left(1 - \frac{\Phi-\phi}{\sinh \phi}\right)\cosh^2 \frac{\phi}{2} \tag{4.117}$$

$$C_B = \frac{k}{\lambda+1}\left(1 + \frac{\Phi-\phi}{\sinh \phi}\right)\sinh^2 \frac{\phi}{2} \tag{4.118}$$

$$C_C = \frac{k}{4(\lambda^2-1)}\left[1 - \frac{(\Phi-\phi)^2}{\sinh^2 \phi}\right]\sinh^2 \phi \tag{4.119}$$

where ϕ is a variable related to x and t by

$$\tanh \frac{\phi}{2} = \left(\frac{\lambda+1}{\lambda-1}\right)^{1/2}\left(\frac{x-u_A t}{x-u_B t}\right)^{1/2} \tag{4.120}$$

If $\Phi>0$, as has been noted above, the lower edge of the boundary region is located at the "unperturbed" boundary position of species B, that is, $x = u_B t$, and the concentration distributions for x between $u_B t$ and $x'(t)$ are given by

$$C_A = 0, \qquad C_C = 0$$

$$C_B = \frac{2k}{\lambda+1}\sinh^2 \frac{\phi'}{2} \tag{4.121}$$

with ϕ' given as the root of equation 4.115. Thus, in this case, the region between the lower edge and $x'(t)$ of the sedimentation boundary is a uniform solution of B only.

If $\Phi < 0$, the lower edge of the boundary is located at the "unperturbed" boundary position of species A, that is, $x = u_A t$, and the concentration distributions between this position and $x'(t)$ are given by

$$C_A = \frac{2k}{\lambda - 1} \cosh^2 \frac{\phi'}{2}$$

$$(4.122)$$

$$C_B = 0, \qquad C_C = 0$$

where ϕ' is the root of equation 4.116. Hence, in this case, the region considered is filled with a uniform solution of A only.

2. Illustrations of the Gilbert-Jenkins Asymptotic Solutions

Inspection of the expression presented above reveals that if the variable ξ defined by equation 4.105 is used as the abscissa, the concentration distribution of each solute species represented by the Gilbert-Jenkins asymptotic solutions is transformed to a "reduced" curve which is independent of time and is governed by three parameters: λ, C_A^0/k, and C_B^0/k. The remaining parameters u_A, u_B, and u_C are absorbed into ξ and λ.

In the upper sections of Figs. 4.10a and 4.10b are shown the "reduced" concentration distributions of A, B, and AB $(=C)$ for two systems in which λ is the same $(=5)$ but the set of values of C_A^0/k and C_B^0/k is different. For these values of the basic parameters, the constant Φ is negative for the system (a) and positive for the system (b). The lower sections of Figs. 4.10a and 4.10b show the corresponding distributions of total refractive index gradients, calculated on the assumption that the three species have the same specific refractive index increment. The dashed lines in these graphs indicate the "unperturbed" boundary positions (expressed in terms of x/t) for the three species.

Of the many interesting features displayed by these theoretical curves the following points are worthy of special mention.

1. It is observed how greatly the position and shape of the gradient curve are affected by the occurrence of a chemical reaction in the system.

2. The gradient curve is separated into two distinct regions, one being infinitely sharp and the other being broad and spreading over a range.

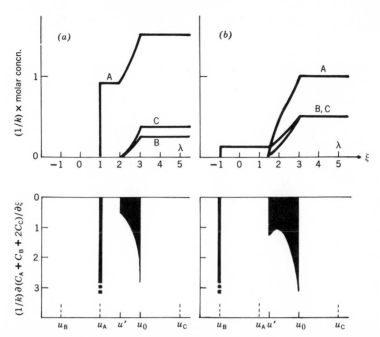

Fig. 4.10. Theoretical distributions of concentrations of individual solute species (A, B, C) and total refractive index gradients in diffusion-free reacting systems of the type $A + B \rightleftharpoons C$ in which $u_B < u_A < u_C$ (a) $C_A^0/k = 1.5$, $C_B^0/k = 0.25$. (b) $C_A^0/k = 1.0$, $C_B^0/k = 0.5$.

3. The spreading boundary is single-peaked or double-peaked, depending on the initial concentrations of A and B and on the equilibrium constant K (or k). In either case, the position of each peak does not correspond to the mobility of any solute species present in the solution. The appearance of two peaks does not mean the existence of two solutes in the spreading boundary. In fact, three solute species coexist in this region.

4. The infinitely sharp boundary moves at a slower rate than the spreading boundary, but this rate does not necessarily equal the mobility of the slowest species B. It may be equal to the rate intrinsic to species A, depending on the initial concentrations of A and B and on the equilibrium constant.

These characteristics of the theoretical gradient curves add another example to the warning that it is extremely hazardous to draw conclusions from sedimentation velocity experiments by formal application of the traditional procedures when a chemical interaction between solutes may be suspected.

C. THE NICHOL-WINZOR TREATMENT IN TERMS OF CONSTITUENT QUANTITIES

Nichol and Winzor[29] have shown that the Gilbert-Jenkins treatment for complex-forming systems of the type $A + B \rightleftarrows AB$ is entirely analogous to that adopted by Johnston and Ogston[30] for systems of two nonreacting solutes whose sedimentation coefficients (or mobilities) depend on the composition of the solution (i.e., the concentrations of the two solute components). In what follows, we give a formal and exact proof of this fact, using, for simplicity, the rectangular-cell approximation to the basic equations.

Suppose that species AB consists of constituent A and constituent B. Then the molar constituent concentration C^A of A in the solution is given as the sum of C_A and C_C, that is,

$$C^A = C_A + C_C \tag{4.123}$$

Similarly, we have for the molar constituent concentration C^B of B

$$C^B = C_B + C_C \tag{4.124}$$

Here, as before, C_A, C_B, and C_C denote the molar concentrations of *solute species* A, B, and AB, respectively. Addition of equations 4.98 and 4.100, followed by use of equations 4.101 and 4.123, yields

$$\frac{\partial C^A}{\partial t} + \frac{\partial}{\partial x}[u_A C^A + (u_C - u_A)C_C] = 0 \tag{4.125}$$

In a similar way, we obtain

$$\frac{\partial C^B}{\partial t} + \frac{\partial}{\partial x}[u_B C^B + (u_C - u_B)C_C] = 0 \tag{4.126}$$

Eliminating C_A and C_B from equations 4.103, 4.123, and 4.124, and solving the resulting quadratic equation for C_C, we obtain

$$C_C = \tfrac{1}{2}\left\{ C^A + C^B + k - \left[(C^A + C^B + k)^2 - 4C^A C^B \right]^{1/2} \right\} \tag{4.127}$$

where, as before, k is the reciprocal of K.

Introduction of C_C from equation 4.127 into equations 4.125 and 4.126 gives

$$\frac{\partial C^A}{\partial t} + \frac{\partial}{\partial x}(u^A C^A) = 0 \tag{4.128}$$

and

$$\frac{\partial C^B}{\partial t} + \frac{\partial}{\partial x}(u^B C^B) = 0 \tag{4.129}$$

where u^A and u^B are the constituent mobilities of constituents A and B defined as

$$u^A = u_A + \frac{(u_C - u_A)}{2C^A}\left\{ C^A + C^B + k - \left[(C^A + C^B + k)^2 - 4C^A C^B\right]^{1/2}\right\}$$

$$\tag{4.130}$$

$$u^B = u_B + \frac{(u_C - u_B)}{2C^B}\left\{ C^A + C^B + k - \left[(C^A + C^B + k)^2 - 4C^A C^B\right]^{1/2}\right\}$$

$$\tag{4.131}$$

Equations 4.128 and 4.129 are formally identical to the transport equations for two nonreacting solutes with the mobilities which are dependent on the composition of the solutes in the solution. Thus it is found that the difficulty associated with the sedimentation analysis of complex-forming systems of the type considered here is essentially of the same kind as the one which is encountered in theoretical treatments of the sedimentation boundaries of multicomponent solutions involving solute-solute interactions.

The formulation given above in terms of constituent quantities suggests an alternative approach to the transport processes of chemically reacting systems. In fact, such an approach has been developed by Nichol and Ogston[31,32] for self-associating systems and complex-forming systems of the types $A + B \rightleftarrows AB$ and $A + B \rightleftarrows C + D$. Their theory mainly purports to elucidate rather qualitatively how the major features of sedimentation or electrophoresis patterns of these reacting systems are influenced by the relative magnitudes of mobilities and initial concentrations of individual solute species. We do not enter into it in the present monograph.

REFERENCES

1. F. J. Reithel, *Adv. Protein Chem.* **3**, 123 (1963).
2. L. W. Nichol, J. L. Bethune, G. Kegeles, and E. L. Hess, in *The Proteins* (H. Neurath, Ed.), Academic Press, New York, 1964, Vol. 2, p. 305.
3. A. Tiselius, *Nova Acta Regiae Soc., Sci. Upsaliensis* **7**, No. 4, 1 (1930).
4. G. G. Belford and R. L. Belford, *J. Chem. Phys.* **37**, 1926 (1962).
5. G. A. Gilbert, *Discuss. Faraday Soc.* **20**, 68 (1955).

6. G. A. Gilbert, *Proc. Roy. Soc.* (*London*) **A250**, 377 (1959).
7. M. S. N. Rao and G. Kegeles, *J. Am. Chem. Soc.* **80**, 5724 (1958).
8. T. Svedberg and K. O. Pedersen, *The Ultracentrifuge*, Oxford University Press, London and New York (Johnson Reprint Corporation, New York), 1940, p. 39 *et seq.*
9. J. L. Bethune and G. Kegeles, *J. Phys. Chem.* **65**, 1761 (1961).
10. G. Kegeles and M. S. N. Rao, *J. Am. Chem. Soc.* **80**, 5721 (1958).
11. V. Massey, W. F. Harrington, and B. S. Hartley, *Discuss. Faraday Soc.* **20**, 24 (1955).
12. G. W. Schwert, *J. Biol. Chem.* **179**, 655 (1949).
13. G. A. Gilbert, *Nature* **186**, 882 (1960).
14. G. A. Gilbert, *Proc. Roy. Soc.* (*London*) **A276**, 354 (1963).
15. L. M. Gilbert and G. A. Gilbert, *Nature* **194**, 1173 (1962).
16. S. N. Timasheff and R. Townend, *J. Am. Chem. Soc.* **83**, 464 (1961).
17. R. Townend and S. N. Timasheff, *J. Am. Chem. Soc.* **82**, 3168 (1960).
18. R. Townend, L. Weinberger, and S. N. Timasheff, *J. Am. Chem. Soc.* **82**, 3175 (1960).
19. J. L. Bethune and G. Kegeles, *J. Phys. Chem.* **65**, 1755 (1961).
20. D. J. Cox, *Arch. Biochem. Biophys.* **129**, 106 (1969).
21. K. E. Van Holde, *J. Chem. Phys.* **37**, 1922 (1962).
22. J. R. Cann, J. G. Kirkwood, and R. A. Brown, *Arch. Biochem. Biophys.* **72**, 37 (1957).
23. P. C. Scholten, *Arch. Biochem. Biophys.* **93**, 568 (1961).
24. T. A. Bak and W. G. Kauman, *Trans. Faraday Soc.* **55**, 1109 (1959).
25. P. C. Scholten and K. J. Mysels, *Trans. Faraday Soc.* **57**, 764 (1961).
26. J. R. Cann and H. R. Bailey, *Arch. Biochem. Biophys.* **93**, 576 (1961).
27. G. A. Gilbert and R. C. Ll. Jenkins, *Nature* **177**, 853 (1956).
28. G. A. Gilbert and R. C. Ll. Jenkins, *Proc. Roy. Soc.* (*London*) **A253**, 420 (1959).
29. L. W. Nichol and D. J. Winzor, *J. Phys. Chem.* **68**, 2455 (1964); see also L. W. Nichol and A. G. Ogston, *ibid.* **69**, 1754 (1965).
30. J. P. Johnston and A. G. Ogston, *Trans. Faraday Soc.* **42**, 789 (1946).
31. L. W. Nichol and A. G. Ogston, *Proc. Roy. Soc.* (*London*) **B163**, 343 (1965).
32. L. W. Nichol and A. G. Ogston, *Proc. Roy. Soc.* (*London*) **B167** 164 (1967).

SEDIMENTATION EQUILIBRIUM IN NONREACTING SYSTEMS

I. INTRODUCTION

A. SEDIMENTATION EQUILIBRIUM

If a solution placed in an ultracentrifuge cell is centrifuged at a constant speed of rotation and at a constant temperature, it will eventually reach a state in which the spatial distribution of any component no longer changes with time This is a state of thermodynamic equilibrium called *sedimentation equilibrium*. Sometimes it is referred to as sedimentation-diffusion equilibrium, a description which derives from the kinetic theory picture in which we visualize forward transport of solutes due to sedimentation being exactly counterbalanced by backward transport due to diffusion at all points in the solution column. During the course of theoretical developments, however, this traditional picture has proved of little use for the formulation of a general and rigorous theory of sedimentation equilibrium phenomena. Thus the term sedimentation-diffusion equilibrium seldom appears in the current literature on ultracentrifugation.

Modern theory of sedimentation equilibrium has developed since Goldberg[1] established its foundation. It is essentially an application of thermodynamics to systems which undergo a field of external force. Thus, it involves no such kinetic quantities as sedimentation and diffusion coefficients and is saved from such complications in the transport theory as the boundary anomaly arising from hydrodynamic interactions between different solute components.

This chapter summarizes and discusses typical theories of sedimentation equilibrium for solutions which consist of chemically nonreacting components. In doing this, emphasis is placed on ways in which the statements to guide thermodynamic analysis of dilute solutions may be derived from fundamental principles governing the sedimentation equilibrium state.

Although its inception dates back to a paper by Tiselius[2] in 1926, theoretical developments on the sedimentation equilibrium of chemically reacting solutes are a relatively new event that has occurred during the last decade. It is still not in a satisfactory stage, but in view of its great significance for the study of proteins in solution, we discuss this subject separately in Chapter 6.

B. APPROACH TO SEDIMENTATION EQUILIBRIUM

Theoretically, the state of sedimentation equilibrium is realized after the rotor is spun at constant speed for an infinitely long interval of time. In reality, after a certain finite time a state is reached in which the distribution of refractive indices or refractive index gradients in the solution column no longer appears to change with time within the precision of the measurement. Measurements taken at this state may be used for studying sedimentation equilibrium behavior of the system under test. For experimentalists it is highly desirable to bring the system to such a state as quickly as possible, mainly for the economy of time. The quicker attainment of equilibrium is even imperative when one has to deal with substances which are relatively unstable. Investigation of the factors which may control the rate of approach to equilibrium thus becomes one of the problems of great practical importance for those who are engaged in ultracentrifugal experiments. In fact, since the early days of the development of the ultracentrifuge, this problem has attracted many investigators, and a variety of ideas have been proposed. Nowadays, one can reach equilibrium within a time of the order of days or even hours, by making use of some of these ideas. Their theoretical foundations are discussed in Chapter 7.

C. BASIC EQUATIONS FOR SEDIMENTATION EQUILIBRIUM

Fundamental equations descriptive of the behavior of a solution at sedimentation equilibrium are the consequence of the condition that the total potential (chemical potential and potentials of external forces) of any component in the solution must be constant throughout the solution column at equilibrium. To derive such equations we impose, for the time being, the restriction that all components either are or may be treated as nonelectrolytes and their partial specific volumes are independent of pressure. The latter condition, that is, incompressibility of the solution, does not seriously limit the practical value of the equilibrium equations to be derived, because the rotor speeds usually employed for sedimentation equilibrium experiments with macromolecular solutes are relatively low so that no great pressure gradient is set up in the cell. The only cases for which its validity becomes debatable are experiments with the meniscus-depletion method (Section II.A.4) and those with the density-gradient method (Section VI), in which the rotor is generally spun at quite high speeds.

As before, let the number of components in a given solution be $q+1$. One of them (labeled 0) is regarded as a solvent and all others are treated as solutes. The condition for sedimentation equilibrium is equivalent to the statement that the gradient of the total potential of any component $i(i=0,1,\ldots,q), X_i$, in the direction of radial distance r must be zero for all r between the air-liquid meniscus r_1 and the cell bottom r_2. For an isothermal incompressible solution at mechanical equilibrium, the X_i is given by equation 1.101. The condition of mechanical equilibrium is automatically satisfied at sedimentation equilibrium. Thus we find that the sedimentation equilibrium state of an incompressible solution is exactly described by the set of equations derived by setting the right-hand side of equation 1.101 equal to zero, that is,

$$(1-\bar{v}_i\rho)\omega^2 r = \sum_{k=1}^{q} \mu_{ik}\frac{dc_k}{dr} \qquad (i=1,2,\ldots,q) \qquad (5.1)^*$$

Here the equation for component 0 has been omitted, because, at thermal and mechanical equilibrium, the condition $X_0=0$ is automatically satisfied if the X_i's for all other components in the system vanish (see equation 1.72). In equations 5.1, ρ is the density of the solution at a particular radial position r, c_i and $\bar{v}_i$ are the c-scale concentration and the partial specific volume of component i at the same radial position, and μ_{ik} is a shorthand notation for the derivative

$$\mu_{ik} = \left(\frac{\partial \mu_i}{\partial c_k}\right)_{T,P,c_m} \qquad (5.2)$$

where μ_i is the chemical potential *per gram* of component i, T is the absolute temperature, and P is the hydrostatic pressure; the subscript c_m has the same meaning as that given in Chapter 1, Section H.

Equations 5.1 are q first-order differential equations to determine q solute concentrations $c_1, c_2, \ldots, c_q$ as functions of r. Integration of these equations requires the following data: (1)[†] ρ and $\bar{v}_i$'s for all solute com-

---

*The partial derivatives $\partial c_k/\partial r$ in equation 1.101 are here replaceable by the ordinary derivatives dc_k/dr, because the equilibrium concentrations are functions of r only.

[†]Note that for incompressible systems under consideration ρ and $\bar{v}_i$'s are independent of P. We also remark that the temperature is held constant here.

ponents as functions of $c_1, c_2, \ldots, c_q$; (2) μ_i's for all solute components as functions of $c_1, c_2, \ldots, c_q$, and P; and (3) P as a function of r. Data (1) and (2) can, in principle, be obtained from separate thermodynamic measurements or appropriate statistical thermodynamic considerations. Instead of (1), we may have data of $\bar{v}_i$ for all $q+1$ components as functions of the solute concentrations $c_1, c_2, \ldots, c_q$, since if we make use of the generally valid relation

$$\sum_{i=0}^{q} \bar{v}_i c_i = 1 \tag{5.3}$$

the density ρ can be expressed as

$$\rho = \frac{1}{\bar{v}_0} + \sum_{i=1}^{q} \left(1 - \frac{\bar{v}_i}{\bar{v}_0}\right) c_i \tag{5.4}$$

At thermal and mechanical equilibrium the pressure distribution in the ultracentrifuge cell is governed by equation 1.69, that is,

$$\frac{dP}{dr} = \rho \omega^2 r \tag{5.5*}$$

In order to determine P as a function of r, this equation must be coupled with equations 5.1, because ρ depends on the solute concentrations.

Equations 5.1 and 5.5 must be supplemented with the boundary conditions for $c_1, c_2, \ldots, c_q$, and P. Those for the solute concentrations are obtained from the physical requirement that the total mass of any solute component must be conserved before and after centrifugation, yielding

$$\int_{r_1}^{r_2} r c_i(r) \, dr = \int_{r_1}^{r_2} r c_i^0 \, dr = \frac{c_i^0}{2}(r_2^2 - r_1^2) \qquad (i = 1, 2, \ldots, q) \tag{5.6}$$

where c_i^0 is the concentration of solute i in the uniform solution before centrifugation, that is, the initial solution. The one for pressure P is obtained from the condition that the pressure at the air-liquid meniscus is maintained at atmospheric pressure P_1 during the entire period of centrifugation. Thus

$$P = P_1 \quad \text{at} \quad r = r_1 \tag{5.7}$$

---

*The partial derivative $\partial P / \partial r$ in equation 1.69 is here replaced by the ordinary derivative dP/dr, because P depends only on r at sedimentation equilibrium.

II. TWO-COMPONENT SYSTEMS

A.   BASIC EQUATIONS AND ASSUMPTIONS

Let us begin with a binary solution which consists of a solvent 0 and a homogeneous solute 1. For this system the set of equations 5.1 reduces to a single equation of the form

$$(1 - \bar{v}_1\rho)\omega^2 r = \mu_{11}\frac{dc_1}{dr} \tag{5.8}$$

Since here we are concerned with nonelectrolyte solutes, the chemical potential μ_1 may be written

$$\mu_1 = (\mu_1^0)_c + \frac{RT}{M_1}\ln(y_1 c_1) \tag{5.9}$$

where $(\mu_1^0)_c$, the reference chemical potential per gram of solute 1, is defined so that the activity coefficient y_1 on the c-concentration scale approaches unity as c_1 tends to zero. The quantity M_1 is the molecular weight of solute 1.

Substitution of equation 5.9 into equation 5.8 yields

$$\frac{M_1(1 - \bar{v}_1\rho)\omega^2 rc}{RT} = \left[1 + c_1\left(\frac{\partial \ln y_1}{\partial c_1}\right)_{T,P}\right]\frac{dc_1}{dr} \tag{5.10}$$

This differential equation for c_1 forms the basis of theoretical treatments of sedimentation equilibrium phenomena in incompressible binary solutions. The data necessary for its integration are y_1 as a function of c_1 and P, and $\bar{v}_1$ and ρ as functions of c_1 (these last two are independent of P for incompressible systems).

We proceed by imposing certain restrictions on terms in equation 5.10. First, it is assumed that the solution is sufficiently dilute with respect to the solute component. Then, since we are concerned with nonelectrolyte solutes, the logarithm of y_1 may be expanded in integral powers of c_1 to give

$$\ln y_1 = M_1 B_1 c_1 + M_1^2 B_2 c_1^2 + \cdots \tag{5.11}$$

The expansion coefficients $B_1, B_2, \ldots$, which all refer to infinite dilution, are functions of T and P. They are equivalent to the nonideality coefficients $B_{11}, B_{111}, \ldots$, which were introduced in Chapter 3, Section

IV.E. Second, it is assumed that y_1 does not depend on P, which is equivalent to assuming the pressure independence of all nonideality coefficients. As is shown in Section II.D, both $\bar{v}_0$ and $\bar{v}_1$ become independent of c_1 under this assumption. If $\bar{v}_0$ is independent of c_1, its inverse can be equated to the density of the pure solvent, ρ_0. Hence equation 5.4 may be written

$$\rho = \rho_0 + (1 - \bar{v}_1 \rho_0) c_1 \qquad (5.12)$$

where $\bar{v}_1$ must be treated as independent of c_1. Thus it is found that if y_1 is independent of P, the density of the solution varies linearly with c_1.

Substitution of equations 5.11 and 5.12 into equation 5.10, followed by expansion in powers of c_1, gives

$$\frac{M_1(1 - \bar{v}_1 \rho_0) \omega^2 r c_1}{RT} = \left[1 + (B_1 M_1 + \bar{v}_1) c_1 \right.$$

$$\left. + (2 B_2 M_1^{\,2} + B_1 M_1 \bar{v}_1 + \bar{v}_1^{\,2}) c_1^{\,2} + \cdots \right] \frac{dc_1}{dr} \qquad (5.13)$$

This provides the basis of thermodynamic analysis of sedimentation equilibrium data for dilute binary solutions of nonelectrolytes in which $\bar{v}_0$ and $\bar{v}_1$ may be treated as independent of pressure and composition. Since no pressure-dependent term is involved, this differential equation for c_1 need not be coupled with the pressure equation 5.5. In what follows, we transform equation 5.13 into statements to guide the evaluation of experimental data.

1. The Integrated Form of Williams et al.

Direct integration of both sides of equation 5.13 with respect to r from the air-liquid meniscus $r = r_1$ to the cell bottom $r = r_2$ and use of equation 5.6 for $i = 1$, that is,

$$\int_{r_1}^{r_2} r c_1(r) dr = \frac{c_1^{\,0}}{2} (r_2^{\,2} - r_1^{\,2}) \qquad (5.14)$$

yields, after some rearrangements of terms,

$$\frac{1}{M_{app}} = \frac{1}{M_1} + \frac{B_1^*}{2} [c_1(r_1) + c_1(r_2)]$$

$$+ \frac{2 B_2^* M_1}{3} \left\{ [c_1(r_1)]^2 + c_1(r_1) c_1(r_2) + [c_1(r_2)]^2 \right\}$$

$$+ \cdots \qquad (5.15)$$

where M_{app} is an apparent molecular weight defined by

$$M_{app} = \frac{2RT}{(1 - \bar{v}_1 \rho_0)(r_2^2 - r_1^2)\omega^2} \frac{[c_1(r_2) - c_1(r_1)]}{c_1^0} \cdot \quad (5.16)$$

and $B_1^*, B_2^*, \ldots$ are defined by

$$B_1^* = B_1 + \frac{\bar{v}_1}{M_1}$$

$$\quad (5.17)$$

$$B_2^* = B_2 + \frac{\bar{v}_1 B_1}{2M_1} + \frac{\bar{v}_1^2}{2M_1^2}$$

Equation 5.15 was derived for the first time by Williams et al.[3] It shows that M_1 and B_1^* can be determined from the ordinate intercept and *initial* slope of a plot of $1/M_{app}$ versus the mean concentration $\bar{c}_1 = \frac{1}{2}[c_1(r_1) + c_1(r_2)]$. It should be noted that the parameter B_1^* is equivalent to B_{11}^* defined by equation 3.224. Thus when the solution is pseudoideal, B_1^* vanishes, and the plot has zero initial slope. For many synthetic polymers examined under theta conditions this type of plot was horizontal over a fairly wide range of the abscissa.

All data required for the application of this method of analysis can be obtained from experiment. The concentrations at the ends of the solution column are most conveniently estimated by extrapolation of Rayleigh fringe patterns at sedimentation equilibrium. The reader is referred to a paper by LaBar[4] for these operations. A somewhat cumbersome procedure must be used[5] when we wish to obtain these concentrations from equilibrium schlieren patterns.

2. Expansion in Powers of c_1^0

Until Williams et al. discovered the method described above, estimate of M_1 and B_1^* had often been made by resort to the following expansion proposed by Wales et al.[6]:

$$\frac{1}{M_{app}} = \frac{1}{M_1} + B_1^* c_1^0 + \cdots \quad (5.18)$$

However, this expansion holds only at the limit of zero rotor speed, and hence should not be applied to data analysis. The verification may be made as follows.

For this purpose we treat the simplest case, in which all B_i^*'s but B_1^* are zero. Then equation 5.13 reduces to

$$\frac{M_1(1 - \bar{v}_1 \rho_0) \omega^2 r c_1}{RT} = (1 + M_1 B_1^* c_1) \frac{dc_1}{dr}$$

which may be integrated to give

$$\ln \theta + \beta \theta = A \xi + k \tag{5.19}$$

where

$$\theta = \frac{c_1}{c_1^{\,0}} \tag{5.20}$$

$$\xi = \frac{r^2 - r_1^{\,2}}{r_2^{\,2} - r_1^{\,2}} \tag{5.21}$$

$$A = \frac{M_1(1 - \bar{v}_1 \rho_0)(r_2^{\,2} - r_1^{\,2}) \omega^2}{2RT} \tag{5.22}$$

$$\beta = B_1^* M_1 c_1^{\,0} \tag{5.23}$$

and k is a constant of integration. In terms of θ and ξ, condition 5.14 may be put in the form

$$\int_0^1 \theta(\xi) d\xi = 1 \tag{5.24}$$

or

$$\int_{\theta_1}^{\theta_2} \theta \left(\frac{d\xi}{d\theta} \right) d\theta = 1 \tag{5.25}$$

with $\theta_1 = c_1(r_1)/c_1^{\,0}$ and $\theta_2 = c_1(r_2)/c_1^{\,0}$. Differentiation of equation 5.19 with respect to θ and substitution of the resulting expression for $d\xi/d\theta$ into equation 5.25 yields

$$\theta_2 - \theta_1 + \frac{\beta}{2} (\theta_2^{\,2} - \theta_1^{\,2}) = A \tag{5.26}$$

On the other hand, it follows from equation 5.19 that

$$\ln \left(\frac{\theta_2}{\theta_1} \right) + \beta(\theta_2 - \theta_1) = A \tag{5.27}$$

These two equations allow the unknown parameters θ_1 and θ_2 to be determined as functions of A and β.

Writing equation 5.19 for $\xi = 0$, we have

$$k = \ln\theta_1 + \beta\theta_1 \tag{5.28}$$

which indicates that the constant k may be determined when θ_1 is obtained as a function of A and β. Insertion of the k so determined back into equation 5.19 completes the solution to the problem under consideration.

Equations 5.26 and 5.27 can be solved analytically for θ_1 and θ_2 only in the case when $\beta = 0$, yielding

$$\theta_1 = \frac{A}{e^A - 1} \tag{5.29}$$

$$\theta_2 = \frac{A}{1 - e^{-A}} \tag{5.30}$$

For small nonzero values of β we may use a series expansion method to solve equations 5.26 and 5.27. The results give for $1/M_{app}$

$$\frac{1}{M_{app}} = \frac{1}{M_1} + Q_1(A)B_1^* c_1^0 - Q_2(A)(B_1^*)^2 M_1 (c_1^0)^2$$

$$+ \text{ higher terms in } c_1^0 \tag{5.31}$$

where

$$Q_1(A) = \frac{A}{2} \coth\frac{A}{2} \tag{5.32}$$

$$Q_2(A) = \left(\frac{A}{2}\right)^2 \tag{5.33}$$

Comparison with equation 5.31 shows that the c_1^0 term in equation 5.18 is incorrect by a factor $Q_1(A)$, which is unity only at $A = 0$, that is, $\omega = 0$. It should be noted that even if all B_i^*'s for $i \geqslant 2$ are assumed to be zero, equation 5.31 contains terms higher than the first power of c_1^0. This feature should be contrasted with equation 5.15, in which $1/M_{app}$ is precisely linear in $\bar{c}_1$ under the same condition.

In place of equation 5.18, use may be made of equation 5.31 for correct evaluation of B_1^* from a plot of $1/M_{app}$ versus c_1^0. For this purpose the experiments must be conducted in such a way that the parameter λ defined by

$$\lambda = \frac{(1 - \bar{v}_1\rho_0)(r_2^2 - r_1^2)\omega^2}{2RT} \tag{5.34}$$

is kept constant for differing initial concentrations, because $Q_1, Q_2, \ldots$ depend on λ through A. The quantity λ has dimensions of reciprocal molecular weight and, as will be seen at many places in the subsequent discussion, it plays a central role in the theory of sedimentation equilibrium.

For a given system the value of λ can be adjusted by the rotor speed ω and the length of the solution column $r_2 - r_1$ (or the volume of a test solution to be placed in the cell). With the current analytical ultracentrifuges, one can work at any ω in the range 2000 to 60,000 rev/min with no attendance of rotor instability. The value of $r_2 - r_1$ can be adjusted between 1 and 0.1 cm or even less with no significant loss in precision of optical measurements of sedimentation behavior. Thus it is possible to vary experimentally the value of λ over an almost 10,000-fold range, depending on the molecular weight of the solute to be investigated. It is due to this experimental flexibility that the sedimentation equilibrium method is capable of studying substances which are so widely different in molecular weight as ranging from small molecules treated in organic chemistry to typical high polymers, both synthetic and biological.

3. Pseudoideal or Very Dilute Solutions

When the solution is very dilute or its nonideality parameters B_i^* are all zero or very small, equation 5.13 may be replaced approximately by

$$\frac{dc_1}{dr} = \frac{M_1(1 - \bar{v}_1 \rho_0)\omega^2 c_1 r}{RT} \tag{5.35}$$

This equation is often called the sedimentation equilibrium equation for ideal binary solutions. However, this convention is not used in the present treatise, since, as was pointed out in Chapter 3, Section IV.E.1, the condition of thermodynamic ideality does not lead to vanishing of all B_i^* coefficients. Instead, we shall refer to it as the equation for very dilute or pseudoideal binary solutions, because it is indeed valid as a good approximation for such solutions.

There are a variety of ways in which equation 5.35 may be compared with experimental data. Here we present some typical examples. For more details the reader should consult Van Holde and Baldwin[7] or Creeth and Pain.[8]

1. Differential Logarithmic Form (Concentration)

Equation 5.35 may be written in the form

$$\frac{d\ln c_1}{d(r^2)} = \frac{M_1(1 - \bar{v}_1\rho_0)\omega^2}{2RT} \qquad (5.36)$$

which indicates that the desired M_1 may be found from the slope of $\ln c_1$ plotted against r^2. This method is suitable for use with the Rayleigh interference optics. However, it faces difficulty when determination of $c_1{}^0$ is not easy, because the value of $c_1{}^0$ is needed for converting observed fringes to c_1. Very recently, an ingenious alternative free from such a difficulty has become available. It is outlined in Appendix A.

2. Differential Form (Concentration Gradient)

Equation 5.35 may be put in the alternate form as

$$\frac{d[(1/r)(d\Delta c_1/dr)]}{d(\Delta c_1)} = \frac{M_1(1 - \bar{v}_1\rho_0)\omega^2}{RT} \qquad (5.37)$$

where

$$\Delta c_1 = c_1(r) - c_1(r_1) \qquad (5.38)$$

Thus the desired value of M_1 may be obtained from the slope of a plot of $(1/r)(d\Delta c_1/dr)$ versus Δc_1.[7] Values of Δc_1 are obtained as the fringe numbers counted from the meniscus when the Rayleigh interference optics is used, while they are calculated by integration of the observed pattern when the schlieren optics is employed. In either case, no determination of $c_1{}^0$ is needed. Despite the convenience in determining Δc_1, the Rayleigh optics is unsuitable for the use of this method, since the less precise procedure of numerical or graphical differentiation becomes necessary to determine $d\Delta c_1/dr$.

3. Differential Logarithmic Form (Concentration Gradient)

Equation 5.35 may be transformed to

$$\frac{d\ln[(1/r)(dc_1/dr)]}{d(r^2)} = \frac{M_1(1 - \bar{v}_1\rho_0)\omega^2}{2RT} \qquad (5.39)$$

which indicates that M_1 may be evaluated from the slope of a plot of $\ln[(1/r)(dc_1/dr)]$ versus r^2. This method, due originally to Lamm,[9] is particularly suitable for schlieren equilibrium data. No integration is necessary, differing from method 2. Marler et al.[10] have pointed out that, mathematically, the Lamm method is essentially equivalent to method 2. However, in practice, somewhat smaller errors may be expected for the Lamm method, in which the values of r chosen as the independent variable may be determined free of error.

When, in any of the three methods of analysis just given, the plot exhibits deviation from linearity, it is an indication of the fact that the second and higher terms in the brackets of equation 5.13 are not negligible under the experimental conditions chosen. In this case, if $B_1^*, B_2^*, \ldots$ are positive, the plot becomes *convex upward*. Nichol et al.[11] discussed equilibrium schlieren patterns of highly nonideal binary solutions, and demonstrated, both theoretically and experimentally, that these higher terms in equation 5.13 lead to the occurrence of a maximum concentration gradient under certain conditions.

If the above-mentioned methods are formally applied to a multi-component solution with c_1 being replaced by the total solute concentration c or the excess refractive index n_c, the respective plots also should become nonlinear. We may then take advantage of the curvatures to obtain information about solute heterogeneity if the nonideality effect is absent or can be eliminated from the plots. A consideration of the problem is deferred to Section V.C.

4. The Meniscus-Depletion Method

The concentration distribution in a very dilute or pseudoideal binary solution at sedimentation equilibrium can be deduced by integration of equation 5.35 followed by determination of the integration constant with the aid of equation 5.14. The result of calculation is

$$c_1(r) = \frac{c_1^0 \lambda M_1}{1 - \exp(-\lambda M_1)} \exp\left[\frac{-\lambda M_1(r_2^2 - r^2)}{r_2^2 - r_1^2} \right] \tag{5.40}$$

where λ is the parameter defined by equation 5.34. The concentrations at the ends of the solution column are represented by

$$c_1(r_1) = \frac{c_1^0 \lambda M_1 \exp(-\lambda M_1)}{1 - \exp(-\lambda M_1)} \tag{5.41}$$

$$c_1(r_2) = \frac{c_1^0 \lambda M_1}{1 - \exp(-\lambda M_1)} \tag{5.42}$$

Hence

$$\frac{c_1(r_2)}{c_1(r_1)} = \exp(\lambda M_1) \tag{5.43}$$

Conventional sedimentation equilibrium experiments are characterized by rotor speeds and solution depths so chosen that the concentration at the cell bottom relative to that at the air-liquid meniscus becomes about 3. Equation 5.43 indicates that such conditions are equivalent to

$$\lambda M_1 \approx 1 \tag{5.44}$$

or, in terms of ω,

$$\omega^2 \approx \frac{2RT}{M_1(1 - \bar{v}_1\rho_0)(r_2^2 - r_1^2)} \tag{5.45}$$

Rotor speeds calculated from this expression, with suitable values taken for the buoyancy factor and the length of the solution column, are relatively low for macromolecular solutes. Thus, the conventional sedimentation equilibrium experiment is often referred to as "low speed" experiment.

If the rotor speed is increased about three times that used in usual low-speed experiments, the ratio $c_1(r_2)/c_1(r_1)$ increases from 3 to 2×10^4, and the concentration at the air-liquid meniscus becomes effectively zero. Under such a condition, the difference in fringe number between a particular radial position r and the air-liquid meniscus, denoted here by $J(r)$, becomes directly proportional to $c_1(r)$ itself. Hence, equation 5.36 may be written

$$\frac{d \ln J(r)}{d(r^2)} = \frac{M_1(1 - \bar{v}_1\rho_0)\omega^2}{2RT} \tag{5.46}$$

Thus the desired M_1 can be evaluated from the slope of a plot of $\ln J$ versus r^2. Since no information about c_1^0 is needed, this method is particularly useful when there is difficulty in the determination of the initial concentration. Figure 5.1 illustrates this method with data for a preparation of 70S ribosomes; $c_1 = 0.5$ mg/ml and $\omega = 3205$ rev/min.[12]

Yphantis[13] was the first to recognize that the need for obtaining c_1^0 can be eliminated by the use of a high rotor speed which makes the concentration at the air-liquid meniscus effectively zero. However, we take note of the fact that Wales et al.[6] had introduced the idea of meniscus depletion in an earlier investigation concerning the evaluation of the number-average molecular weights of polydisperse solutes by sedimentation equilibrium experiments. Thus the method based on equation 5.46 may be termed the

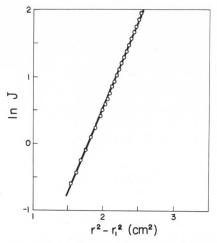

Fig. 5.1. Values of ln J (J: fringe number) as a function of r^2 (r: radial distance) from an experiment of the meniscus-depletion type with a sample of 70S ribosomes[12]; $c_0 = 0.5$ mg/ml, $\omega = 3205$ rev/min. Only data corresponding to more than 0.5 fringe are recorded. The line gives $M_1 = 2.65 \times 10^6$.

Wales-Yphantis method. However, in the current literature, it is generally referred to as the *meniscus-depletion method* or sometimes the *high-speed sedimentation equilibrium method*.

The rotor speed of 3205 rev/min used in obtaining the data of Fig. 5.1 is rather low from the viewpoint of experimentalists. The fact that nonetheless the condition of meniscus depletion was realized suggests that the term "high speed" does not necessarily convey the correct meaning of "meniscus depletion" experiments. Actually, in the experiment which led to Fig. 5.1, the molecular weight of the solute was so high that the value of the product λM_1 became much larger than unity in spite of the relatively low speed employed. Thus it seems more proper to define the meniscus-depletion method in terms not of high rotor speeds but of large values of a dimensionless parameter λM_1, that is,

$$\lambda M_1 \gg 1 \tag{5.47}$$

In recent years, this method has acquired a great popularity among biochemists as well as molecular biologists. The reason is not only its above-mentioned merit but also such practical advantages that it needs only a very limited amount of sample (we may use concentrations of about 0.01%, an order of magnitude lower than in conventional low-speed experiments), and that because of the centrifugal fractionation obtainable at high rotor speeds, it can be of effective use for analyzing paucidisperse systems which characterize most biochemical preparations.

These merits, however, are somewhat offset by several shortcomings. First, this method is restricted to very dilute or pseudoideal solutions. It should be noted that lowering concentration is attended by a loss of precision of the measurement. Second, high rotor speeds (more correctly, high values of λM_1) produce very high concentration gradients in the region near the cell botton, which may give rise to unresolved fringes and thus preclude accurate measurements of $J(r)$. Third, in the same circumstances, a substantial portion of the solute is swept down to the cell bottom. Thus even if the initial concentration is sufficiently low, relatively high concentrations are set up in the vicinity of the cell bottom, and nonideality effects may not become entirely negligible over the solution column. Then the plot of $\ln J$ versus r^2 exhibits a curvature, and we must confront the problem of which portion of the plot to choose to calculate M_1 from the slope. Theoretically, we should approach the air-liquid meniscus as closely as possible in order to be free from nonideality effects. However, in practice, the data at very low concentrations are sensitive to baseline errors. Thus we will have to compromise in determining the desired slope. Finally, when relatively small solutes have to be treated, the concentration at the air-liquid meniscus may not be depressed to zero even at the highest rotor speed available at present.[4, 14]

Due to these disadvantages, errors of a few percent in the evaluated molecular weights by the meniscus-depletion method are unavoidable. Creeth and Pain[8] state that this method should be used when it has been verified that the particular instrument and drive unit are capable of giving an adequate performance in the speed range and solution conditions chosen.

B. VIRIAL EXPANSION FOR SEDIMENTATION EQUILIBRIUM

Statistical thermodynamics allows derivation of the expansions for the osmotic pressure π and the excess turbidity τ_e of a binary solution in powers of the solute concentration.[15-17] If, as in the system treated above, the partial specific volumes and also the refractive index increments of the components are independent of pressure and composition, these expansions are given by

$$\frac{\pi}{RTc_1{}^0} = \frac{1}{M_1} + \left(\frac{B_1^*}{2}\right)c_1{}^0 + \text{ higher terms in } c_1{}^0 \qquad (5.48)$$

$$\frac{Hc_1{}^0}{\tau_e} = \frac{1}{M_1} + B_1^* c_1{}^0 + \text{ higher terms in } c_1{}^0 \qquad (5.49)$$

where B_1^* is the nonideality parameter defined before, and H is a constant known as the light-scattering factor. These series in c_1^0 are called the virial expansion for osmotic pressure and the virial expansion for light scattering, respectively. In each of them, the coefficients for $c_1^0, (c_1^0)^2, \ldots$ are referred to as the second virial coefficient, the third virial coefficient, and so on. In this treatise, the osmotic pressure second virial coefficient is denoted by $B_{OS},$* and the light-scattering second virial coefficient by B_{LS}. It is seen that B_{LS} is just equal to B_1^* and is twice as large as B_{OS}. This relationship does not hold when the partial specific volumes and/or the refractive index increments of the components are either pressure dependent or composition dependent.[15] Under pseudoideality where $B_1^* = 0$, both B_{OS} and B_{LS} vanish simultaneously, and when either of these coefficients becomes zero, B_1^* vanishes. Therefore, the necessary and sufficient condition for a binary solution of the type under consideration to be pseudoideal is the vanishing of either B_{OS} or B_{LS}.

Up through the second terms, equations 5.15 and 5.49 are similar in form, except for the fact that the c_1^0 in the latter is replaced by the mean concentration $\bar{c}_1 = \frac{1}{2}[c_1(r_1) + c_1(r_2)]$ in the former. Because of this difference it does not seem proper to call equation 5.15 the virial expansion for sedimentation equilibrium. In this treatise, equation 5.31 will be referred to as such, and the coefficient for c_1^0 in this equation, that is, the *sedimentation equilibrium second virial coefficient*, is denoted by the symbol B_{SE}. Thus

$$B_{SE} = B_1^* \left(\frac{\lambda M_1}{2} \right) \coth \left(\frac{\lambda M_1}{2} \right) \qquad (5.50)$$

where the relationship $A = \lambda M_1$ has been used.

Unlike B_{OS} and B_{LS}, the quantity B_{SE} depends on the external conditions of the experiment through the parameter λ. This difference is associated with the fact that in osmotic pressure or light-scattering measurements we deal with solutions of uniform concentration, whereas in sedimentation experiments there is set up a distribution of concentration gradients in the test solution. Finally, it should be noted that B_{SE} also vanishes under pseudoideality.

C. DETERMINATION OF CHEMICAL POTENTIALS

Our considerations so far in this chapter have centered on the methods that may be utilized to evaluate the molecular weight of a solute from

---

*In polymer physical chemistry, the osmotic pressure second virial coefficient is usually denoted by A_2.

sedimentation equilibrium data for dilute binary solutions. In fact, since the beginning of the development of the ultracentrifuge it has been the molecular weight determination that drew the primary interest of most of the investigators who were engaged in sedimentation equilibrium experiments. However, there are other thermodynamic quantities whose magnitude may be available by this type of equilibrium experiment. Typical of them is the chemical potential.

From the structure of equation 5.8 it is readily recognized that measurements of c_1 and dc_1/dr at sedimentation equilibrium allow the computation of μ_{11}, provided the value of $\bar{v}_1 \rho$ at the specified c_1 is known from separate density measurements. When μ_{11} is obtained in this way for a series of c_1 (at constant temperature), integration of the resulting curve of μ_{11} versus c_1 yields the chemical potential of the solute μ_1 as a function of c_1. Also from the experimental determination of $\mu_{11}(c_1)$ it is possible to calculate $\mu_{01}(c_1)$ by making use of the Gibbs-Duhem relation, though this process requires data for M_0, M_1, and $\rho(c_1)$. Integration of $\mu_{01}(c_1)$ with respect to c_1 then gives the dependence of the chemical potential of the solvent on the concentration c_1.

This possibility of determining the chemical potentials of components in binary solutions from sedimentation equilibrium measurements was early recognized by Pedersen[18] and Drucker,[19] but it was not elaborated until Young and associates[20] developed a rigorous theory of binary solutions in the way discussed in the next section. Wales[21] was probably the first investigator who applied the above-mentioned idea to thermodynamic analysis of concentrated polymer solutions. He calculated, from sedimentation equilibrium data for methyl ethyl ketone solutions of polystyrene and polyvinyl acetate, the apparent osmotic pressure second virial coefficients (defined as $(1/c_1)[(\pi/Rc_1T)-(1/M_1)]$) up to polymer concentrations of about 20% and found them to agree closely with the values derived from direct osmotic pressure measurements.

Very recently, Scholte[22] presented a further confirmation of the idea, performing more extensive measurements with polystyrene in toluene and in cyclohexane up to as high a polymer concentration as about 80 wt%. His results for the toluene solutions of polystyrene are illustrated in Fig. 5.2, where the values of $-\Delta\hat{\mu}_0/RT$ ($\Delta\hat{\mu}_0$ is the difference between $M_0\mu_0$ in the solution and that in the pure solvent at a fixed temperature) are compared to those derived by previous investigators from other equilibrium experiments.

One might hesitate to employ the sedimentation equilibrium method for thermodynamic studies of concentrated polymer solutions, on the consideration that because the viscosity of such solutions may be very high one would have to wait an exceedingly long interval of time for the state of

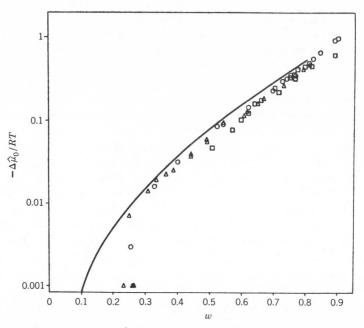

Fig. 5.2. Relation between $-\Delta\hat{\mu}_0/RT$ and w (weight fraction of solute) for the system polystyrene-touluene. ———, from sedimentation equilibrium by Scholte[22]; $\bigcirc$, from vapor pressure data of Bawn et al. (25, 60, and 80°C) [*Trans Faraday Soc.* **46**, 677 (1950)]; $\square$, from vapor pressure data of Baughan (20°C) [*Trans. Faraday Soc.* **44**, 495 (1948)]; $\triangle$, from vapor pressure data of Schmoll and Jenckel (20–60°C) [*Z. Elektrochem.* **60**, 756 (1956)].

equilibrium to be reached. In fact, Wales[21] observed the attainment of equilibrium after a few weeks of centrifugation in his work quoted above. However, as has been demonstrated by Scholte,[22] this kind of difficulty can be largely obviated by use of a short solution column (see Section IV.D). Hence, at present, even concentrated polymer solutions may be examined rather routinely by the sedimentation equilibrium method.

D.   CORRECTION FOR PRESSURE EFFECTS

The theory described above has introduced, for its development, two important assumptions that the partial specific volumes of the components and the activity coefficient of the solute are independent of pressure P. Although these assumptions are probably appropriate under the conditions of conventional low-speed experiment, it is of great theoretical interest to extend the theory to the case to which they are inapplicable. Such work was undertaken by Young et al.[20]

For the treatment of pressure-dependent systems it is important to choose concentration variables which are independent of P. Here we express the concentration of the solute (component 1) on the molality scale and designate it by m_1. If we denote the chemical potential *per mole* of component 1 by $\hat{\mu}_1$, the total potential per mole of the same component in the ultracentrifuge cell is represented by $\hat{\mu}_1(m_1, P, T) - \frac{1}{2}M_1\omega^2 r^2$. The condition that at sedimentation equilibrium the total potential of the solute is constant throughout the cell therefore gives

$$d\hat{\mu}_1 - M_1\omega^2 r\, dr = 0 \qquad (T = \text{constant}) \tag{5.51}$$

This equation must be coupled with the pressure equation 5.5, which may be written

$$dP = \rho\omega^2 r\, dr \tag{5.52}$$

Because

$$d\hat{\mu}_1 = \left(\frac{\partial\hat{\mu}_1}{\partial m_1}\right)_{T,P} dm_1 + \left(\frac{\partial\hat{\mu}_1}{\partial P}\right)_{T,m_1} dP \qquad (T = \text{constant}) \tag{5.53}$$

and because, from thermodynamics,

$$\left(\frac{\partial\hat{\mu}_1}{\partial P}\right)_{T,m_1} = M_1\bar{v}_1 \tag{5.54}$$

equation 5.51, after substitution of equation 5.52, may be put in the form*

$$\left(\frac{\partial\hat{\mu}_1}{\partial m_1}\right)_P \frac{dm_1}{dr} = M_1[1 - \bar{v}_1(m_1, P)\rho(m_1, P)]\omega^2 r \tag{5.55}$$

We write $\hat{\mu}_1$ as

$$\hat{\mu}_1 = (\hat{\mu}_1{}^0)_m + RT\ln[m_1\gamma_1(m_1, P)] \tag{5.56}$$

where $(\hat{\mu}_1{}^0)_m$, the reference potential per mole of the solute, is defined so that $\gamma_1(m_1, P)$, the practical activity coefficient on the molality scale of the solute, tends to unity as m_1 approaches zero. Insertion of equation 5.56 and consideration that the reference potential does not depend on m_1 transform equation 5.55 into

$$\left\{\frac{\partial\ln[m_1\gamma_1(m_1, P)]}{\partial m_1}\right\}_P \frac{dm_1}{dr} = \frac{M_1\omega^2 r}{RT}[1 - \bar{v}_1(m_1, P)\rho(m_1, P)] \tag{5.57}$$

---

*Since here we are concerned with an isothermal system, the subscript T is omitted from all derivatives which appear below.

Equations 5.57 and 5.5, a set of differential equations of the first order, must be solved simultaneously for the independent variables m_1 and P as functions of r. In order to perform the real integration, it is necessary to know the forms of $\hat{\mu}_1$, $\bar{v}_1$, and ρ as functions of m_1 and P. The boundary conditions for m_1 and P are provided by equations 5.14 and 5.7, respectively. For the present purpose, it is convenient to cast equation 5.14 into the form*

$$\int_{r_1}^{r_2} \frac{\rho(m_1, P)m_1(r)}{1000 + M_1 m_1(r)} r\, dr = \frac{(r_2^2 - r_1^2)}{2} \frac{\rho(m_1^0, P_1)m_1^0}{1000 + M_1 m_1^0} \qquad (5.58)$$

by using the following interrelation between c_1 and m_1:

$$c_1 = \frac{\rho M_1 m_1}{1000 + M_1 m_1} \qquad (5.59)$$

The general solutions to the set of equations just given cannot be obtained, but, as has been shown by Young et al.,[20] it is possible to derive an equation which allows the computation of the activity of the solute in a binary solution of molality $(m_1)_\alpha$ relative to its activity in a solution of molality $(m_1)_\beta$, both at the same pressure P''.

This process starts with the introduction of equations 5.54 and 5.56 into the identity

$$\left[\frac{\partial}{\partial P} \left(\frac{\partial \hat{\mu}_1}{\partial m_1} \right)_P \right]_{m_1} = \left[\frac{\partial}{\partial m_1} \left(\frac{\partial \hat{\mu}_1}{\partial P} \right)_{m_1} \right]_P$$

Then

$$\left\{ \frac{\partial}{\partial P} \left[\frac{\partial \ln(m_1 \gamma_1)}{\partial m_1} \right]_P \right\}_{m_1} = \frac{M_1}{RT} \left(\frac{\partial \bar{v}_1}{\partial m_1} \right)_P \qquad (5.60)$$

Integration from $P = P'$ to $P = P''$ with m_1 fixed gives

$$\left[\frac{\partial \ln(m_1 \gamma_1)}{\partial m_1} \right]_{P''} - \left[\frac{\partial \ln(m_1 \gamma_1)}{\partial m_1} \right]_{P'} = \frac{M_1}{RT} \int_{P'}^{P''} \left(\frac{\partial \bar{v}_1}{\partial m_1} \right)_P dP \qquad (5.61)$$

*m_1^0 denotes the molality of the solute in the original solution.

Further integration with respect to m_1 from $(m_1)_\alpha$ to $(m_1)_\beta$ at constant P'' yields

$$\ln\left[\frac{(m_1)_\beta\gamma_1((m_1)_\beta, P'')}{(m_1)_\alpha\gamma_1((m_1)_\alpha, P'')}\right] = \int_{(m_1)_\alpha}^{(m_1)_\beta}\left[\frac{\partial\ln(m_1\gamma_1)}{\partial m_1}\right]_{P'}dm_1$$

$$+\frac{M_1}{RT}\int_{(m_1)_\alpha}^{(m_1)_\beta}\left[\int_{P'}^{P''}\left(\frac{\partial\bar{v}_1}{\partial m_1}\right)_P dP\right]dm_1 \quad (5.62)$$

So far both m_1 and P' have been treated as independent of each other. We now allow P' to vary with m_1 in such a way that P' is equal to P which satisfies the differential equations 5.57 and 5.5. The relation between P and m_1 determined from these equations may be termed the "equilibrium line" on the (P, m_1) plane.

The first integral on the right-hand side of equation 5.62 becomes, with substitution for $[\partial\ln(m_1\gamma_1)/\partial m_1]_{P'}$ from equation 5.57 with $P = P'$,

$$\int_{(m_1)_\alpha}^{(m_1)_\beta}\left[\frac{\partial\ln(m_1\gamma_1)}{\partial m_1}\right]_{P'}dm_1$$

$$= \left(\frac{\omega^2 M_1}{2RT}\right)(r_\beta^2 - r_\alpha^2) - \left(\frac{\omega^2 M_1}{RT}\right)\int_{r_\alpha}^{r_\beta}\bar{v}_1(m_1, P')\rho(m_1, P')r\,dr \quad (5.63)$$

where r_α and r_β denote the values of r at which m_1 becomes equal to $(m_1)_\alpha$ and $(m_1)_\beta$, respectively.

The double integral in equation 5.62 can be evaluated to yield

$$\int_{(m_1)_\alpha}^{(m_1)_\beta}\int_{P'}^{P''}\left(\frac{\partial\bar{v}_1}{\partial m_1}\right)_P dP\,dm_1$$

$$= \omega^2\left\{\int_{r_\alpha}^{r_\beta}[\bar{v}_1(m_1, P') - \bar{v}_1((m_1)_\beta, P')]\rho(m_1, P')r\,dr\right.$$

$$\left.+ \int_{P'(r_\alpha)}^{P''}[\bar{v}_1((m_1)_\alpha, P) - \bar{v}_1((m_1)_\beta, P)]dP\right\} \quad (5.64$$

Introduction of equations 5.63 and 5.64 into equation 5.62 and rearrangement complete the desired relation:

$$\ln\left[\frac{(m_1)_\beta\gamma_1((m_1)_\beta, P'')}{(m_1)_\alpha\gamma_1((m_1)_\alpha, P'')}\right]$$

$$=\left(\frac{\omega^2 M_1}{2RT}\right)(r_\beta^2 - r_\alpha^2) - \left(\frac{\omega^2 M_1}{RT}\right)\int_{r_\alpha}^{r_\beta}\bar{v}_1((m_1)_\beta, P')\rho(m_1, P')r\,dr$$

$$+\left(\frac{M_1\omega^2}{RT}\right)\int_{P'(r_\alpha)}^{P''}[\bar{v}_1((m_1)_\alpha, P) - \bar{v}_1((m_1)_\beta, P)]\,dP \quad (5.65)$$

For the special case in which P'' is chosen equal to $P'(r_\alpha)$, the last term in equation 5.65 vanishes, and we obtain, after changing the pressure notation from P' to P,

$$\ln\left[\frac{(m_1)_\beta\gamma_1((m_1)_\beta, P_\alpha)}{(m_1)_\alpha\gamma_1((m_1)_\alpha, P_\alpha)}\right] = \left(\frac{\omega^2 M_1}{2RT}\right)(r_\beta^2 - r_\alpha^2)$$

$$-\left(\frac{\omega^2 M_1}{RT}\right)\int_{r_\alpha}^{r_\beta}\bar{v}_1((m_1)_\beta, P)\rho(m_1, P)r\,dr \quad (5.66)$$

Here P_α is the abbreviation of $P(r_\alpha)$, that is, the pressure found at the position along the equilibrium line where the molality of the solute is $(m_1)_\alpha$. If we take the position r_α as the air-liquid meniscus and the position r_β as an arbitrary point r in the cell, equation 5.66 may be written

$$\ln\left[\frac{m_1(r)\gamma_1(m_1(r), P_1)}{m_1(r_1)\gamma_1(m_1(r_1), P_1)}\right] = \left(\frac{\omega^2 M_1}{2RT}\right)(r^2 - r_1^2)$$

$$-\left(\frac{\omega^2 M_1}{RT}\right)\int_{r_1}^{r}\bar{v}_1(m_1(r), P(r'))\rho(m_1(r'), P(r'))r'\,dr' \quad (5.67)$$

This relation allows the computation of the ratio of the activities of the solute in binary solutions of different molalities at atmospheric pressure from the following data: $\bar{v}_1$ and ρ as functions of m_1 and P, and the distribution of m_1 and P in the solution column at sedimentation equilibrium.

Finally, we verify that equation 5.57 reduces to equation 5.10 when the pressure dependence of $\bar{v}_0$ and $\bar{v}_1$ is absent. This should be the case, because equation 5.10 was derived under such an assumption. With the partial specific volumes being independent of pressure P, it follows that ρ becomes independent of P and that c_0 becomes a unique function of c_1. Hence ρ depends only on c_1. Then from equation 5.59 it is found that m_1 also depends on c_1 only. If this consequence is combined with the relation $\hat{\mu}_1 = (\hat{\mu}_1{}^0)_m + RT\ln(m_1\gamma_1) = M_1(\mu_1{}^0)_c + RT\ln(c_1 y_1)$, it follows that

$$\left[\frac{\partial \ln(m_1\gamma_1)}{\partial m_1}\right]_P \frac{dm_1}{dr} = \left[\frac{\partial \ln(c_1 y_1)}{\partial c_1}\right]_P \frac{dc_1}{dr} \tag{5.68}$$

Introduction of this relation into equation 5.57 yields

$$\frac{M_1(1-\bar{v}_1\rho)\omega^2 r c_1}{RT} = \left[1 + c_1\left(\frac{\partial \ln y_1}{\partial c_1}\right)_{T,P}\right]\frac{dc_1}{dr} \tag{5.69}$$

which agrees with equation 5.10. Note that the term $\bar{v}_1\rho$ in this equation may be allowed to vary with c_1, although it must be treated as independent of P.

In Section II.A it was stated that if y_1 is independent of P, both $\bar{v}_0$ and $\bar{v}_1$ in incompressible binary solutions become independent of c_1. This statement can be deduced in the following way. If the solution is incompressible, an argument similar to that described above allows equation 5.60 to be rewritten in terms of c_1 and y_1 as

$$\left[\frac{\partial}{\partial P}\left(\frac{\partial \ln y_1}{\partial c_1}\right)_P\right]_{c_1} = \frac{M_1}{RT}\left(\frac{\partial \bar{v}_1}{\partial c_1}\right)_P$$

or

$$\left[\frac{\partial}{\partial c_1}\left(\frac{\partial \ln y_1}{\partial P}\right)_{c_1}\right]_P = \frac{M_1}{RT}\left(\frac{\partial \bar{v}_1}{\partial c_1}\right)_P \tag{5.70}$$

This proves that if y_1 does not depend on P, $\bar{v}_1$ is independent of c_1. The independence of $\bar{v}_0$ from c_1 follows from this consequence combined with the Gibbs-Duhem relation.

III. THREE-COMPONENT SYSTEMS

A. BASIC EQUATIONS

A great variety of three-component systems may be conceived as the objects of ultracentrifugal analysis, but we do not intend here to discuss their sedimentation equilibrium behavior in an exhaustive manner. Rather, we shall restrict ourselves to a special type of ternary solution, in which, in addition to a macromolecular solute and its solvent, there coexists a second component composed of small molecules. This added component may be either a liquid compound (a second solvent) or a solid compound (a second solute). It is labeled here as component 2, with the main solvent as component 0 and the macromolecular solute as component 1. We ask whether it is possible to determine the molecular weight of component 1 from sedimentation equilibrium measurements on such a ternary solution.

As in Section II, we again impose the condition that all components in the solution either are or may be treated as nonelectrolytes. Also it is assumed that the solution is incompressible. Then the equations basic to the sedimentation equilibria of components 1 and 2 are given by equations 5.1 with $q = 2$. However, for the present discussion, it is convenient to start with the equations in which the concentrations are expressed in terms of molalities. It is a simple matter to show that they are given by*

$$M_1(1 - \bar{v}_1\rho)\omega^2 r = \left(\frac{\partial\hat{\mu}_1}{\partial m_1}\right)_{P,m_2}\frac{dm_1}{dr} + \left(\frac{\partial\hat{\mu}_1}{\partial m_2}\right)_{P,m_1}\frac{dm_2}{dr} \qquad (5.71)$$

$$M_2(1 - \bar{v}_2\rho)\omega^2 r = \left(\frac{\partial\hat{\mu}_2}{\partial m_1}\right)_{P,m_2}\frac{dm_1}{dr} + \left(\frac{\partial\hat{\mu}_2}{\partial m_2}\right)_{P,m_1}\frac{dm_2}{dr} \qquad (5.72)$$

where m_i is the molality of component $i(i = 1, 2)$, and $\hat{\mu}_i$ denotes the chemical potential *per mole* of component i. If use is made of the identity

$$\left(\frac{\partial\hat{\mu}_1}{\partial m_2}\right)_{P,m_1} = \left(\frac{\partial\hat{\mu}_2}{\partial m_1}\right)_{P,m_2} \qquad \text{(at constant } T) \qquad (5.73)$$

*We omit affixing the subscript T to the derivatives which appear below, since we are concerned here with isothermal solutions.

the set of differential equations given above may be transformed to

$$2rA_1 = \left(\frac{\partial\hat{\mu}_1}{\partial m_1}\right)_{P,m_2}\frac{dm_1}{dr} + \left(\frac{\partial\hat{\mu}_2}{\partial m_1}\right)_{P,m_2}\frac{dm_2}{dr} \qquad (5.74)$$

$$2rA_2 = \left(\frac{\partial\hat{\mu}_2}{\partial m_1}\right)_{P,m_2}\frac{dm_1}{dr} + \left(\frac{\partial\hat{\mu}_2}{\partial m_2}\right)_{P,m_1}\frac{dm_2}{dr} \qquad (5.75)$$

where

$$A_i = \frac{M_i}{2}(1 - \bar{v}_i\rho)\omega^2 \qquad (i = 1,2) \qquad (5.76)$$

A dimensionless factor Γ, usually called the "binding coefficient," is defined by

$$\Gamma = -\frac{(\partial\hat{\mu}_2/\partial m_1)_{P,m_2}}{(\partial\hat{\mu}_2/\partial m_2)_{P,m_1}} \qquad (5.77)$$

It can be shown easily that this expression may be rewritten

$$\Gamma = \left(\frac{\partial m_2}{\partial m_1}\right)_{P,\hat{\mu}_2} \qquad (5.78)$$

Making use of equation 5.77 and solving equations 5.74 and 5.75 for dm_1/dr and dm_2/dr, we obtain

$$\frac{dm_1}{dr} = \frac{2r(A_1 + \Gamma A_2)}{(\partial\hat{\mu}_1/\partial m_1)_{P,m_2} - \Gamma^2(\partial\hat{\mu}_2/\partial m_2)_{P,m_1}} \qquad (5.79)$$

$$\frac{dm_2}{dr} = \frac{2r\{\Gamma A_1 + A_2[(\partial\hat{\mu}_1/\partial m_1)_{P,m_2}/(\partial\hat{\mu}_2/\partial m_2)_{P,m_1}]\}}{(\partial\hat{\mu}_1/\partial m_1)_{P,m_2} - \Gamma^2(\partial\hat{\mu}_2/\partial m_2)_{P,m_1}} \qquad (5.80)$$

Analogous to equation 5.56 we may write for $\hat{\mu}_1$ and $\hat{\mu}_2$

$$\hat{\mu}_1 = (\hat{\mu}_1{}^0)_m + RT\ln(m_1\gamma_1) \qquad (5.81)$$

$$\hat{\mu}_2 = (\hat{\mu}_2{}^0)_m + RT\ln(m_2\gamma_2) \qquad (5.82)$$

where $(\hat{\mu}_i{}^0)_m$ is the reference chemical potential per mole of component i appropriate to the molality scale and γ_i is the practical activity coefficient of that component on the same concentration scale. Thus γ_i has the limiting property

$$\lim_{\substack{m_1 \to 0 \\ m_2 \to 0}} \gamma_i = 1 \qquad (i = 1, 2) \tag{5.83}$$

Differentiation of equations 5.81 and 5.82 gives

$$\left(\frac{\partial \hat{\mu}_1}{\partial m_1} \right)_{P, m_2} = RT(m_1^{-1} + \beta_{11}) \tag{5.84}$$

$$\left(\frac{\partial \hat{\mu}_2}{\partial m_2} \right)_{P, m_1} = RT(m_2^{-1} + \beta_{22}) \tag{5.85}$$

where β_{11} and β_{22} are defined, respectively, by

$$\beta_{11} = \left(\frac{\partial \ln \gamma_1}{\partial m_1} \right)_{P, m_2} \tag{5.86}$$

$$\beta_{22} = \left(\frac{\partial \ln \gamma_2}{\partial m_2} \right)_{P, m_1} \tag{5.87}$$

which are functions of m_1, m_2, and P. Substitution of equations 5.81 and 5.82 into equation 5.77 yields

$$\Gamma = -\frac{m_2 \beta_{21}}{1 + m_2 \beta_{22}} \tag{5.88}$$

where

$$\beta_{21} = \left(\frac{\partial \ln \gamma_2}{\partial m_1} \right)_{P, m_2} \tag{5.89}$$

which is also a function of m_1, m_2, and P.

Equations 5.79 and 5.80, in which equations 5.84, 5.85, and 5.88 are inserted, along with the pressure equation 5.5, form the basis from which theoretical statements to guide the evaluation of sedimentation equilibrium data for ternary solutions may be derived. These equations are supplemented with the boundary conditions presented in Section I.C. The actual integration requires the thermodynamic quantities β_{11}, β_{12}, β_{22}, $\bar{v}_1$, and $\bar{v}_2$ to be given as functions of m_1, m_2, and P, but for incompressible systems the last two quantities must be treated as independent of P. By a similar argument to that given for binary solutions, the pressure dependence of the activity coefficients γ_1 and γ_2 (hence that of β_{11}, β_{12}, and β_{22}) may be neglected at rotor speeds which are employed in conventional low-speed sedimentation experiments. When this neglect is permissible, the partial specific volumes of all components become independent of composition of the solution, as can be verified in a manner similar to that described in Section II.D. In what follows, we proceed by confining ourselves to the case in which all these assumptions are applicable.

1. Effects of Binding

We restrict the equilibrium equations 5.79 and 5.80 to the system in which component 1 is so dilute that terms of the first and higher powers of m_1 are negligible in comparison to m_2. Then

$$\frac{dm_1}{d\xi} = C\left[M_1(1 - \bar{v}_1\rho^0) + \Gamma^0 M_2(1 - \bar{v}_2\rho^0)\right]m_1 \qquad (5.90)$$

$$\frac{dm_2}{d\xi} = C\frac{M_2(1 - \bar{v}_2\rho^0)m_2}{1 + \beta_{22}{}^0 m_2} \qquad (5.91)$$

where C and ξ are defined by

$$C = \frac{(r_2{}^2 - r_1{}^2)\omega^2}{2RT} \qquad (5.92)$$

$$\xi = \frac{r^2 - r_1{}^2}{r_2{}^2 - r_1{}^2} \qquad (5.93)$$

and the superscript 0 refers to the limit $m_1 = 0$. Thus ρ^0 and $\beta_{22}{}^0$ are functions of m_2 only (the dependence of these quantities on P need not be considered here because of the assumptions made above).

For practical purposes it is convenient to have equations in which concentrations are expressed on the c-scale. To this end we note that

$$m_i = \frac{(1000/M_i)c_i\bar{v}_0}{1 - \bar{v}_1 c_1 - \bar{v}_2 c_2}$$

and that, at $m_1 = 0$, there holds the relation (see Section II.D)

$$\left(\frac{\partial \ln(m_2\gamma_2)}{\partial m_2}\right)_P \frac{dm_2}{d\xi} = \left(\frac{\partial \ln(c_2 y_2)}{\partial c_2}\right)_P \frac{dc_2}{d\xi}$$

if the solution is incompressible. Then equations 5.90 and 5.91 are transformed to

$$\frac{d\ln c_1}{d\xi} + \frac{c_2\bar{v}_2}{1 - \bar{v}_2 c_2}\frac{d\ln c_2}{d\xi} = C\left[M_1(1 - \bar{v}_1\rho^0) + \Gamma^0 M_2(1 - \bar{v}_2\rho^0)\right] \quad (5.94)$$

$$\frac{d\ln c_2}{d\xi} = C\frac{M_2(1 - \bar{v}_2\rho^0)}{1 + c_2\alpha_{22}{}^0} \quad (5.95)$$

where

$$\alpha_{22}{}^0 = \left(\frac{\partial\ln y_2}{\partial c_2}\right)^0_{P,c_1} \quad (5.96)$$

$$\rho^0 = \rho_0 + (1 - \bar{v}_2\rho_0)c_2 \quad (5.97)$$

with ρ_0 being the density of component 0 in the pure state.

Integration of equation 5.95, with the boundary condition and $\alpha_{22}{}^0$ given as a function of c_2, yields the concentration distribution of component 2 at sedimentation equilibrium. Inasmuch as its expression does not contain quantities which are related to component 1, the same concentration distribution of component 2 ought to be observed in the "blank" experiment, in which only components 0 and 2 are centrifuged at the same compositions and the same external conditions as in the "solution" experiment. Thus if a given solution is sufficiently dilute with respect to component 1, the equilibrium distribution of component 2 is not disturbed, at least to a first approximation, by the presence of component 1. However, the equilibrium distribution of component 1 is affected by the presence of component 2 through the factor Γ^0 and also through the c_2-dependence of ρ^0.

If the coexistence of component 1 has no thermodynamic influence on component 2, the derivative $(\partial\hat{\mu}_2/\partial m_1)_{P,m_2}$ should be zero and hence Γ may be zero, since derivative $(\partial\hat{\mu}_2/\partial m_2)_{P,m_1}$ generally does not vanish. Therefore, the factor Γ^0 may be regarded as a measure of thermodynamic interaction between components 1 and 2. Thus the term multiplied by Γ^0 in equation 5.94 is associated with the extent to which the equilibrium distribution of component 1 is influenced by the thermodynamic interaction it would experience from component 2. In general, this interaction should disturb the equilibrium distribution of component 2, but equation 5.95 indicates that such an effect is negligible if the concentration of component 1 is very low.

When $\Gamma^0 = 0$, equation 5.94 reduces to

$$\frac{d\ln c_1}{d\xi} + \frac{c_2\bar{v}_2}{1 - \bar{v}_2 c_2}\frac{d\ln c_2}{d\xi} = CM_1(1 - \bar{v}_1\rho^0) \qquad (5.98)$$

If component 2 undergoes negligible redistribution, a condition that would be met if the densities of components 0 and 2 were not very much different, equation 5.98 becomes

$$\frac{d\ln c_1}{d\xi} = CM_1(1 - \bar{v}_1\rho^0) \qquad (5.99)$$

with ρ^0 now given by

$$\rho^0 = \rho_0 + (1 - \bar{v}_2\rho_0)c_2^0 \qquad (5.100)$$

Here c_2^0 is the c-concentration of component 2 in the initial uniform solution. Equation 5.99 is identical in form to the sedimentation equilibrium equation for very dilute or pseudoideal binary solutions (see equation 5.36). Thus, if $\Gamma^0 = 0$ and redistribution of component 2 is negligible, component 1 in a ternary solution very dilute with respect to it would behave as if it were centrifuged in a single solvent of uniform density ρ^0, and its molecular weight can be determined by a variety of methods as described in Section II.A.3.

When $\Gamma^0 = 0$ but component 2 undergoes significant redistribution, substitution of the equilibrium distribution of component 2 premeasured by a "blank" experiment allows equation 5.98 to be integrated to compute c_1 as a function of ξ. In this case, an equilibrium distribution of component 1 is established in a prescribed density gradient of a mixture of components 0 and 2. If densities of the last two components and the rotor speed are chosen suitably, it is possible to set up a density gradient in which the buoyancy factor $1 - \bar{v}_1\rho^0$ vanishes at a certain point in the solution column. Then, at equilibrium, component 1 should find itself only in a narrow band around such a particular point. Meselson et al.[23] were the first to recognize that this type of sedimentation equilibrium experiment would be a very powerful tool for characterizing macromolecular substances of biological interest, such as DNA, in terms of their partial specific volumes. In view of its great appeal both to biochemists and molecular biologists, a detailed account of the density-gradient equilibrium sedimentation is presented in Section V.

Next, when Γ^0 is nonzero but no redistribution of component 2 occurs, equation 5.94 assumes the form

$$\frac{d\ln c_1}{d\xi} = C\left[M_1(1 - \bar{v}_1\rho^0) + \Gamma^0 M_2(1 - \bar{v}_2\rho^0)\right] \tag{5.101}$$

with ρ^0 given by equation 5.100. Equation 5.101 differs from equation 5.99 only in the respect that M_1 in the latter is replaced by a quantity M_1^* defined as

$$M_1^* = M_1 + \frac{\Gamma^0 M_2(1 - \bar{v}_2\rho^0)}{1 - \bar{v}_1\rho^0} \tag{5.102}$$

Therefore, in this case, plots of $\ln c_1$ versus r^2 also follow a straight line, but the slope allows only the computation of M_1^*, not of M_1. Since there exists no experimental means of estimating Γ^0, we are unable to proceed further. This difficulty, first pointed out by Lansing and Kraemer,[24] persists also when component 2 undergoes redistribution.

Equation 5.101, identical in form to the sedimentation equilibrium equation for a very dilute binary solution in which a solute of molecular weight M_1^* is dissolved in a solvent of density ρ^0 (see equation 5.36), indicates that when component 2 is subject to no redistribution, the thermodynamic interaction between components 1 and 2 simply changes the molecular weight of component 1 from its true value M_1 to the value M_1^* given by equation 5.102. Some authors describe this change as caused by "binding" (or "solvation") of component 2 on component 1, and regard the factor Γ^0 as the strength of the binding. Although the factor Γ is called "binding coefficient" on the basis of such a description, there exists no positive reason to accept the meaning of Γ implied by such a name. A nonzero value of this coefficient does not necessarily mean a chemical binding of molecules of component 1 with molecules of component 2, but it is simply an expression of the fact that some thermodynamic interaction exists between these two components.

2. The Lamm Theory for Electrolyte Solutions

Up to this point we have proceeded with the assumption that solutes may be treated as nonelectrolytes. Thus it might be considered that charged macromolecules, such as proteins and synthetic polyelectrolytes, would have to be precluded from the discussions developed so far. This is not the case, however, if these molecules are studied in the presence of an excess of a low-molecular-weight electrolyte, that is, the so-called supporting electrolyte. The reason is that the added electrolyte will swamp out the long-range interactions between macroions. However, since the addition of a supporting electrolyte to a polyelectrolyte solution makes the system a three-component one, a consideration of the binding effect becomes necessary for molecular weight determination of polyelectrolytes from sedimentation equilibrium experiments on such solutions. The first theoretical treatment of this problem was given by Lamm[25] in 1944.

Let us consider a ternary solution which contains a polyelectrolyte PX_z and a supporting electrolyte BX in pure water. It is assumed that these electrolytes ionize completely according to the schemes:

$$PX_z \rightarrow P^{z+} + zX^-$$

$$BX \rightarrow B^+ + X^-$$

Here z is the number of ionizable sites on one polyelectrolyte molecule, and B^+ and X^- represent, respectively, a univalent cation and a univalent anion. For simplicity, we assume that the solution is incompressible and that the mean activity coefficient on the molality scale of each electrolyte component is equal to unity. The latter assumption is not very realistic, and may be removed if desired, but it does not affect the essential points of the results derived below.

We label the neutral PX_z as component 1 and the neutral BX as component 2. The sedimentation equilibrium equations for these components are given by equations 5.71 and 5.72, which may be rewritten

$$RT\left(\frac{d\ln a_1}{dr}\right) = 2A_1 r \tag{5.103}$$

$$RT\left(\frac{d\ln a_2}{dr}\right) = 2A_2 r \tag{5.104}$$

by substituting for $\hat{\mu}_i$

$$\hat{\mu}_i = (\hat{\mu}_i^0)_m + RT\ln a_i \tag{5.105}$$

where a_i is the activity of component i. According to electrochemistry, we may write

$$a_1 = (\gamma_{\pm})_1^{z+1}(m_{P^{z+}})(m_{X^-})^z \tag{5.106}$$

$$a_2 = (\gamma_{\pm})_2^2(m_{B^+})(m_{X^-}) \tag{5.107}$$

where $(\gamma_{\pm})_i$ is the mean activity coefficient on the molality scale of component i, and $m_{P^{z+}}$, m_{B^+}, and m_{X^-} are the molalities of ionic species P^{z+}, B^+, and X^-, respectively. Under the assumption that the mean activity coefficient of each electrolyte component is unity, substitution of equations 5.106 and 5.107 into equations 5.103 and 5.104 yields

$$RT\left(\frac{d\ln m_1}{dr} + z\frac{d\ln m_{X^-}}{dr}\right) = 2A_1 r \tag{5.108}$$

$$RT\left(\frac{d\ln m_2}{dr} + \frac{d\ln m_{X^-}}{dr}\right) = 2A_2 r \tag{5.109}$$

where account has been taken of the fact that $m_{P^{z+}}$ and m_{B^+} are equal to the molalities of the neutral components m_1 and m_2, respectively. The condition of electric neutrality is written

$$zm_{P^{z+}} + m_{B^+} = m_{X^-}$$

or

$$zm_1 + m_2 = m_{X^-} \tag{5.110}$$

Equations 5.108 and 5.109 may be solved simultaneously, with the aid of equation 5.110, to yield expressions for dm_1/dr and dm_2/dr. If the solution is dilute in both polyelectrolyte and low-molecular-weight electrolyte, the results may be simplified to[3]

$$\frac{d\ln c_1}{d(r^2)} = \frac{2A_1\left(\dfrac{z}{2}\dfrac{M_2}{M_1}\dfrac{c_1}{c_2} + 1\right) - A_2 z}{RT\left[2 + z\dfrac{M_2}{M_1}\dfrac{c_1}{c_2}(1+z)\right]} \tag{5.111}$$

$$\frac{dc_2}{d(r^2)} = \frac{A_2\left(\dfrac{M_1}{M_2}\dfrac{c_2}{c_1} + z + z^2\right) - A_1 z}{RT\left[2 + z\dfrac{M_2}{M_1}\dfrac{c_1}{c_2}(1+z)\right]}\frac{M_2}{M_1}c_1 \tag{5.112}$$

From equation 5.111 it follows that if c_2 is nonzero,

$$\lim_{c_1\to 0}\frac{d\ln c_1}{d(r^2)} = \frac{(1-\bar{v}_1\rho^0)\omega^2}{2RT}M_1^* \tag{5.113}$$

where M_1^* stands for

$$M_1^* = M_1 - \left(\frac{z}{2}\right) M_2 \left(\frac{1 - \bar{v}_2 \rho^0}{1 - \bar{v}_1 \rho^0}\right) \tag{5.114}$$

and ρ^0 is the density of the solution in the absence of the polyelectrolyte solute. Equation 5.113 indicates that if redistribution of the low-molecular-weight electrolyte does not occur (so that ρ^0 may be treated as constant), a plot of $\ln c_1$ versus r^2 becomes linear, as is the case with a very dilute binary solution of a nonelectrolyte solute (see Section II.A.3), but its slope gives an apparent molecular weight M_1^* which differs from the true molecular weight M_1 of the polyelectrolyte component by the amount $-(zM_2/2)(1 - \bar{v}_2\rho^0)/(1 - \bar{v}_1\rho^0)$. The difference between M_1^* and M_1 is usually called the *residual charge effect* or the *secondary charge effect*.

Thus, only when $(z/2)M_2$ is sufficiently small compared to M_1 or when the buoyancy factor of the supporting electrolyte is close to zero, the experimentally determined M_1^* may be equated to M_1 with a trivial error. The former condition is equivalent to a very small charge-to-weight ratio of the polyelectrolyte. It is usually satisfied for proteins near the isoelectric point.

The term $(z/2)(M_2/M_1)(c_1/c_2)$ in equation 5.111 is usually referred to as the *primary charge effect*. It depends on the concentration of the polyelectrolyte relative to that of the supporting electrolyte. Thus this effect can be eliminated either by diluting the solution to zero polyelectrolyte concentration or by increasing the concentration of the supporting electrolyte. This latter operation is the usual addition of an excess of a supporting electrolyte to polyelectrolyte solutions. It is important to comprehend that such an operation can be of effective use only for elimination of the primary charge effect.

The question then arises as to what should be done to eliminate the residual or secondary charge effect so that the true molecular weight of a polyelectrolyte may be determined from sedimentation equilibrium experiments. In fact, from this question there has evolved a series of theoretical investigations which are discussed in the next section.

Comparison of equations 5.114 and 5.102 shows that the ternary solution under consideration has a binding coefficient Γ^0 equal to $-z/2$. This implies that the residual charge effect may be regarded as being due to an apparent binding of the added supporting electrolyte on the polyelectrolyte solute. Close examination of the derivation given above reveals that this apparent binding has its root not in a special thermodynamic interaction between the two electrolytes but in the requirement that the condition of electric neutrality be satisfied in any volume element of the solution.

When the solution is "salt free," that is, $c_2 = 0$, equation 5.111 reduces to

$$\frac{d\ln c_1}{d(r^2)} = \frac{(1 - \bar{v}_1 \rho_0)\omega^2}{2RT} \frac{M_1}{z+1} \tag{5.115}$$

This equation indicates that a plot of $\ln c_1$ versus r^2 is linear, but its slope now yields an apparent molecular weight which is smaller than M_1 by a factor $(1 + z)^{-1}$. Thus, the sedimentation equilibrium experiments with charged solutes in pure water lead to a pronounced underestimation of molecular weight even if the number of ionizable sites per solute is only one or two. This important fact was first recognized by Tiselius.[2] The significance of Lamm's disclosure is that Tiselius' effect can be removed by the addition of an excess of a supporting electrolyte but we are still left with the residual charge effect even after such an operation.

B.   THE CASASSA-EISENBERG THEORY

Efforts to find a way by which the residual charge effect could be eliminated were initiated by Scatchard[26] in 1946, and were continued by a number of investigators.[27-36] Thus, Vrij,[30] and independently Casassa and Eisenberg,[32-34] showed that this effect, or more generally the binding effect, can be removed by a conceptual procedure which involves a modified definition of components on the basis of Scatchard's idea.[26] Subsequently, Eisenberg[35] verified that results of greater generality can be derived by a rigorous thermodynamic method which does not resort to the artificial redefinition of components. Then Casassa and Eisenberg,[36] in an important review article published in 1964, elaborated this new idea to a very general and elegant shape.

Some readers may have a preference for the approach along Scatchard's idea, because it is simpler in mathematical context. For a summary of this approach the review articles written by Creeth and Pain[8] and Adams[37] seem to be pertinent references among many others. In this treatise, we wish to give a description of the 1964 method of Casassa and Eisenberg, because it indeed provides an excellent opportunity for comprehension of thermodynamic analysis of multicomponent systems.

Although the original Casassa-Eisenberg theory is concerned with a very general multicomponent system, we restrict it here to a ternary solution which consists of a solvent (component 0), a macromolecular solute (component 1), and a low-molecular-weight solute (component 2). The theory described below is equally applicable to both charged and uncharged systems, so that we do not specify, before moving to the actual discussion, whether or not the solutes are ionizable in the solvent. It is hoped that the present summary can be a useful guide for a study of the more complete and general Casassa-Eisenberg paper.

1. Osmotic Equilibrium

Let us dialyze our ternary solution against a solution (dialyzate) which contains components 0 and 2 at a predetermined composition. It is assumed that the volume of the "outer" dialyzate is so large that its composition does not undergo any appreciable change when either component 0 or 2 diffuses into or out of the "inner" solution through a semipermeable membrane of the dialysis bag. The dialysis is performed with the experimental setup in which the pressure of the dialyzate phase is held constant at a given value of P'. Also, the whole system is maintained at a constant temperature T.

The system reaches a thermodynamic equilibrium state, called osmotic equilibrium, when the following relations are satisfied:

$$(\hat{\mu}_0)_{\text{inner}} = (\hat{\mu}_0)_{\text{outer}} = \text{constant} \tag{5.116}$$

$$(\hat{\mu}_2)_{\text{inner}} = (\hat{\mu}_2)_{\text{outer}} = \text{constant} \tag{5.117}$$

The second relation in each of these comes from the condition that the composition and pressure of the outer dialyzate phase are held constant. The quantity $\hat{\mu}_i$ denotes, as before, the chemical potential *per mole* of component i. The set of relations above is equivalent to

$$d(\hat{\mu}_0)_{\text{inner}} = 0 \tag{5.118}$$

$$d(\hat{\mu}_2)_{\text{inner}} = 0 \tag{5.119}$$

We denote by π the osmotic pressure set up in the inner solution, and by $d\pi$ the change that occurs when dn_1 moles of component 1 are added to it. Writing the Gibbs-Duhem relation for this change and considering conditions 5.118 and 5.119, we obtain

$$V\left(\frac{\partial \pi}{\partial n_1} \right)_{\hat{\mu}} = n_1\left(\frac{\partial \hat{\mu}_1}{\partial n_1} \right)_{\hat{\mu}} \tag{5.120}*$$

where V is the volume of the inner solution, and the subscript $\hat{\mu}$ indicates that differentiation with respect to n_1 is carried out at fixed $\hat{\mu}_0$ and $\hat{\mu}_2$.

---

*To all the derivatives which appear below we omit attaching the subscript T, since we are not concerned with effects caused by variation of temperature.

Equation 5.120 may be rewritten

$$V_m\left(\frac{\partial \pi}{\partial m_1}\right)_{\hat{\mu}} = m_1\left(\frac{\partial \hat{\mu}_1}{\partial m_1}\right)_{\hat{\mu}} \tag{5.121}$$

where V_m is the volume of the solution containing 1000 g of component 0.

Since $\hat{\mu}_1$ and $\hat{\mu}_2$ are functions of m_1, m_2, pressure P, and temperature T, we have at constant T

$$d\hat{\mu}_1 = \overline{V}_1 dP + RT(a_{11} dm_1 + a_{12} dm_2) \tag{5.122}$$

$$d\hat{\mu}_2 = \overline{V}_2 dP + RT(a_{21} dm_1 + a_{22} dm_2) \tag{5.123}$$

where $\overline{V}_i$ is the partial molar volume of component i, and a_{ij} stands for

$$a_{ij} = \left[\frac{\partial \ln(m_i \gamma_i)}{\partial m_j}\right]_{P, m_{k \neq j}} \tag{5.124}$$

with γ_i being the activity coefficient of component i on the molality scale. Solving equations 5.122 and 5.123 for dm_1 and dm_2, we obtain

$$RT|\Delta| dm_1 = \left(d\hat{\mu}_1 - \overline{V}_1 dP\right) a_{22} - \left(d\hat{\mu}_2 - \overline{V}_2 dP\right) a_{12} \tag{5.125}$$

$$RT|\Delta| dm_2 = -\left(d\hat{\mu}_1 - \overline{V}_1 dP\right) a_{21} + \left(d\hat{\mu}_2 - \overline{V}_2 dP\right) a_{11} \tag{5.126}$$

where

$$|\Delta| = a_{11} a_{22} - a_{12} a_{21} \tag{5.127}$$

Let these equations be restricted to the state of osmotic equilibrium where equations 1.118 through 1.120 hold. Then, P in these equations may be replaced by $P' + \pi$. Hence

$$RT|\Delta| m_1 = a_{22}\left(\frac{\partial \pi}{\partial m_1}\right)_{\hat{\mu}}\left[V_m - m_1\left(\overline{V}_1 - \overline{V}_2 \frac{a_{12}}{a_{22}}\right)\right] \tag{5.128}$$

$$-RT|\Delta| m_1\left(\frac{\partial m_2}{\partial m_1}\right)_{\hat{\mu}} = a_{21}\left(\frac{\partial \pi}{\partial m_1}\right)_{\hat{\mu}}\left[V_m - m_1\left(\overline{V}_1 - \overline{V}_2 \frac{a_{11}}{a_{21}}\right)\right] \tag{5.129}$$

Elimination of $(\partial \pi / \partial m_1)_{\hat{\mu}}$ yields

$$\left(\frac{\partial m_2}{\partial m_1}\right)_{\hat{\mu}} = -\left(\frac{a_{21}}{a_{22}}\right)\left[V_m - m_1\left(\overline{V}_1 - \overline{V}_2\frac{a_{11}}{a_{21}}\right)\right]\Bigg/\left[V_m - m_1\left(\overline{V}_1 - \overline{V}_2\frac{a_{12}}{a_{22}}\right)\right]$$

$$(5.130)$$

This indicates how components 1 and 2 interact with each other in the inner solution at osmotic equilibrium. Though not indicated explicitly, each term in equations 5.128 through 5.130 refers to a particular pressure given by $P' + \pi$.

Let the density of the inner solution be denoted by ρ. Then it follows that

$$\left(\frac{\partial \rho}{\partial m_1}\right)_{\hat{\mu}} = \left(\frac{\partial \rho}{\partial m_1}\right)_{P, m_2} + \left(\frac{\partial \rho}{\partial m_2}\right)_{P, m_1}\left(\frac{\partial m_2}{\partial m_1}\right)_{\hat{\mu}} + \left(\frac{\partial \rho}{\partial P}\right)_{m_1, m_2}\left(\frac{\partial P}{\partial m_1}\right)_{\hat{\mu}} \quad (5.131)$$

If the solution is incompressible,* the last term on the right-hand side vanishes, and we obtain

$$\left(\frac{\partial \rho}{\partial m_1}\right)_{\hat{\mu}} = \left(\frac{\partial \rho}{\partial m_1}\right)_{P, m_2} + \left(\frac{\partial \rho}{\partial m_2}\right)_{P, m_1}\left(\frac{\partial m_2}{\partial m_1}\right)_{\hat{\mu}} \quad (5.132)$$

By proper use of equations 5.129 and 5.130, equation 5.132 may be put in the form

$$\frac{(\partial \rho/\partial m_1)_{\hat{\mu}}}{(\partial \pi/\partial m_1)_{\hat{\mu}}} = \left(\frac{1}{RT|\Delta|m_1}\right)\Bigg\{V_m\left[a_{22}\left(\frac{\partial \rho}{\partial m_1}\right)_{P, m_2} - a_{21}\left(\frac{\partial \rho}{\partial m_2}\right)_{P, m_1}\right]$$

$$- m_1\left[(\overline{V}_1 a_{22} - \overline{V}_2 a_{12})\left(\frac{\partial \rho}{\partial m_1}\right)_{P, m_2} + (\overline{V}_2 a_{11} - \overline{V}_1 a_{21})\left(\frac{\partial \rho}{\partial m_2}\right)_{P, m_1}\right]\Bigg\}$$

$$(5.133)$$

*We can develop the argument without this assumption.[36]

2. The Sedimentation Equilibrium Equation for the Macromolecular Component on the c-Concentration Scale

In terms of molalities, the density of a ternary solution, ρ, can be written

$$\rho = \frac{1}{V_m}(1000 + M_1 m_1 + M_2 m_2) \tag{5.134}$$

where, as before, V_m denotes the volume of the solution containing 1000 g of component 0. It follows from this expression that

$$\left(\frac{\partial \rho}{\partial m_i}\right)_{P, m_{j \neq i}} = \frac{M_i}{V_m}(1 - \bar{v}_i \rho) \qquad (i = 1, 2) \tag{5.135}$$

where $\bar{v}_i$ is the partial specific volume of component i. By virtue of this relation the quantity A_i defined by equation 5.76 may be written

$$A_i = \omega^2 \left(\frac{V_m}{2}\right)\left(\frac{\partial \rho}{\partial m_i}\right)_{P, m_{j \neq i}} \tag{5.136}$$

Introduction of equation 5.136 and use of the relation

$$\left(\frac{\partial \hat{\mu}_i}{\partial m_j}\right)_{P, m_{k \neq j}} = RT a_{ij} \tag{5.137}$$

allow equations 5.79 and 5.80 to be written

$$\frac{dm_1}{dr} = \frac{\omega^2 r V_m}{RT|\Delta|}\left[a_{22}\left(\frac{\partial \rho}{\partial m_1}\right)_{P, m_2} - a_{12}\left(\frac{\partial \rho}{\partial m_2}\right)_{P, m_1}\right] \tag{5.138}$$

$$\frac{dm_2}{dr} = \frac{\omega^2 r V_m}{RT|\Delta|}\left[a_{11}\left(\frac{\partial \rho}{\partial m_2}\right)_{P, m_1} - a_{21}\left(\frac{\partial \rho}{\partial m_1}\right)_{P, m_2}\right] \tag{5.139}$$

with $|\Delta|$ defined by equation 5.217. These equations are exact, and can be applied even when the system is compressible and also when the activity coefficients of the components depend on pressure.[36]

The c-scale concentration of component 1 is related to m_1 by the equation

$$c_1 = \frac{M_1 m_1}{V_m} \tag{5.140}$$

Note that V_m is a function of m_1, m_2, P, and T. Thus if the solution is incompressible, we have at constant temperature

$$\frac{dc_1}{dr} = \left(\frac{\partial c_1}{\partial m_1}\right)_{P,m_2} \frac{dm_1}{dr} + \left(\frac{\partial c_1}{\partial m_2}\right)_{P,m_1} \frac{dm_2}{dr} \tag{5.141}$$

It follows from equation 5.140 that

$$\frac{V_m}{M_1}\left(\frac{\partial c_1}{\partial m_1}\right)_{P,m_2} = 1 - \frac{\overline{V}_1}{V_m}m_1, \qquad \frac{V_m}{M_1}\left(\frac{\partial c_1}{\partial m_2}\right)_{P,m_1} = -\frac{\overline{V}_2}{V_m}m_1 \tag{5.142}$$

Hence equation 5.141 gives, after substitution for dm_1/dr and dm_2/dr from equations 5.138 and 5.139,

$$\frac{RT|\Delta|V_m}{M_1\omega^2 r}\frac{dc_1}{dr} = V_m\left[a_{22}\left(\frac{\partial\rho}{\partial m_1}\right)_{P,m_2} - a_{12}\left(\frac{\partial\rho}{\partial m_2}\right)_{P,m_1}\right]$$

$$-m_1\left[\left(\overline{V}_1 a_{22} - \overline{V}_2 a_{12}\right)\left(\frac{\partial\rho}{\partial m_1}\right)_{P,m_2} + \left(\overline{V}_2 a_{11} - \overline{V}_1 a_{21}\right)\left(\frac{\partial\rho}{\partial m_2}\right)_{P,m_1}\right] \tag{5.143}$$

Combining equations 5.133 and 5.143 and using equation 5.140, we arrive at a surprisingly simple relation:

$$\left(\frac{\partial\pi}{\partial m_1}\right)_{\hat{\mu}}\frac{d\ln c_1}{dr} = r\omega^2\left(\frac{\partial\rho}{\partial m_1}\right)_{\hat{\mu}} \tag{5.144}$$

The molality m_1 can be eliminated if both sides are multiplied by $(\partial m_1/\partial c_1)_{\hat{\mu}}$. Thus

$$\left(\frac{\partial\pi}{\partial c_1}\right)_{\hat{\mu}}\frac{d\ln c_1}{dr} = r\omega^2\left(\frac{\partial\rho}{\partial c_1}\right)_{\hat{\mu}} \tag{5.145}$$

This is the final result of the Casassa-Eisenberg treatment for the ternary solution considered. Although the derivation above assumed incompressibility of the solution, it can be shown that exactly the same equation is obtained even if this assumption is removed.[36] However, in what follows, we will proceed with this assumption, since it not only simplifies the treatment but also pertains to relatively low pressure gradients set up in conventional low-speed equilibrium experiments. Furthermore, it is assumed that the activity coefficients on the c-scale of all components are independent of pressure. Except for special cases, this assumption is also sufficient at relatively low pressures.

3. Actual Processes for the Determination of M_1

Under the above-mentioned assumptions, the derivatives $(\partial \pi / \partial c_1)_{\hat{\mu}}$ and $(\partial \rho / \partial c_1)_{\hat{\mu}}$ become independent of pressure. Hence, their values to be used in equation 5.145 may be derived from dialysis experiments performed under ordinary pressure, although, strictly speaking, use must be made of the values which correspond to the hydrostatic pressure $P^{(r)}$ existing at position r in the solution column at sedimentation equilibrium.

We now make the solution very dilute with respect to component 1. Subject to the above-mentioned assumptions, the osmotic virial expansion for such a solution may be written

$$\frac{\pi}{RTc_1} = \frac{1}{M_1} + A_2^{(0)}c_1 + O(c_1^2) \tag{5.146}$$

where the osmotic second virial coefficient $A_2^{(0)}$ depends only on c_2, the concentration of component 2 in the osmotically equilibrated solution. Its possible dependence on temperature T is not indicated, because we are not concerned here with effects caused by variation of T. Substitution of equation 5.146 into equation 5.145 gives

$$\frac{\omega^2}{2RT}\left(\frac{\partial \rho}{\partial c_1}\right)_{\hat{\mu}}\left(\frac{d\ln c_1}{d(r^2)}\right)^{-1} = \frac{1}{M_1} + 2A_2^{(0)}c_1 + \cdots \tag{5.147}$$

At the limit of vanishingly small c_1, this series tends to

$$\left(\frac{2RT}{\omega^2 M_1}\right)\frac{d\ln c_1}{d(r^2)} = \left(\frac{\partial \rho}{\partial c_1}\right)_{\hat{\mu}}^{0} \tag{5.148}$$

Here $(\partial \rho / \partial c_1)_{\hat{\mu}}^{0}$ denotes the value of this derivative at $c_1 = 0$. It is a function of c_2 for the incompressible solution considered here.

The process of evaluating $(\partial \rho / \partial c_1)_{\hat{\mu}}^{0}$ from dialysis experiments may be as follows. Let a given ternary solution be placed in a dialysis bag, which is in turn immersed in a large volume of a mixture of components 0 and 2 (i.e., dialyzate) whose composition is predetermined. The whole system is placed in a thermostated bath. The external pressure may be left unspecified, inasmuch as the solution and the dialyzate are assumed to be incompressible. After osmotic equilibrium has been reached, we measure the density of the solution and determine the concentration c_1 in it. Repetition of similar experiments by maintaining the dialyzate constant but by changing the amount of component 1 to be placed in the dialysis bag yields a plot of ρ versus c_1 for a fixed value of c_2', where c_2' denotes the c-scale concentration of component 2 in the dialyzate. Evaluation of its initial slope yields $(\partial\rho/\partial c_1)_{\hat{\mu}}^0$ which corresponds to the specified c_2' value. Similar determinations at other fixed values of c_2' finally yield a plot of $(\partial\rho/\partial c_1)_{\hat{\mu}}^0$ versus c_2' (at a given temperature). In the limit of $c_1 = 0$ the concentration c_2 of component 2 in the dialysis bag tends to c_2', so that this plot may be regarded as the functional relation between $(\partial\rho/\partial c_1)_{\hat{\mu}}^0$ and c_2, that is, the datum to be substituted into equation 5.148.

Although $(\partial\rho/\partial c_1)_{\hat{\mu}}$ at fixed c_2' can be obtained as a function of c_1, it would be forbiddingly difficult to determine this derivative at fixed c_2 as a function of c_1, since c_2 may change as c_1 varies. The same difficulty exists in the evaluation of $(\partial\pi/\partial c_1)_{\hat{\mu}}$ at fixed c_2. It is these derivatives as functions of c_1 and c_2 (not c_2') that we need in considering the behavior of equation 5.145 at concentrations of component 1 which are not too low. Thus, actual application of equation 5.145 to solutions concentrated in component 1 is almost prohibitively difficult.

Equation 5.148 refers to a particular radial position r in the solution column at sedimentation equilibrium. Therefore, the derivative $(\partial\rho/\partial c_1)_{\hat{\mu}}^0$ in this equation must be evaluated pointwise with the c_2 value existing at that level. In the limit of vanishingly small c_1, the equilibrium distribution of component 2 in the cell may be taken to be identical to that observed for the corresponding "blank" experiment. Thus we may substitute into equation 5.148 the values of $(\partial\rho/\partial c_1)_{\hat{\mu}}^0$ corresponding to the observed $[c_2(r)]_{\text{blank}}$. Only if negligible redistribution of component 2 occurs, then $[c_2(r)]_{\text{blank}}$ is simply replaced by c_2^0, the concentration of component 2 in the original uniform solution. In this case, a plot of $\ln c_1$ versus r^2 (at infinite dilution of macromolecular component 1) should be linear. The desired M_1 then can be derived from its slope and the separately determined value of $(\partial\rho/\partial c_1)_{\hat{\mu}}^0$ corresponding to the given c_2^0. When component 2 undergoes appreciable redistribution, it is advantageous to plot $\ln c_1$ against $I(r)$, where $I(r)$ is defined by

$$I = \frac{\omega^2}{RT} \int_{r_1}^{r} \left(\frac{\partial\rho}{\partial c_1}\right)_{\hat{\mu}}^0 r\, dr$$

According to equation 5.148, the resulting plot should be linear and give M_1 from its slope. The values of $(\partial\rho/\partial c_1)_{\hat{\mu}}^0$ necessary to compute this integral may be obtained by the procedure described above.

These results are a triumph of the Casassa-Eisenberg theory, proving that there is a practical way in which the true molecular weight of a macromolecular solute in a ternary solution can be determined from sedimentation equilibrium measurements, regardless of whether or not the solute interacts with a second low-molecular-weight solute (or solvent). The theory has introduced no assumption into the nature of the solute components, except for the one that only one of them is diffusible through the dialysis bag. Therefore the procedure described is equally applicable to both charged and uncharged systems. The only point to note is that it has been established on the neglect of the pressure dependence of partial specific volumes and activity coefficients of all components. Thus its limitation may become significant when it is applied to high-speed equilibrium experiments.

Though not shown here, the Casassa-Eisenberg procedure may also apply to systems which contain more than one kind of diffusible low-molecular-weight solute. Typical of such systems are protein solutions, in which, in addition to a given protein preparation and a supporting strong electrolyte, there are weak electrolytes added to buffer the solution.

There are actual systems to which direct application of the Casassa-Eisenberg method is not feasible. Such systems are, for example, ones in which dialysis of the added low-molecular-weight solutes is difficult or in which the density gradient set up by the added solutes prevents establishment of a stable distribution of the macromolecular component, especially in the region of low concentrations.

IV. POLYMER SOLUTIONS

A. BASIC ASSUMPTIONS AND EQUATIONS

The Casassa-Eisenberg theory can be written for general multicomponent solutions which contain arbitrary numbers of macromolecular components and diffusible small solutes.[36] Derivations of such general forms involve too complex a formalism of less practical interest, and are not discussed in this treatise. Instead, we are concerned here with a special type of multicomponent system, in which an arbitrary number (say q) of macromolecular solutes differing only in molecular weight exist at sufficiently low concentrations. Thus these solutes have the same partial specific volume and the same specific refractive index increment. It is generally accepted, and confirmed experimentally, that ordinary synthetic homopolymers belong to this class of macromolecules. The theory described in this section, therefore, will be of primary interest to polymer physical chemists.

Before moving toward the actual formulation, two points are worthy of note. First, copolymers must be precluded, because this class of macromolecules is generally heterogeneous not only in molecular weight but also in partial specific volume and specific refractive index increment. Second, although the ensuing discussion assumes a single liquid for the solvent component, this assumption is not always imperative. When a "mixed" solvent, that is, a mixture of two or more liquids having different affinities to a given polymeric substance, is used, as is often done in experimental studies of polymers, the theory described below is still valid if it is applied to the solution that has been dialyzed against the mixed solvent.

For simplicity of mathematical developments it is assumed that the solution is incompressible and that the practical activity coefficients, y_i, on the c-scale concentration of all solutes do not depend on pressure. Then it follows that the partial specific volumes of all components including the solvent are independent of both pressure and composition. It is further assumed that all the solutes either are or may be treated as nonionizable in the given solvent and that their concentrations are sufficiently low. Then, the chemical potential per gram of solute $i(i = 1, 2, \ldots, q), \mu_i$, may be written

$$\mu_i = (\mu_i^0)_c + \frac{RT}{M_i} \ln(y_i c_i) \tag{5.149}$$

and the logarithm of y_i may be expanded in Taylor's series as

$$\ln y_i = M_i \sum_{k=1}^{q} B_{ik} c_k + M_i^2 \sum_{k=1}^{q} \sum_{m=1}^{q} B_{ikm} c_k c_m + \cdots \tag{5.150}$$

Here M_i is the molecular weight of solute i and c_i is the c-scale concentration of the same solute. Under the assumption made above for y_i, all the nonideality coefficients appearing in equation 5.150 can be treated as independent of P.

Substitution of equation 5.150 into equations 5.1 and consideration of the condition that all solutes have the same partial specific volume (say $\bar{v}$) give for sedimentation equilibrium of the solution

$$\frac{M_i(1 - \bar{v}\rho)\omega^2 r c_i}{RT} = \frac{dc_i}{dr} + M_i c_i \sum_{k=1}^{q} B_{ik} \frac{dc_k}{dr} + O(c_i c_k c_m) \tag{5.151}$$

$$(i = 1, 2, \ldots, q)$$

From the condition that $\bar{v}$ is independent of composition and the Gibbs-Duhem relation the partial specific volume, $\bar{v}_0$, of the solvent component may be equated to the reciprocal density of the pure solvent, $1/\rho_0$. Thus equation 5.4 gives

$$\rho = \rho_0 + (1 - \bar{v}\rho_0) \sum_{k=1}^{q} c_k \qquad (5.152)$$

Insertion of this relation into equations 5.151 yields, after rearrangements of terms,

$$\lambda M_i c_i = \frac{dc_i}{d\xi} + M_i c_i \sum_{k=1}^{q} B_{ik} \frac{dc_k}{d\xi} + \bar{v} \sum_{k=1}^{q} c_k \frac{dc_i}{d\xi} + O(c_i c_k c_m) \qquad (5.153)$$

$$(i = 1, 2, \ldots, q)$$

Here λ is a parameter defined by

$$\lambda = \frac{(1 - \bar{v}\rho_0)(r_2^2 - r_1^2)\omega^2}{2RT} \qquad (5.154)$$

and ξ is the previously defined reduced cell coordinate, that is,

$$\xi = \frac{r^2 - r_1^2}{r_2^2 - r_1^2} \qquad (5.155)$$

Equations 5.153 are supplemented by the conservation-of-mass conditions, equations 5.6, which may be written in terms of ξ as

$$\int_0^1 c_i(\xi)d\xi = c_i^0 \qquad (i = 1, 2, \ldots, q) \qquad (5.156)$$

As before, the total solute concentration in the initial solution is denoted by c_0. Thus

$$\sum_{i=1}^{q} c_i^0 = c_0 \qquad (5.157)$$

Equations 5.153 and 5.156 form the basis of analysis of sedimentation equilibrium data for dilute solutions of homopolymers.

B. VIRIAL EXPANSION FOR SEDIMENTATION EQUILIBRIUM

If both sides of equations 5.153 are integrated from $\xi=0$ to $\xi=1$ and conditions 5.156 are applied, then

$$\lambda M_i c_i^0 = (c_i)_{\xi=1} - (c_i)_{\xi=0} + \sum_{k=1}^{q} M_i B_{ik} \int_0^1 c_i \frac{dc_k}{d\xi} d\xi$$

$$+ \bar{v} \sum_{k=1}^{q} \int_0^1 c_k \frac{dc_i}{d\xi} d\xi + O(c_i c_k c_m)$$

$$(i = 1, 2, \ldots, q)$$

This relation is summed over all solute components and the resultant expression is divided by c_0 to yield, after rearrangements,

$$\lambda \sum_{i=1}^{q} M_i g_i = \frac{(c)_{\xi=1} - (c)_{\xi=0}}{c_0}$$

$$+ c_0 \sum_{i=1}^{q} \sum_{k=1}^{q} (M_i B_{ik} + \bar{v}) g_i g_k \int_0^1 \theta_i \frac{d\theta_k}{d\xi} d\xi + O(c_0^2) \qquad (5.158)$$

where

$$g_i = \frac{c_i^0}{c_0} \qquad (5.159)$$

$$(c)_{\xi=0} = c(r_1) = \sum_{i=1}^{q} (c_i)_{\xi=0}$$

$$(c)_{\xi=1} = c(r_2) = \sum_{i=1}^{q} (c_i)_{\xi=1} \qquad (5.160)$$

$$\theta_i = \frac{c_i}{c_i^0} \qquad (5.161)$$

The quantity g_i represents the weight fraction of solute i in the sample, and the quantity θ_i is the local concentration of solute i at sedimentation equilibrium relative to its value in the initial solution. At low speeds employed for conventional sedimentation experiments the variation of θ_i with distance in the cell may be moderate; in other words, the magnitude of θ_i may remain at the order of unity over the solution column. In what follows, we shall assume that this condition is obeyed by all solute components in the solution.

We define an apparent molecular weight M_{app} by

$$M_{app} = \frac{(c)_{\xi=1} - (c)_{\xi=0}}{c_0 \lambda} \tag{5.162}$$

This relation is a natural generalization of equation 5.16. It is possible to determine M_{app} as a function of c_0 from a series of experiments at differing initial concentrations. The primary purpose here is to derive theoretical relations which permit thermodynamic information about the system to be deduced from this kind of experimental data.

Equations 5.162 and 3.173 allow equation 5.158 to be rewritten

$$M_{app} = M_w - \frac{c_0}{\lambda} \sum_{i=1}^{q} \sum_{k=1}^{q} (M_i B_{ik} + \bar{v}) g_i g_k \int_0^1 \theta_i \frac{d\theta_k}{d\xi} d\xi + O(c_0^2) \tag{5.163}$$

Expression of equations 5.153 in terms of θ_i gives

$$\lambda M_i \theta_i = \frac{d\theta_i}{d\xi} + c_0 \left(M_i \theta_i \sum_{k=1}^{q} B_{ik} g_k \frac{d\theta_k}{d\xi} \right.$$

$$\left. + \bar{v} \sum_{k=1}^{q} g_k \theta_k \frac{d\theta_i}{d\xi} \right) + O(c_0^2) \tag{5.164}$$

$$(i = 1, 2, \ldots, q)$$

This form suggests a solution of $\theta_i(\xi)$ in powers of c_0 as

$$\theta_i(\xi) = \theta_{i0}(\xi) + c_0 \theta_{i1}(\xi) + c_0^2 \theta_{i2}(\xi) + \cdots \tag{5.165}$$

where the $\theta_{ij}(\xi)$'s are unknown functions of ξ to be determined. Substitution of this expansion into equations 5.164, followed by comparison of terms of equal powers in c_0, yields the following set of differential equations for $\theta_{ij}(\xi)$:

$$\lambda M_i \theta_{i0} = \frac{d\theta_{i0}}{d\xi} \tag{5.166}$$

$$\lambda M_i \theta_{i1} = \frac{d\theta_{i1}}{d\xi} + M_i \theta_{i0} \sum_{k=1}^{q} B_{ik} g_k \frac{d\theta_{k0}}{d\xi} + \bar{v} \sum_{k=1}^{q} g_k \theta_{k0} \frac{d\theta_{i0}}{d\xi}$$

$$\cdots \tag{5.167}$$

In a similar way, we obtain from equations 5.156 the following set of conditions for $\theta_{ij}(\xi)$:

$$\int_0^1 \theta_{i0}(\xi) d\xi = 1 \qquad (i = 1, 2, \ldots, q) \tag{5.168}$$

$$\int_0^1 \theta_{ij}(\xi) d\xi = 0 \qquad (i = 1, 2, \ldots, q; \quad j \geqslant 1) \tag{5.169}$$

Solution of equation 5.166 with equation 5.168 gives

$$\theta_{i0}(\xi) = \frac{\lambda M_i \exp(\lambda M_i \xi)}{\exp(\lambda M_i) - 1} \tag{5.170}$$

After equation 5.170 has been introduced into the right-hand terms, equation 5.167 may be integrated to obtain $\theta_{i1}(\xi)$, where the constant of integration should be determined by equation 5.169 for $j = 1$. But only the function $\theta_{i0}(\xi)$ is needed for the approximation considered below.

Now we insert equation 5.165 into equation 5.163, with θ_{i0} obtained above, and perform the integration indicated. The result may be rearranged to give

$$\frac{1}{M_{app}} = \frac{1}{M_w}$$

$$+ \frac{c_0}{M_w^2} \sum_{i=1}^{q} \sum_{k=1}^{q} \frac{\lambda\{1 - \exp[-\lambda(M_i + M_k)]\} M_i^2 M_k^2 g_i g_k}{[1 - \exp(-\lambda M_i)][1 - \exp(-\lambda M_k)](M_i + M_k)} B_{ik}^*$$

$$+ \text{higher terms in } c_0 \tag{5.171}$$

where

$$B_{ik}^* = B_{ik} + \frac{\bar{v}}{M_k} \tag{5.172}$$

It should be recalled that, in Chapter 3, Section IV.E.1, a solution was defined as thermodynamically pseudoideal if B_{ik}^* is zero for all pairs of i and k.

For binary solutions which consist of solvent 0 and solute 1, equation 5.171 reduces to

$$\frac{1}{M_{app}} = \frac{1}{M_1} + B_{11}^* \left(\frac{\lambda M_1}{2} \right) \coth \left(\frac{\lambda M_1}{2} \right) c_0$$

$$+ \text{higher terms in } c_0 \tag{5.173}$$

It is easily verified that this equation agrees with equation 5.31 to terms in c_0. Equation 5.171, which is the virial expansion for sedimentation equilibrium of polydisperse polymer solutions, dictates that the M_w of the solute and the sedimentation equilibrium second virial coefficient B_{SE} may be evaluated from measurements of M_{app} as a function of c_0. Here B_{SE} is a quantity defined by

$$B_{SE} = \frac{1}{M_w^2} \sum_{i=1}^{q} \sum_{k=1}^{q} \frac{\lambda \{ 1 - \exp[-\lambda(M_i + M_k)] \} M_i^2 M_k^2 g_i g_k}{[1 - \exp(-\lambda M_i)][1 - \exp(-\lambda M_k)](M_i + M_k)} B_{ik}^* \tag{5.174}$$

It is to be noted that this coefficient depends on external experimental conditions through the parameter λ and vanishes under pseudoideality.

The virial expansions for osmotic pressure π and excess turbidity τ_e of a multicomponent solution of the type considered here are as follows[38-40]:

$$\frac{\pi}{RTc_0} = \frac{1}{M_n} + \frac{c_0}{2} \sum_{i=1}^{q} \sum_{k=1}^{q} B_{ik}^* g_i g_k + O(c_0^2) \tag{5.175}$$

$$\frac{Hc_0}{\tau_e} = \frac{1}{M_w} + \frac{c_0}{M_w^2} \sum_{i=1}^{q} \sum_{k=1}^{q} B_{ik}^* M_i M_k g_i g_k + O(c_0^2) \tag{5.176}$$

From these it is seen that both osmotic pressure second virial coefficient B_{OS} (the coefficient for c_0 in equation 5.175) and light-scattering second virial coefficient B_{LS} (the coefficient for c_0 in equation 5.176) simultaneously vanish when the solution is pseudoideal, but that, in general, they are not simply related, unlike the case of binary solutions for which we have noted the relationship $B_{LS} = 2B_{OS}$ (Section II.B). Neither B_{OS} nor B_{LS} is equal to B_{SE}, but it is possible to derive a relation which allows an approximate estimate of B_{LS} from the measurements of B_{SE} and the z-average molecular weight M_z.

For this purpose we expand the right-hand side of equation 5.174 in powers of λ and consider the definition of B_{LS}. Then there results

$$B_{SE} = B_{LS}\left[1 + \left(\frac{\lambda^2}{12}\right)F_1 + \left(\frac{\lambda^4}{720}\right)F_2 + \cdots\right] \qquad (5.177)$$

where

$$F_1 = \frac{\displaystyle\sum_{i=1}^{q}\sum_{k=1}^{q} B_{ik}^* M_i^2 M_k^2 g_i g_k}{\displaystyle\sum_{i=1}^{q}\sum_{k=1}^{q} B_{ik}^* M_i M_k g_i g_k} \qquad (5.178)$$

$$F_2 = \frac{\displaystyle\sum_{i=1}^{q}\sum_{k=1}^{q} B_{ik}^*(M_i M_k)^2(M_i M_k - M_i^2 - M_k^2) g_i g_k}{\displaystyle\sum_{i=1}^{q}\sum_{k=1}^{q} B_{ik}^* M_i M_k g_i g_k} \qquad (5.179)$$

For conventional low-speed sedimentation equilibrium experiments, for which λM_w may be of the order of unity, the second and higher terms in the brackets of equation 5.177 may be treated as small corrections to the leading term. In this case, approximate estimates of $F_1, F_2, \ldots$ would be sufficient for the calculation of B_{SE}. As a first approximation we assume that B_{ik}^* is independent of i and k. In this connection, it is worth recalling that B_{ik}^* has the symmetric property $B_{ik}^* = B_{ki}^*$ (see Chapter 3, Section IV.E.1). With B_{ik}^* being the same for all pairs of i and k, it follows that

$$F_1 = (M_z)^2$$

$$F_2 = (M_z M_{z+1})^2 - 2(M_z)^2 M_{z+1} M_{z+2} \qquad (5.180)$$

$$\cdots$$

Thus equation 5.177 becomes approximately

$$B_{SE} = B_{LS} \left\{ 1 + \tfrac{1}{12}(\lambda M_z)^2 + \tfrac{1}{720}(\lambda M_z)^4 \left[\frac{(M_{z+1})^2}{M_z} - 2\frac{M_{z+1}M_{z+2}}{(M_z)^2} \right] + \cdots \right\} \tag{5.181}$$

Unless the sample is so polydisperse that terms higher than $(\lambda M_z)^2$ have to be included, this expression may be replaced by

$$B_{LS} = \frac{B_{SE}}{1 + (\lambda M_z)^2/12} \tag{5.182}$$

which shows that B_{LS} can be estimated (though not exactly) from the measurements of B_{SE} and M_z.

The treatment above is unsatisfactory because of the assumption made for B_{ik}^*. Recently, Billick[41] has examined how much correction should be applied for equation 5.182 if, as is the case with actual polymer solutions, B_{ik}^* varies with M_i and M_k. For this purpose he assumed the form which had been proposed by Casassa,[42] that is,

$$(B_{ik}^* M_i M_k)^{1/3} = \tfrac{1}{2} B_0^{1/3} \left[M_i^{(2-\alpha)/3} + M_k^{(2-\alpha)/3} \right] \tag{5.183}$$

where B_0 and α are constants characteristic of a given polymer-solvent pair. Values of α seldom exceed 0.25 even in very good solvents. Billick further assumed that the molecular weight distribution is represented by the Schulz-Zimm form (see Chapter 3, Appendix B). It was shown that for a polymer sample not too heterogeneous in molecular weight, that is, with M_z/M_w smaller than, say, 2, the inclusion of the molecular weight dependence of B_{ik}^* would not change the value calculated from equation 5.182 by more than 5%, and in most cases by less than that.

In early stages of the development of sedimentation equilibrium theory, it was anticipated that a quantity equivalent to M_{app} defined by equation 5.162 should provide the weight-average molecular weight of a dissolved polydisperse substance. However, it soon became apparent that this did not hold for polymer solutes. For example, Mosimann[43] and others[44,45] found that the values of M_{app} for such solutes depended on the solvent used, the initial concentration of the test solution, and even the rotor speed and the solution depth, but they were unable to describe the cause. Furthermore, the actually measured concentration or concentration gradient curves at sedimentation equilibrium sometimes exhibited marked deviations from those expected by the theory available at that time. It was eventually recognized that these discrepancies arose because the theory had not taken into account thermodynamic nonideality of the solution. A partial solution was offered by Schulz[46] and independently by Wales et al.[47] Later, Wales[48,49] and also Goldberg[1] discussed these earlier ideas in great detail. All of these workers started with the assumption that the nonideality coefficients B_{ik}^* are independent of molecular weight and thus can be replaced by a single parameter B. This assumption seemed adequate from the theoretical and experimental information which was then available. However, recent theoretical studies in polymer physical chemistry have proved its inadequacy, and the earlier parameter B now must be replaced by an array of B_{ik}^*, although the dependence of B_{ik}^* on the molecular weights M_i and M_k is as yet not established. It was a relatively recent event that the expression for B_{SE} given by equation 5.174 was derived by taking the array of B_{ik}^* into consideration.[5,50]

Equation 5.171 explains all the discrepancies observed by Mosimann and other earlier workers. Besides the obvious dependence on the initial concentration as seen directly, the right-hand side of this equation depends on the solvent through B_{ik}^*'s and also on both the rotor speed and the solution depth through the parameter λ.

Although the third and higher terms in the braces of equation 5.181 can be made small in comparison to unity by an appropriate choice of the parameter λ, we still must be concerned with the effects of solute heterogeneity on these terms. Utiyama et al.[51] have examined this problem with mixtures of two narrow-distribution samples of polystyrene (whose M_w's were 1.97×10^6 and 1.62×10^5) in methyl ethyl ketone at 25°C. The results of their computations are presented in Table 5.1, where the values of the second and third terms in equation 5.181 at a fixed value (1.10) of λM_w are compared. It is seen that the contributions of the third term are quite small except for the mixture $M_{0.2}$, which was the most heterogeneous of the mixtures examined (having an M_z/M_w ratio of as large as about 3). In Fig. 5.3 are depicted the experimental data of Utiyama et al.[51] for the least heterogeneous mixture $M_{0.8}$, plotted in the form of $1/M_{app}$ versus $[1 + (\lambda M_z)^2/12]c_0$. The result is consistent with the fact that for this mixture the third term in question actually makes an insignificant contribution to B_{SE}. On the other hand, Fig. 5.4, which shows the data for the mixture $M_{0.2}$ plotted in the form of $1/M_{app}$ versus λ^2 with c_0 as a parameter, indicates that the plotted points follow a curve which is convex upward and that the curvature becomes pronounced as c_0 is increased. It seems reasonable to attribute this trend primarily to the contribution of the third term, but we should not overlook the possibility that terms higher than the first power of c_0 in the virial expansion for $1/M_{app}$ may give rise to a similar effect.

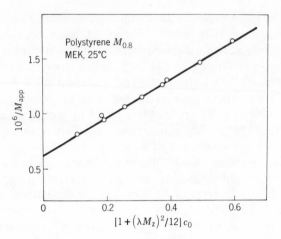

Fig. 5.3. Variation of $1/M_{app}$ with $[1+(\lambda M_z)^2/12]c_0$ for a binary mixture, $M_{0.8}$, of almost monodisperse samples of polystyrene in methyl ethyl ketone at 25°C. [51]

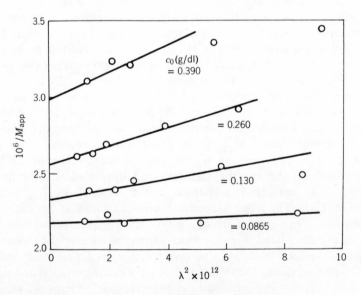

Fig. 5.4. Dependence of $1/M_{app}$ at fixed c_0 on λ^2 for a binary mixture, $M_{0.2}$, of almost monodisperse samples of polystyrene in methyl ethyl ketone at 25°C. [51]

TABLE 5.1. Comparison of the Second and Third Terms in Equation 5.181 for Blends of Two Monodisperse Samples of Polystyrene Having M of 1.97×10^6 and 1.62×10^5, Respectively[a]

Blend[b]	M_w $\times 10^{-6}$	M_z $\times 10^{-6}$	M_{z+1} $\times 10^{-6}$	M_{z+2} $\times 10^{-6}$	Second term	Third term
$M_{0.8}$	1.61	1.93	1.97	1.97	0.143	-0.0044
$M_{0.6}$	1.21	1.87	1.96	1.97	0.236	-0.0129
$M_{0.4}$	0.89	1.77	1.95	1.97	0.388	-0.0394
$M_{0.2}$	0.50	1.50	1.92	1.97	0.882	-0.2839

[a] $\lambda M_w = 1.10$ for all blends.

[b] Subscripts to M represent weight fractions of the higher molecular weight component.

C. OTHER EXPANSIONS FOR $1/M_{app}$

It was shown for binary solutions that the reciprocal apparent molecular weight can be expanded not only in powers of the initial concentration but also in the form of equation 5.15. The theory described in the preceding section proved that the former type of expansion was possible also for heterogeneous polymer solutions. Can an expansion like equation 5.15 be formulated for these solutions?

An answer to this question has been offered recently by Fujita,[52] and subsequently, Deonier and Williams[53] have discussed the significance of Fujita's approach in comparison with earlier work by Van Holde and Williams.[54]

Fujita's theory is based on two simplifying assumptions that the term $\bar{v}/M_k$ may be ignored compared with B_{ik} and that the array of B_{ik}'s may be replaced by a single parameter B. The first assumption is equivalent to approximating the solution density in the sedimentation equilibrium equations by the solvent density ρ_0. Subject to these assumptions, the reciprocal apparent molecular weight can be represented by[52]

$$\frac{1}{M_{app}} = \frac{1}{M_w} + \frac{B}{2}[c(r_1) + c(r_2)](1 + \Delta) + \cdots \qquad (5.184)$$

where Δ stands for

$$\Delta = \frac{\int_0^1 M_{\mathrm{d}}(dc^2/d\xi)d\xi}{\int_0^1 M_{\mathrm{w}}(dc^2/d\xi)d\xi} \tag{5.185}$$

with M_{d} defined by

$$M_{\mathrm{d}} = M_{wr} - M_{\mathrm{w}} \tag{5.186}$$

The quantity M_{wr} is the weight-average molecular weight of the solutes found at a radial position r in the solution column, that is,

$$M_{wr} = \frac{\sum_{i=1}^q M_i c_i}{c} \tag{5.187}$$

with c being the total solute concentration at the same position.

For a monodisperse solute, M_{d} is identically zero, and hence Δ vanishes regardless of the value of B. This suggests that deviation of Δ from zero may arise primarily from solute heterogeneity, and not so much from thermodynamic nonideality. On this idea, Fujita calculated Δ by use of the solutions to equations 5.153 with only the first term on the right-hand side retained, and obtained

$$\Delta = \frac{(\lambda M_{\mathrm{w}})^2}{12}\left[\left(\frac{M_z}{M_{\mathrm{w}}}\right)^2 - \frac{M_z}{M_{\mathrm{w}}}\right] + O(\lambda^4 M_{\mathrm{w}}^4) \tag{5.188}$$

For a numerical check, a solute which has the most probable distribution of molecular weights (see Chapter 3, Appendix B) is considered, and a value of unity is assumed for λM_{w}. Then the leading term of equation 5.188 becomes 0.0625. Because the magnitude of λM_{w} chosen is typical of conventional low-speed experiments, it may be concluded that Δ generally remains as small as 0.1 or less if the solute is not too heterogeneous in molecular weight and the solution is not too nonideal. However, it is too crude an approximation to discard the factor Δ entirely from equation 5.184. Thus the conclusion from Fujita's formulation may be stated as follows. An expansion like equation 5.15 does not hold for heterogeneous polymer solutions, but a plot of $1/M_{\mathrm{app}}$ versus the mean concentration $\bar{c}$ allows the determination of M_{w} and $B(1+\Delta)$ from its ordinate intercept and initial slope, respectively. The experimental data for use of this method must be obtained with λ kept at a constant value for differing initial concentration, since Δ depends on λ. To calculate B from the slope the value of M_z for the sample must be known by a separate means. These inconveniences are not contained in the corresponding plot for a binary solution.

The theory given above is restrictive because of the two assumptions made at the onset of its development. However, Deonier and Williams[53] verified that the factor B in equation 5.184 may be replaced by the light-scattering second virial coefficient B_{LS} in the approximation that errors higher than $(\lambda M_z)^2/12$ may be ignored in comparison with unity.

As early as 1953 Van Holde and Williams,[54] recognizing that a refined concentration variable, instead of the initial concentration c_0, ought to be used for extrapolation of $1/M_{app}$ to infinite dilution, proposed the expansion

$$\frac{1}{M_{app}} = \frac{1}{M_w} + Bc_0^* Q + \cdots \tag{5.189}$$

where c_0^* is the new concentration defined by

$$c_0^* = \frac{\left(\int_0^1 (M_{wr}^{app})^2 c^2 d\xi\right)\left(\int_0^1 c \, d\xi\right)}{\left(\int_0^1 M_{wr}^{app} c \, d\xi\right)^2} \tag{5.190}$$

and Q is a correction factor defined by

$$Q = \frac{1 + B\int_0^1 (M_{wr}^{app})^3 c^3 d\xi / \int_0^1 (M_{wr}^{app})^2 c^2 d\xi + \cdots}{1 + B\int_0^1 (M_{wr}^{app})^2 c^2 d\xi / \int_0^1 M_{wr}^{app} c \, d\xi + \cdots} \tag{5.191}$$

Here M_{wr}^{app}, an apparent molecular weight of the solutes found at a position r in the solution column, is defined by

$$M_{wr}^{app} = \frac{1}{\lambda c} \frac{dc}{d\xi} \tag{5.192}$$

Unlike the M_{wr} defined by equation 5.187, this quantity can be evaluated from experimental data. Thus the concentration c_0^* is calculable for each sedimentation equilibrium experiment. Deonier and Williams[53] showed (the proof is omitted here) that the product $c_0^* Q$ can be transformed to $\bar{c}(1 + \Delta)$, where Δ is given by equation 5.185. Hence, Fujita's theory is mathematically equivalent to the earlier theory of Van Holde and Williams. However, this mathematical equivalence does not mean the practical equivalence of the two theories. The concentration variable of Fujita, $\bar{c}$, is more convenient for practical use than that of Van Holde and Williams, c_0^*, especially in the present-day situation where sedimentation equilibrium data are usually obtained by the Rayleigh interference method (so that solute concentration, rather than its gradient, is measured as a function of r).

D. THE SHORT COLUMN METHOD

Although its proof is deferred to Chapter 7, the rate of approach to sedimentation equilibrium is greatly accelerated by a reduction in volume of the sample solution placed in the ultracentrifuge cell. This idea was first introduced at Uppsala,[55] and later elaborated by Van Holde and Baldwin[7] to a form very useful for the experimental worker. Nowadays it is common practice to use solution columns 2.5 to 3.0 mm in length in conventional sedimentation equilibrium experiments. Some authors[51,56] used columns as short as 1 mm or less. The dimensionless parameter λM_w is made smaller than unity for such short solution columns. The second and higher terms in the braces of equation 5.181 then become negligibly small, and B_{SE} may be equated to B_{LS} with a trivial error. However, such a theoretical simplification is partially offset by a lowering in precision of the determination of M_{app}, because, as λM_w is lowered, the concentration difference between the ends of the solution column diminishes. For practical application of the short column method, therefore, the concentration gradient at the midpoint of the column is used as primary data. The schlieren optical system is usually employed for the measurement. Hence no great accuracy is expected for the results. Probably, this type of experiment is most useful as a survey method when a large number of samples are to be examined or a variety of conditions are to be studied. The multichannel cells designed by Yphantis[57] will be of great use in such circumstances.

Under the same assumptions as used in the preceding sections it can be shown that if the solution is pseudoideal, an apparent molecular weight M_{app}^* defined by

$$M_{app}^* = (\lambda c_0)^{-1} \left(\frac{dc}{d\xi} \right)_{\xi = 1/2} \tag{5.193}$$

is related to the parameter λ and the average molecular weights of the sample by

$$M_{app}^* = M_w \left[1 - \frac{(\lambda M_w)^2}{24} \frac{M_z M_{z+1}}{M_w^2} + O(\lambda^4 M_w^4) \right] \qquad (5.194)$$

The difference between the point $\xi = \frac{1}{2}$ and the midpoint of the solution column is approximately equal to $(r_2 - r_1)^2/4(r_1 + r_2)$. For example, if $r_1 = 6.8$ cm and $r_2 = 7.0$ cm, this amounts to about 0.0007 cm. Thus, under usual conditions of sedimentation equilibrium experiments, the difference is negligible, and M_{app}^* may be evaluated from the concentration gradient measured at the midpoint of the column. Equation 5.194 allows estimate of the error which will be committed if the M_{app}^* thus obtained is equated to M_w. For example, if the sample has a most probable distribution of molecular weight, for which $M_z/M_w = \frac{3}{2}$, $M_{z+1}/M_z = \frac{4}{3}$, and so on, and if the experiment is done with conditions such that λM_w becomes 0.5, we will obtain an M_{app}^* value which is about 2% lower than M_w.

When, as is usually the case with polymer solutes, the solution is nonideal, it is required that data for M_{app}^* be determined at a series of differing initial concentrations and then extrapolated to infinite dilution. Equation 5.194 is valid for this extrapolated M_{app}^*.

V. DETERMINATIONS OF AVERAGE MOLECULAR WEIGHTS AND MOLECULAR WEIGHT DISTRIBUTION

A. INTRODUCTION

As early as 1928, in his celebrated thesis dealing with the settling of spherical colloid particles under gravity, Rinde[58] derived an integral equation which relates the distribution of radii of the particles to the concentration or concentration gradient distribution at sedimentation equilibrium and proposed several methods for solving the equation. In this paper we find the beginning of an extensive literature concerning the sedimentation equilibrium experiment as a tool for investigating heterogeneity of macromolecular solutes in size (molecular weight) and in density (partial specific volume).

In this section, confining ourselves to polymeric substances which are heterogeneous only in molecular weight, we discuss typical methods proposed, and actually tested in many cases, for the evaluation of average molecular weights and molecular weight distributions from sedimentation equilibrium measurements. In doing this, it is assumed, unless otherwise stated, that the system is incompressible and that the partial specific volumes of all polymeric solutes not only have the same value but also are independent of the composition. Furthermore, the discussion is restricted to solutions for which the contributions higher than the first power of individual solute concentrations or total solute concentration need not be considered. Thus, strictly speaking, all the theoretical relations given below should be applied to data which have been extrapolated, in one way or another, to infinite dilution. In practice, the necessary extrapolation is not always easy, because polymer solutions are generally markedly nonideal. The difficulty is relaxed if the experiment can be performed under theta conditions, because the data taken at a moderately low initial concentration may be regarded with accuracy as those at infinite dilution.

It is a relatively new disclosure that the sedimentation equilibrium experiment with a density gradient, initiated by Meselson et al.,[23] can be of effective use for a quantitative detection of heterogeneity in density of macromolecular solutes. The discussion of this subject is deferred to Section VI.

B. INTEGRAL EQUATIONS FOR MOLECULAR WEIGHT DISTRIBUTION

The total solute concentration, $c(r)$, at a radial position r is the sum of $c_i(r)$ over all solute components (numbered $1, 2, \ldots, q$, as before). Hence, in terms of equations 5.159 and 5.161, it may be written

$$c(r) = c_0 \sum_{i=1}^{q} g_i \theta_i(\xi) \qquad (5.195)$$

where c_0 is the initial concentration of the solution. Since we are concerned with the formulation correct to the first power of c_0, equation 5.170 for $\theta_{i0}(\xi)$ may be substituted for $\theta_i(\xi)$ in equation 5.195. Thus

$$c(\xi) = c_0 \sum_{i=1}^{q} \frac{\lambda M_i g_i}{\exp(\lambda M_i) - 1} \exp(\lambda M_i \xi) \qquad (5.196)$$

Differentiation with respect to ξ gives

$$\frac{dc(\xi)}{d\xi} = c_0 \sum_{i=1}^{q} \frac{(\lambda M_i)^2 g_i}{\exp(\lambda M_i) - 1} \exp(\lambda M_i \xi) \qquad (5.197)$$

These relations are fundamentals for analysis of molecular weight hetero-geneity of macromolecular solutes.

If, as is usually the case with ordinary polymer samples, q is essentially infinite, the weight fraction g_i may be regarded as a continuous function of M_i, and the sum in each of the relations above may be replaced by the integral over the range of M from 0 to ∞. Thus for solutes polydisperse in molecular weight

$$c(\xi) = c_0 \int_0^\infty \frac{\lambda M g(M)}{\exp(\lambda M) - 1} \exp(\lambda M \xi) dM \qquad (5.198)$$

$$\frac{dc(\xi)}{d\xi} = c_0 \int_0^\infty \frac{\lambda^2 M^2 g(M)}{\exp(\lambda M) - 1} \exp(\lambda M \xi) dM \qquad (5.199)$$

where $g(M)$ represents the distribution of molecular weights on a weight basis. Equations 5.198 and 5.199 are mathematically equivalent integral equations of the Fredholm type for an unknown function $g(M)$. The quantities on the left-hand sides are given as functions of ξ, over the range $0 < \xi < 1$, from measurements of the refractive index distribution (by the Rayleigh interference method) or the refractive index gradient distribution (by the schlieren method), provided the solute molecules all have the same specific refractive index increment. This optical condition is obeyed accurately by a homologous series of homopolymers. Hence, as long as we are concerned with a macromolecular solute of this type, the quantities $c(\xi)$ and c_0 in the equations above may be replaced by $n_c(\xi)$ and n_c^0, respectively.

1. The Lansing-Kraemer Expressions for Average Molecular Weights

From equations 5.198 and 5.199 it is possible to derive expressions which allow the computation of average molecular weights from the measurement of c or $dc/d\xi$ as a function of ξ at a single rotor speed. The derivation was given for the first time by Lansing and Kraemer.[59]

In the first place, it follows from equation 5.198 that

$$M_w = \frac{(c)_{\xi=1} - (c)_{\xi=0}}{\lambda c_0} = \frac{1}{\lambda c_0} \int_0^1 \frac{dc}{d\xi} d\xi \qquad (5.200)$$

Thus the weight-average molecular weight of a polymeric material may be determined by measuring the concentration difference between the cell bottom and the air-liquid meniscus or the total area under the $dc/d\xi$ versus ξ curve between these ends of the solution column.

Next, equation 5.199 yields

$$M_z = \frac{1}{\lambda^2 c_0 M_w}\left[\left(\frac{dc}{d\xi}\right)_{\xi=1} - \left(\frac{dc}{d\xi}\right)_{\xi=0}\right] \qquad (5.201)$$

which indicates that the z-average molecular weight of the material may be calculated from the difference in $dc/d\xi$ between the ends of the solution column.

If equation 5.199 is differentiated with respect to ξ and the result at $\xi=1$ is subtracted from that at $\xi=0$, it is found that M_{z+1} can be expressed by

$$M_{z+1} = \frac{1}{\lambda^3 c_0 M_w M_z}\left[\left(\frac{d^2c}{d\xi^2}\right)_{\xi=1} - \left(\frac{d^2c}{d\xi^2}\right)_{\xi=0}\right] \qquad (5.202)$$

Proceeding to higher derivatives, we obtain, in general,

$$M_{z+p} = \frac{1}{\lambda^{p+2} c_0 M_w M_z \cdots M_{z+p-1}}\left[\left(\frac{d^{p+1}c}{d\xi^{p+1}}\right)_{\xi=1} - \left(\frac{d^{p+1}c}{d\xi^{p+1}}\right)_{\xi=0}\right] \qquad (5.203)$$

where $p > 1$.

Equation 5.198 may be integrated to yield

$$M_n = \frac{c_0}{\lambda}\left\{\int_0^1\left[\int_{-\infty}^\xi c(\xi')d\xi'\right]d\xi\right\}^{-1} \qquad (5.204)$$

This relation indicates that it is necessary for the determination of the number-average molecular weight M_n to know the function $c(\xi)$ over the range $-\infty < \xi < 1$. However, the actual values of $c(\xi)$ can be obtained only in the region between the air-liquid meniscus and the cell bottom, that is, for values of ξ from zero to unity. Thus equation 5.204 remains of theoretical interest unless there is available a means of estimating the behavior of $c(\xi)$ outside the region of actual measurement. This point may be made clearer if equation 5.204 is recast in an alternate form as

$$M_n^{-1} = \frac{\lambda}{c_0}[F(\lambda) + K] \qquad (5.205)$$

where

$$F(\lambda) = \int_0^1 \left[\int_0^\xi c(\xi') d\xi' \right] d\xi \qquad (5.206)$$

$$K = \int_{-\infty}^0 c(\xi) d\xi \qquad (5.207)$$

It is this additive constant K that we are unable to calculate directly from experiment.

Wales and his associates[6,49] discussed various approximations to K, and came to the idea that if the rotor speed is chosen so high that the concentration at the air-liquid meniscus may be depressed effectively to zero, then the constant K becomes negligibly small and the desired M_n can be obtained unambiguously. Though not stated explicitly in his paper, Yphantis[13] was unquestionably influenced by this earlier idea in his development of the meniscus-depletion method described in Section II.A.4. At present, no other idea but that of Wales et al. is known for the estimate of K.

A fairly high precision can be expected for the determinations of M_w and M_z by equations 5.200 and 5.201 if use is made of Rayleigh interference data for the former and of schlieren data for the latter. The evaluation of M_{z+1} by equation 5.202 requires that the slopes of the concentration gradient curve at the ends of the solution column be measured. The accuracy of schlieren data does not warrant sufficient precision for these measurements. For the curvatures of the gradient curve at the ends more than a moderate precision cannot be expected even in very favorable circumstances. Thus, with the current optical systems, the maximum number of average molecular weights that can be estimated by use of the Lansing-Kraemer method will be three or four at most. Despite such a practical limitation the sedimentation equilibrium method, among many other methods now available for molecular weight determination, is the only one which allows average molecular weights higher than M_w to be evaluated. This unique feature is associated with the power of the ultracentrifuge that is capable of fractionating a substance into molecules of different weight.

2. The Approach to $g(M)$ by the Method of Moments

The kth moment, ν_k, of a molecular weight distribution $g(M)$ about $M = 0$ is defined by

$$\nu_k = \int_0^\infty M^k g(M) dM \qquad (k = -1, 0, 1, \ldots) \qquad (5.208)$$

It is a simple matter to show that the values of successive ν_k's can be expressed in terms of average molecular weights as follows:

$$\nu_{-1} = M_n^{-1}, \quad \nu_0 = 1, \quad \nu_1 = M_w, \quad \nu_2 = M_w M_z, \quad \nu_3 = M_w M_z M_{z+1}, \quad \cdots$$

$$(5.209)$$

Thus the moment of $g(M)$ of any positive integral order can, in principle, be evaluated if either the $c(\xi)$ versus ξ curve or the $dc(\xi)/d\xi$ versus ξ curve is obtained with infinitely high precision, because then all the average molecular weights higher than M_n may be determined by the Lansing-Kraemer equations. According to statistical mathematics, the form of $g(M)$ can then be determined exactly, provided $g(M)$ satisfies certain mathematical restrictions.

Actually, infinite precision is not attainable. Hence, one must be content with finding an approximate shape of $g(M)$ from a finite number of moments of lower order. This process generally involves assuming a reasonable function for $g(M)$ which contains as many adjustable parameters as there are given moments and then determining those parameters in such a way that the assigned moment values are reproduced. The central problem in an application of this method is the choice of a "reasonable" function for $g(M)$. When for some kinetic or statistical reasons it is possible to know in advance what particular form the $g(M)$ of a given material assumes, the above-mentioned procedure yields an exact determination of the molecular weight distribution, provided as many moments are known as the number of parameters. Often two-parameter functions, such as those summarized in Chapter 3, Appendix B, are selected for $g(M)$, and the combination of M_n (determined from osmotic pressure measurements) and M_w (from light-scattering or sedimentation equilibrium measurements) or M_w and M_z (from sedimentation equilibrium measurements) is taken to evaluate the parameters contained. The results so obtained are of limited value, yielding at best certain approximations to $g(M)$. Wales et al.[6] described a fitting procedure in which $g(M)$ was expressed as a finite sum of Laguerre's polynomials with adjustable parameters. Herdan[60] criticized it, and proposed another, similar method. At any rate, since the number of moments determinable experimentally is only three or four, minute details of the distribution functions cannot be disclosed whatever elaborate mathematical techniques may be used. In this connection, it is quite instructive that Koningsveld and Tuijnman[61] showed distinctly different distribution functions which give the same values for the first three or four moments.

3. The Integral Equation Approach to $g(M)$

Since the classic contribution of Rinde,[58] every worker appears to have had a continued belief that direct inversion of the integral equation 5.198 or 5.199 should provide the most reliable form of $g(M)$ when the data to be substituted for $c(\xi)$ or $dc/d\xi$ are considered to be accurate. Thus various methods have been proposed for the inversion of these equations. To mention a few, Wales[48] attempted expressing the function

$$\frac{\lambda Mg(M)}{(1-e^{-\lambda M})}$$

in a series form as

$$\exp(-bM)\sum_{j=1}^{n} d_j M^j$$

and evaluating the parameters involved from the coefficients in a polynomial which fits the observed concentration or concentration gradient distribution curve. Donnelly[62] described a method in which the observed curve is represented in terms of functions that can be treated by Laplace transforms. Provencher [63] has presented still another approach of similar nature. Despite these efforts, the actual results obtained were not necessarily as satisfactory as might be expected. For example, the calculated $g(M)$ curves showed negative values in some regions of M.[6] However, none of the previous authors have been able to describe the real cause of such physically intolerable behavior. Probably, many of them were of the opinion that the failures of the integral equation approach might be due to the fault or the incompleteness of the approximation schemes employed for inversion of the integral equations.

Eventually, in a recent paper, Lee[64] stated that the reported unsatisfactory results should have been attributed to an intrinsic property of the problem. According to him, the problem of inferring $g(M)$ from either equation 5.198 or 5.199 is "ill posed" in Hadamard's sense; that is, these integral equations have the mathematical property which leads to significantly different forms of $g(M)$ from concentration or concentration gradient distribution curves that differ so slightly as to be indetectable by the experiment. This important conclusion suggests the need for the development of a new approach which differs in idea from previous investigators. In fact, Gehatia and Wiff[65] have shown very recently that if the upper limit of the integral in equation 5.198 or 5.199 is replaced by a certain finite adjustable value, the "ill posed" character of the original equations can be eliminated and that good approximations to $g(M)$ are derivable by application of Tikhonov's regularizing functions. Although their theory is not discussed here because of its very sophisticated mathematical context, it appears to point to a new area of research in the theory of sedimentation equilibrium.

C. SENSITIVITY OF SEDIMENTATION EQUILIBRIUM TO SOLUTE POLYDISPERSITY IN MOLECULAR WEIGHT

It was shown in Section II.A.3 that plots of $\ln c$ versus r^2 for a pseudoideal binary solution at sedimentation equilibrium give a straight line whose slope is proportional to the molecular weight of the solute. On the basis of this fact, it has long been believed, and is still believed by some workers, that solute polydispersity in molecular weight M should manifest itself quite sensitively as an upward curvature of this type of plot. Thus when this plot is virtually linear, one very often concludes that the solute must be essentially homogeneous in M. However, there are known many actual cases in which values of $\ln c$ vary linearly with r^2 within the precision of measurement, despite the existence of other evidence which shows that the sample is undoubtedly heterogeneous. Such observations may be accounted for, at least in part, by a thermodynamic nonideality of the solution which generally tends to bend the plot downward. But before attempting such an interpretation it is worthwhile to examine theoretically what type of representation of sedimentation equilibrium data for a pseudoideal solution is the most sensitive to solute polydispersity in M. A partial solution to this problem has been offered by Fujita and Williams.[66]

For a pseudoideal solution, any one of the following observations may be taken as an indication of the molecular weight heterogeneity of the solute: (1) curvature of $\ln c$ versus ξ; (2) variation of M_{wr}^{app} with ξ; and (3) deviation of $I(\xi)$ versus $c(\xi)$ from linearity. Here

$$M_{wr}^{app} = \frac{1}{\lambda}\left(\frac{d\ln c}{d\xi}\right) \tag{5.210}$$

$$I(\xi) = \int_0^\xi c(\xi')d\xi' \tag{5.211}$$

To discuss these criteria in quantitative terms, Fujita and Williams proposed three measures of deviation, in order, as

$$\Delta_1 = \tfrac{1}{2}[\ln c(1) + \ln c(0)] - \ln c(\tfrac{1}{2}) \tag{5.212}$$

$$\Delta_2 = \frac{2[M_{wr}^{app}(1) - M_{wr}^{app}(0)]}{M_{wr}^{app}(0) + M_{wr}^{app}(1)} \tag{5.213}$$

$$\Delta_3 = \left\{ \frac{[c(1) - c(0)](dI/dc)_{\xi=0}}{I(1)} \right\} - 1 \qquad (5.214)$$

Figure 5.5 provides an illustration of the geometrical meaning of these measures of deviation.

The general expressions obtained by Fujita and Williams give the following results if the Schulz-Zimm form is assumed for $g(M)$ (see Chapter 3, Appendix B for this particular distribution function):

$$\Delta_1 = \frac{(\lambda M_w)^2}{8(h+1)} \left[1 - \frac{(\lambda M_w)^2}{8} \frac{2h+3}{(h+1)^2} + O(\lambda^4 M_w{}^4) \right] \qquad (5.215)$$

$$\Delta_2 = \frac{(\lambda M_w)}{h+1} \left[1 - \frac{(\lambda M_w)^2}{6} \frac{h+2}{(h+1)^2} + O(\lambda^4 M_w{}^4) \right] \qquad (5.216)$$

$$\Delta_3 = \frac{(\lambda M_w)}{2(h+1)} \left[1 + \frac{(\lambda M_w)}{6} \frac{h+2}{h+1} - \frac{(\lambda M_w)^2}{12} \frac{h+2}{(h+1)^2} + O(\lambda^4 M_w{}^4) \right] \qquad (5.217)$$

For example, let us consider the case in which $\lambda M_w = 1$ and $h = 1$; this value of h corresponds to the most probable distribution for which $M_w/M_n = 2$, $M_z/M_w = \frac{3}{2}$, and so on. Then

$$\Delta_1 = 0.053, \qquad \Delta_2 = 0.437, \qquad \Delta_3 = 0.297$$

These values give not only the relative order of the three measures of deviation but also something about their absolute magnitudes. For the value of λM_w chosen, which, as has been mentioned repeatedly in preceding sections, is typical of conventional low-speed sedimentation equilibrium experiments, the logarithm of $c(1)/c(0)$ may be about unity, so that the height designated A in Fig. 5.5a becomes about 0.5. If this value is compared to $\Delta_1 = 0.053$, it is found that the plot of $\ln c$ versus r^2 has a slight upward curvature. Obviously, the curvature becomes smaller as h is increased, that is, the solute tends to be less heterogeneous. Probably, such small upward curvatures are partly or sometimes almost entirely canceled by opposing nonideality effects. This explains why, in many actual cases, substantially linear relations are obtained between $\ln c$ and r^2 even if the solutes are considered to be fairly polydisperse in molecular weight.

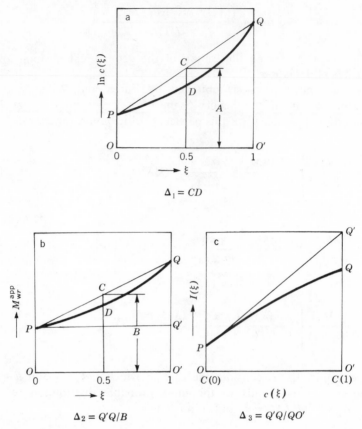

Fig. 5.5. Illustration of three measures of deviation for a discussion of the effects of polydispersity on sedimentation equilibrium behavior.

The relatively large magnitudes of Δ_2 and Δ_3 indicated above suggest that plots of M_{wr}^{app} versus r^2 and those of I versus c, especially the former, would be more suitable than the conventional $\ln c$ versus r^2 for detection of solute polydispersity from sedimentation equilibrium measurements. However, the sensitivities of these kinds of plot are also lowered by thermodynamic nonideality.

D.   THE VARIABLE λ METHOD

The difficulty associated with the determination of M_n from experiments at single rotor speeds motivated Fujita[67] to search for an alternate method which would permit the computation of average molecular weights including M_n from sedimentation equilibrium measurements. He showed that such a method, herein referred to as the variable λ method, can be developed on the basis of the idea previously expressed, though implicitly, by Wales and associates.[6] Recently, Scholte[68] has elaborated it further and checked the validity of some of its results with measurements on typical polymers.

The variable λ method takes advantage of the fact that the equilibrium concentration or concentration gradient at a fixed level in the solution column depends on the parameter λ. Fujita selected $\xi = \frac{1}{2}$ as such a fixed point and defined a function $q(\lambda)$ by

$$q(\lambda) = \frac{(dc/d\xi)_{\xi=1/2}}{\lambda c_0} \tag{5.218}$$

Then equation 5.199 yields for this function

$$q(\lambda) = \int_0^\infty K(\lambda M) M g(M) dM \tag{5.219}$$

where

$$K(x) = \frac{x/2}{\sinh(x/2)} \tag{5.220}$$

It is readily concluded from equations 5.219 and 5.220 that $q(\lambda)$ approaches zero as λ increases indefinitely. The rate of approach depends on the behavior of $g(M)$ in the region of small values of M. Generally speaking, it is slower as the sample contains more low-molecular-weight fractions.

1. Expressions for Average Molecular Weights

Equation 5.219 allows the derivation of relations by which successive average molecular weights including M_n can be computed from data for $q(\lambda)$.

First, we note that $K(x)$ tends to unity as x approaches zero. Application of this fact to equation 5.219 yields

$$q(0) = \lim_{\lambda \to 0} q(\lambda) = \int_0^\infty M g(M) dM = M_w \tag{5.221}$$

Thus M_w can be evaluated by extrapolating the curve for $q(\lambda)$ back to $\lambda = 0$. Since the kernel $K(\lambda M)$ is an even function of its argument, it is more advantageous to plot $q(\lambda)$ against λ^2, rather than λ, in order to facilitate the extrapolation.

Next, integration of equation 5.219 with respect to λ^2 over the range from zero to infinity gives

$$\int_0^\infty q(\lambda)d(\lambda^2) = 33.66 \int_0^\infty \frac{g(M)}{M} dM \qquad (5.222)$$

In terms of M_n, this equation may be written

$$M_n = 33.66 \left[\int_0^\infty q(\lambda)d(\lambda^2) \right]^{-1} = 16.83 \left[\int_0^\infty \lambda q(\lambda)d\lambda \right]^{-1} \qquad (5.223)$$

Thus M_n can be determined by evaluating the area under the curve of $q(\lambda)$ versus λ^2 or of $\lambda q(\lambda)$ versus λ. In passing, we note that

$$\int_0^\infty q(\lambda)d\lambda = \frac{\pi^2}{2} = 4.935 \qquad (5.224)$$

This relation may be utilized to check the reliability of experimentally determined curves for $q(\lambda)$.

Differentiation of equation 5.219 with respect to λ^2 and then integration of the resulting expression with respect to λ from zero to infinity gives

$$\int_0^\infty \frac{dq(\lambda)}{d(\lambda^2)} d\lambda = -0.1732 \int_0^\infty M^2 g(M) dM \qquad (5.225)$$

which may be expressed in terms of M_w and M_z as

$$M_z = -\frac{5.774}{M_w} \int_0^\infty \frac{dq(\lambda)}{d(\lambda^2)} d\lambda \qquad (5.226)$$

Similarly, it is possible to express the higher average molecular weights $M_{z+1}, M_{z+2}, \ldots,$ in terms of $q(\lambda)$. For example,

$$M_{z+1} = -\frac{24}{M_w M_z} \left[\frac{dq(\lambda)}{d(\lambda^2)} \right]_{\lambda=0} \qquad (5.227)$$

Since the derivative $dq/d(\lambda^2)$ may not be determined very accurately from an experimentally given $q(\lambda)$, much reliance cannot be placed upon the values of M_z and M_{z+1}, especially the latter, calculated from these formulas. The expressions for the higher average molecular weights contain derivatives of higher orders, that is, $d^2q/d(\lambda^2)^2$, $d^3q/d(\lambda^2)^3$, and so on. Hence, the values to be derived for these averages are even less reliable. In actual cases, M_{z+1} will be the higest of average molecular weights for which fairly accurate values may be obtained by the λ method.

2. Experimental Determination of the Function $q(\lambda)$

The practical value of the variable λ method depends on whether it is possible to determine experimentally the form of the function $q(\lambda)$ over the entire range of positive λ. Values of λ can be varied by changing either the rotor speed or the length of the solution column. Because the cell length is of the order of 1 cm or less, the maximum range of λ that may be attainable experimentally is limited by the maximum operational speed of the rotor. It is generally observed that when a macromolecular solution is centrifuged at speeds usually employed in sedimentation velocity experiments, almost all solutes are swept down to the region near the cell bottom, and the substantial portion of the solution column is left with the solvent alone, unless the sample contains a measurable amount of low-molecular-weight fractions or the buoyancy factor of the solute is too close to zero. This observed fact implies that by operational speeds of the current ultracentrifuges one may reach the region of λ where $q(\lambda)$ becomes effectively zero. Accurate evaluation of M_n and M_z by the present method depends primarily on how correctly one can determine the behavior of $q(\lambda)$ in this region.

For the determination of M_w and M_{z+1} it is sufficient if the behavior of $q(\lambda)$ at small values of λ is known with accuracy so that $q(0)$ and $(dq/d\lambda^2)_{\lambda=0}$ may be extrapolated. Experimentally, very low speeds of rotation may not be advantageous to obtain small λ, because they are attended by a loss in resolving power of the centrifuge and also by a fluctuation of rotor speed which would give rise to blurring schlieren or Rayleigh interference patterns. Thus, desired small values of λ should be acquired by reduction of volume of the test solution. Scholte[68] used solution columns of about 1.5 mm to determine data in the vicinity of $\lambda = 0$.

3. Extensions of the Method

The original theory of Fujita was written in the form convenient for use with schlieren data. A similar theory suitable for Rayleigh interference data can be developed on the basis of equation 5.198. In this case the integral equation for $g(M)$ is

$$p(\lambda) = \int_0^\infty K(\lambda M) g(M) \, dM \tag{5.228}$$

where

$$p(\lambda) = \frac{(c)_{\xi=1/2}}{c_0} \tag{5.229}$$

From equation 5.228 it can be shown that the first three average molecular weights are expressed in terms of $p(\lambda)$ as follows:

$$M_n = 4.935 \left[\int_0^\infty p(\lambda) \, d\lambda \right]^{-1} \tag{5.230}$$

$$M_w = -5.774 \int_0^\infty \frac{dp(\lambda)}{d(\lambda^2)} \, d\lambda \tag{5.231}$$

$$M_z = -\frac{24}{M_w} \left[\frac{dp(\lambda)}{d(\lambda^2)} \right]_{\lambda=0} \tag{5.232}$$

It is seen that when $p(\lambda)$ is used as the experimental quantity, it must be determined over the entire range of positive λ for the calculations not only of M_n but also of M_w. Thus, as far as M_w is concerned, the method based on $q(\lambda)$ is more convenient for practical use than that based on $p(\lambda)$.

The variable λ method can be formulated by use of concentrations or concentration gradients at two fixed levels in the solution column.[68] To this end we choose the concentrations at $\xi = \frac{1}{4}$ and $\xi = \frac{3}{4}$ and define dimensionless quantities $p_{1/4}(\lambda)$ and $p_{3/4}(\lambda)$ by

$$p_{1/4}(\lambda) = \frac{(c)_{\xi=1/4}}{c_0} \tag{5.233}$$

$$p_{3/4}(\lambda) = \frac{(c)_{\xi=3/4}}{c_0} \tag{5.234}$$

Then it readily follows from equation 5.198 that

$$\frac{2}{\lambda} \left[p_{3/4}(\lambda) - p_{1/4}(\lambda) \right] = \int_0^\infty \left[\cosh\left(\frac{\lambda M}{4} \right) \right]^{-1} M g(M) \, dM \tag{5.235}$$

Hence

$$M_w = \lim_{\lambda \to 0} \frac{2}{\lambda} \left[p_{3/4}(\lambda) - p_{1/4}(\lambda) \right] \tag{5.236}$$

which shows that M_w may be estimated if the concentrations at $\xi = \frac{1}{4}$ and $\xi = \frac{3}{4}$ are measured as functions of λ in the vicinity of $\lambda = 0$. The following relation may also be derived easily:

$$M_z = \lim_{\lambda \to 0} \frac{48}{\lambda^2 M_w} \left[2 - p_{1/4}(\lambda) - p_{3/4}(\lambda) \right] \tag{5.237}$$

A similar formulation can be developed in terms of the concentrations at the air-liquid meniscus ($\xi = 0$) and the cell bottom ($\xi = 1$), but the results are not shown here.[68, 69] Note that these concentrations, especially at the cell bottom, may not be measured as accurately as those in the central region of the solution column.

Next, we define dimensionless quantities $q_{1/4}(\lambda)$ and $q_{3/4}(\lambda)$ by

$$q_{1/4}(\lambda) = \frac{(dc/d\xi)_{\xi=1/4}}{\lambda c_0} \tag{5.238}$$

$$q_{3/4}(\lambda) = \frac{(dc/d\xi)_{\xi=3/4}}{\lambda c_0} \tag{5.239}$$

Equation 5.199 then yields

$$\tfrac{1}{2} \left[q_{1/4}(\lambda) + q_{3/4}(\lambda) \right] = \int_0^\infty K\left(\frac{\lambda M}{2} \right) M g(M) \, dM \tag{5.240}$$

where

$$K\left(\frac{\lambda M}{2} \right) = \frac{\lambda M/4}{\sinh(\lambda M/4)} \tag{5.241}$$

From equation 5.240 the following expressions are easily derivable:

$$M_w = \lim_{\lambda \to 0} \tfrac{1}{2} \left[q_{1/4}(\lambda) + q_{3/4}(\lambda) \right] \tag{5.242}$$

$$M_w M_z M_{z+1} = - \lim_{\lambda \to 0} 48 \frac{d}{d(\lambda^2)} \left[q_{1/4}(\lambda) + q_{3/4}(\lambda) \right] \tag{5.243}$$

Equation 5.199 also yields

$$\tfrac{1}{2}[q_{3/4}(\lambda) - q_{1/4}(\lambda)] = \int_0^\infty L\left(\frac{\lambda M}{2}\right) Mg(M)\,dM \qquad (5.244)$$

with

$$L\left(\frac{\lambda M}{2}\right) = \frac{\lambda M/4}{\cosh(\lambda M/4)} \qquad (5.245)$$

Hence

$$M_z = \frac{2}{M_w} \lim_{\lambda \to 0} \frac{d}{d\lambda}[q_{3/4}(\lambda) - q_{1/4}(\lambda)] \qquad (5.246)$$

Finally, M_n can also be expressed in terms of $q_{1/4}(\lambda)$ as

$$M_n = 5.31\left[\int_0^\infty \lambda q_{1/4}(\lambda)\,d\lambda\right]^{-1} \qquad (5.247)$$

or in terms of $q_0(\lambda)$ $[=(dc/d\xi)/\lambda c_0$ at $\xi = 0]$ as

$$M_n = 2.40\left[\int_0^\infty \lambda q_0(\lambda)\,d\lambda\right]^{-1} \qquad (5.248)$$

4. Experimental Tests

The variable λ method was first used by Osterhoudt and Williams[69] in a study of molecular weight heterogeneity of synthetic polymers. They used data for $p_0(\lambda)$, $p_{1/2}(\lambda)$, and $p_1(\lambda)$ to compute M_w, M_z, and M_{z+1}. However, since they restricted measurements to relatively small λ, the problem of calculating M_n from this type of experiment was left unexplored.

Recently, Scholte[68] undertook more extensive experiments, with pseudoideal solutions of linear polyethylene, polystyrene, and polycaprolactam, to check the validity of Fujita's original method as well as its subsequent modifications described above. The data were taken with the aid of the schlieren method. In what follows, his experimental results on polystyrene are presented in some detail in order to show the accuracy and limitations of the variable λ method.

Scholte's sample of polystyrene had $M_n = 20.7 \times 10^4$ (by osmotic pressure) and $M_w = 34.6 \times 10^4$ (by light scattering). The solvent used was cyclohexane at 35°C (theta temperature). Experiments were conducted at 11 rotor speeds ranging from 3,397 to 24,630 rev/min and at single polymer concentration of 3.845×10^{-3} g/ml. These rotor speeds corresponded to λ values varying from 1.54×10^{-6} to 81.39×10^{-6}. The length of the solution column was kept as short as about 1.5 mm, with glycerol used as a bottom liquid.

Table 5.2 gives the values of M_w, M_z, and M_{z+1} for the polystyrene determined from the concentration gradient curves at the four lowest rotor speeds by means of the Lansing-Kraemer equations 5.200, 5.201, and 5.202. Scholte states that the values given for M_{z+1} are very inaccurate because of the small slopes of the gradient curves. It is to be noted that the M_w values obtained are in excellent agreement with that from light-scattering measurements.

TABLE 5.2. Average Molecular Weights of Scholte's Polystyrene from Measurements at a Single Rotor Speed (Solvent: Cyclohexane; Temperature: 35°C)

Rotor speed rev/min	$\lambda \times 10^6$	$M_w \times 10^{-4}$	$M_z \times 10^{-4}$	$M_{z+1} \times 10^{-4}$
3397	1.54	34.5	65.2	157
4059	2.20	33.9	55.6	48.0
4908	3.21	34.0	51.9	—
5784	4.46	34.0	50.2	—
		Av. 34.1	56.0	100

Figure 5.6 shows the values of $q_0(\lambda)$, $q_{1/4}(\lambda)$, and $q_{1/2}(\lambda)$ plotted against λ. The areas under the three curves are 4.85, 2.59, and 1.58, respectively, which compare well to the theoretically expected values of 4.93, 2.64, and 1.645.[68] Extrapolation of the curve for $q_{1/2}(\lambda)$ to $\lambda = 0$ gives $M_w = 33.4 \times 10^4$. It can be verified that either $q_0(\lambda)$ or $q_{1/4}(\lambda)$ [$q_{3/4}(\lambda)$ and $q_1(\lambda)$ as well] should approach M_w in the limit of $\lambda = 0$, as is actually illustrated in this graph.

In Fig. 5.7 are shown the curves of $\lambda q_{1/2}(\lambda)$, $\lambda q_{1/4}(\lambda)$, and $\lambda q_0(\lambda)$. The areas under these curves are 82.5×10^{-6}, 24.3×10^{-6}, and 10.3×10^{-6} in order from top. If substituted into equations 5.223, 5.247, and 5.248, these figures yield 20.4×10^4, 21.9×10^4, and 23.4×10^4 for M_n. The third value is less accurate because of the smaller value of the area under the curve. The first value is in close agreement with the value of 20.7×10^4 that has been derived from osmometry. However, this agreement seems to be somewhat

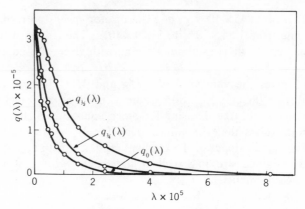

Fig. 5.6. Dependence of q_0, $q_{1/4}$, and $q_{1/2}$ on λ for a sample of polystyrene in cyclohexane at 35°C.[68]

fortuitous, because, as seen in Fig. 5.7, there is a certain arbitrariness in extrapolating the curve for $\lambda q_{1/2}(\lambda)$ to the asymptotic limit.

Figure 5.8 shows the values of $\frac{1}{2}[q_{1/4}(\lambda) + q_{3/4}(\lambda)]$ and $\frac{1}{2}[q_{3/4}(\lambda) - q_{1/4}(\lambda)]$ as functions of λ. Extrapolation of the former quantity to $\lambda = 0$ gives $M_w = 34.0 \times 10^4$. The initial slope of the plot for the latter quantity yields $M_w M_z = 18.6 \times 10^{10}$ when equation 5.246 is applied. Thus we obtain $M_z = 54.7 \times 10^4$. If we replot $\frac{1}{2}[q_{1/4}(\lambda) + q_{3/4}(\lambda)]$ against λ^2 and evaluate the initial slope, a value of 146×10^{15} is obtained for $M_w M_z M_{z+1}$ by means of equation 5.243. Hence M_{z+1} is 78.5×10^4. This value of M_{z+1} is the least accurate among the average molecular weights obtained.

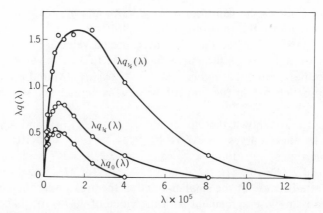

Fig. 5.7. Dependence of $\lambda q_{1/2}$, $\lambda q_{1/4}$, and λq_0 on λ for the same system as in Fig. 5.6.[68]

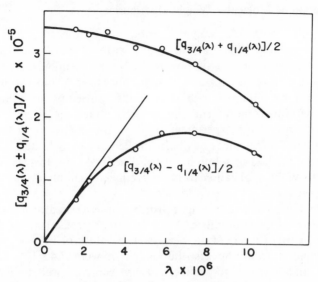

Fig. 5.8. Variation of $\frac{1}{2}(q_{1/4}+q_{3/4})$ and $\frac{1}{2}(q_{3/4}-q_{1/4})$ with λ for the same system as in Fig. 5.6.[68] The straight line shows the initial slope of the lower curve.

Table 5.3 summarizes Scholte's results on the three polymers studied. Except for polycaprolactam, the values of M_n obtained are in fair agreement with those derived from other equilibrium experiments. The poor value of M_n for polycaprolactam is attributable to the fact that the trailing portions of $\lambda q_0(\lambda)$, $\lambda q_{1/4}(\lambda)$, and $\lambda q_{1/2}(\lambda)$ for this polymer had to be inferred by a rather drastic extrapolation.

TABLE 5.3. Scholte's Results from Variable λ Experiments

Polymer	Polystyrene	Polyethylene	Polycaprolactam
Solvent	Cyclohexane	Biphenyl	85% formic acid with 2 N of KCl
Temperature	35°C	123.2°C	25°C
$M_n \times 10^{-4}$	21.2 (20.7)[a]	7.90 (9.0)[a]	2.7 (1.83)[b]
$M_w \times 10^{-4}$	33.7 (34.6)[c]	12.6 (13.0)[c]	3.2 (3.0)[c]
$M_z \times 10^{-4}$	54.7	29.0	4.4
$M_{z+1} \times 10^{-4}$	78.5	72.0	6.1

[a] By osmometry.

[b] By end-group titration.

[c] By light-scattering photometry.

5. Calculation of Molecular Weight Distribution

To determine molecular weight distributions by the λ method we may solve the integral equation 5.219 for $g(M)$, with $q(\lambda)$ obtained experimentally over the entire range of positive λ under pseudoideal conditions. Probably, this integral equation is not "ill posed," and its numerical solutions can be obtained with no essential difficulty on a high-speed computer. However, more generally, we may attack the original equation 5.197 numerically with experimental values for $dc/d\xi$ at a number of different ξ and λ. This ideas has been worked out by Scholte.[70] Details of his numerical method of solution are not described here. Instead, we quote some results of its application to the polyethylene sample discussed in the preceding section.

The calculated molecular weight distribution consisted of a major peak located at an M of about 10×10^4 and a minor peak at an M of about 100×10^4. No negative $g(M)$ appeared. The values of M_n, M_w, M_z, and M_{z+1} computed from the distribution curve were 7.4×10^4, 12.5×10^4, 31.0×10^4, and 95×10^4, respectively. These compare well to the corresponding values given in Table 5.3 Scholte[70] also examined the resolving power of his method with a mixture of three narrow-distribution samples of polystyrene.

VI. SEDIMENTATION EQUILIBRIUM IN A DENSITY GRADIENT

A. INTRODUCTION

Since the first description of its theory by Meselson et al.[23] and its classic application to an analysis of the DNA replication mechanism by Meselson and Stahl,[71] the method of sedimentation equilibrium in a density gradient, or more simply the density-gradient sedimentation equilibrium method, has received a great deal of attention from biochemists and molecular biologists, and it is now regarded as one of the indispensable physical techniques for the investigations of such biological substances as DNA, RNA, and synthetic polynucleotides.[72, 73] This section surveys recent developments in the theory of this method. For more details of the subject the reader is referred to an excellent review article by Vinograd and Hearst.[72]

B. BASIC EQUATIONS AND DEFINITIONS

The system considered here is a ternary solution which consists of a solvent (component 0—water in general), a very dilute (homogeneous) macromolecular solute (component 1), and a moderately concentrated low-molecular-weight solute of high density (component 2). A typical example is composed of water, DNA, and cesium chloride. The concentration and density of component 2 and the rotor speed are adjusted in such a way that an appreciable density gradient is set up in the solution column at sedimentation equilibrium. Then there may appear in the solution a particular point at which the buoyancy factor for component 1 becomes zero. In this case, the macromolecules automatically collect around this particular point, since the effective centrifugal force acting on them vanishes there. This tendency, however, is opposed by thermal motion. Thus there is established an equilibrium distribution of component 1 in a band of a certain finite width. The theory involved here aims at deriving equations which transform the data for the position and width of this band to statements to guide the evaluation of molecular weight and density of the macromolecular component.

The analysis starts with equations 5.79 and 5.80. First, we focus attention on the macromolecular component, and define new quantities M_s and $\bar{v}_s$ by the relations

$$M_s = M_1(1 + \Gamma') \tag{5.249}$$

$$\bar{v}_s = \frac{\bar{v}_1 + \Gamma' \bar{v}_2}{1 + \Gamma'} \tag{5.250}$$

where Γ' is related to the previously defined "binding" coefficient Γ (see equation 5.77 or 5.78) by

$$\Gamma' = \left(\frac{M_2}{M_1}\right)\Gamma \tag{5.251}$$

This new binding coefficient is often referred to as the binding coefficient on a weight basis. Using equations 5.249, 5.250, and 5.76, we obtain

$$A_1 + \Gamma A_2 = \left(\frac{\omega^2}{2}\right)M_s(1 - \bar{v}_s \rho) \tag{5.252}$$

Thus it follows from equation 5.79 that dm_1/dr vanishes at the position at which $\bar{v}_s$ becomes equal to the reciprocal of the solution density ρ. We define this point as the *band position*, and denote its position by r^0. Then

$$\frac{1}{\rho(r^0)} = \bar{v}_s(r^0) = \left(\frac{\bar{v}_1 + \Gamma' \bar{v}_2}{1 + \Gamma'}\right)_{r = r^0} \tag{5.253}$$

This is called the *buoyancy condition*.

In order to find the expression for the concentration of component 1 in the band around the position $r = r^0$, it is necessary to integrate equations 5.79 and 5.80 simultaneously under appropriate boundary conditions. Since this task is formidably difficult, we introduce the same assumption as in Section III.A.1; that is, terms of order m_1 may be neglected in comparison to those of order m_2. This is consistent with the usual experimental conditions in which the solution is sufficiently dilute in component 1 but moderately concentrated in component 2. Then equations 5.79 and 5.80 are simplified to

$$\frac{dm_1}{dr} = \left(\frac{\omega^2 r}{RT}\right) M_s^0 (1 - \bar{v}_s^0 \rho^0) m_1 \tag{5.254}$$

$$\frac{dm_2}{dr} = \frac{(\omega^2 r / RT) M_2 (1 - \bar{v}_2^0 \rho^0) m_2}{1 + m_2 \beta_{22}^0} \tag{5.255}$$

where the superscript 0 refers to the limit $m_1 = 0$; ρ^0 and β_{22}^0 are therefore functions of m_2 and pressure P. Equations 5.254 and 5.255 are essentially the same as equations 5.90 and 5.91.

Before moving to the integration of equation 5.254, we expand ρ^0, M_s^0, and $\bar{v}_s^0$, all being dependent on r through their dependence on m_2, in Taylor's series about $r = r^0$:

$$\rho^0 = \rho^0(r^0) + \left(\frac{d\rho^0}{dr}\right)_{r=r^0} (r - r^0) + \cdots \tag{5.256}$$

$$M_s^0 = M_s^0(r^0) + \left(\frac{dM_s^0}{dr}\right)_{r=r^0} (r - r^0) + \cdots \tag{5.257}$$

$$\bar{v}_s^0 = \bar{v}_s^0(r^0) + \left(\frac{d\bar{v}_s^0}{dr}\right)_{r=r^0} (r - r^0) + \cdots \tag{5.258}$$

Introduction of these expressions into equation 5.254, followed by integration, yields

$$\ln\left[\frac{m_1(r)}{m_1(r^0)}\right] = -\frac{\omega^2 r^0}{2RT} M_s^0(r^0) \bar{v}_s^0(r^0) \left(\frac{d\rho^0}{dr} + \frac{\rho^0}{\bar{v}_s^0} \frac{d\bar{v}_s^0}{dr}\right)_{r=r^0} (r - r^0)^2$$

$$+ O\left[(r - r^0)^3\right] \tag{5.259}$$

Thus, in the approximation that terms higher than $(r-r^0)^2$ may be ignored,

$$m_1(r) = m_1(r^0) \exp\left[\frac{-(r-r^0)^2}{2\sigma^2}\right] \tag{5.260}$$

where

$$\sigma^2 = \frac{RT/(\omega^2 r^0)}{M_s^{\,0}(r^0)\bar{v}_s^{\,0}(r^0)(d\rho/dr)_{r=r^0}^{\text{eff}}} \tag{5.261}$$

$$M_s^{\,0}(r^0) = M_1[1 + \Gamma'^0(r^0)] \tag{5.262}$$

$$\left(\frac{d\rho}{dr}\right)_{r=r^0}^{\text{eff}} = \left(\frac{d\rho^0}{dr}\right)_{r=r^0} + \left(\frac{\rho^0}{\bar{v}_s^{\,0}}\frac{d\bar{v}_s^{\,0}}{dr}\right)_{r=r^0} \tag{5.263}$$

with the buoyancy condition

$$\frac{1}{\rho^0(r^0)} = \bar{v}_s^{\,0}(r^0) = \left(\frac{\bar{v}_1^{\,0} + \Gamma'^0 \bar{v}_2^{\,0}}{1 + \Gamma'^0}\right)_{r=r^0} \tag{5.264}$$

Equation 5.260 indicates that the molality of component 1 distributes in a Gaussian form about the band position r^0, with the standard deviation σ given by equation 5.261. Equations 5.260 through 5.264 are formally identical to those derived by Hearst and Vinograd,[74] who used the assumption of independent solvated macromolecules, which may be stated as follows:

$$\left(\frac{\partial \hat{\mu}_1}{\partial m_1}\right)_{\hat{\mu}_2} = \left(\frac{\partial \hat{\mu}_1}{\partial m_1}\right)_{m_2} + \left(\frac{\partial \hat{\mu}_1}{\partial m_2}\right)_{m_1}\left(\frac{\partial m_2}{\partial m_1}\right)_{\hat{\mu}_2} \simeq \frac{RT}{m_1} \tag{5.265}$$

If no thermodynamic interaction acts between components 1 and 2, that is, $\Gamma'^0 = 0$, and if the partial specific volumes $\bar{v}_1$ and $\bar{v}_2$ are constant, equations 5.261, 5.263, and 5.264 reduce to simpler forms as

$$\sigma^2 = \frac{(RT/\omega^2 r^0)}{M_1 \bar{v}_1 (d\rho/dr)_{r=r^0}^{\text{eff}}} \tag{5.266}$$

$$\left(\frac{d\rho}{dr}\right)_{r=r^0}^{\text{eff}} = \left(\frac{d\rho^0}{dr}\right)_{r=r^0} \tag{5.267}$$

$$\frac{1}{\rho^0(r^0)} = \bar{v}_1 \tag{5.268}$$

These are the expressions obtained by Meselson et al.[23] in their pioneering paper.

C. DETERMINATIONS OF M_1 AND $\bar{v}_1$

Equations 5.266 to 5.268 show that if Γ'^0 is zero and if $\bar{v}_1$ and $\bar{v}_2$ are constant, the molecular weight M_1 and the partial specific volume $\bar{v}_1$ of the macromolecular component can be determined from the measurements of σ^2 and r^0, provided there is available a means of estimating $\rho^0(r^0)$ and $(d\rho^0/dr)_{r=r^0}$ from separate experiments. In fact, methods, either theoretical or empirical, useful for this purpose have been worked out, as is shown in the next three sections.

If Γ'^0 is nonzero, this type of sedimentation experiment does not allow the unambiguous determination of M_1 and $\bar{v}_1$, because there is no experimental means of estimating $(d\rho/dr)_{r=r^0}^{\text{eff}}$ and Γ'^0. However, if $\bar{v}_1^0$, $\bar{v}_2^0$, and Γ'^0 vary so slowly with distance r, at least over a short interval about the band position, that $\bar{v}_s^0$ may be considered substantially constant, the second term on the right-hand side of equation 5.263 may be dropped, and the following expression for M_1 is obtained from equations 5.261 through 5.264:

$$M_1 = \frac{(RT/\omega^2 r^0)[1 - \bar{v}_2^0 \rho^0(r^0)]}{\sigma^2(\bar{v}_1^0 - \bar{v}_2^0)(d\rho^0/dr)_{r=r^0}} \tag{5.269}$$

This allows the determination of M_1 to be made even if Γ'^0 is nonzero.

For systems in which the pressure dependence of $\bar{v}_1$, $\bar{v}_2$, and activity coefficients of the components is negligible, the corresponding theory can be formulated in terms of the c-scale concentrations.[75] Thus, in place of equation 5.260, we obtain

$$c_1(r) = c_1(r^0) \exp\left[\frac{-(r - r^0)^2}{2\sigma^2}\right] \tag{5.270}$$

and the various terms in equation 5.261 for σ^2 are redefined as follows:

$$M_s^0(r^0) = M_1[1 + \lambda'^0(r^0)] \tag{5.271}$$

$$\left(\frac{d\rho}{dr}\right)_{r=r^0}^{\text{eff}} = \left(\frac{d\rho^0}{dr}\right)_{r=r^0} - \left[\frac{1 - \bar{v}_2^0 \rho^0}{\bar{v}_1^0 + \lambda'^0 \bar{v}_2^0}\left(\frac{d\lambda'^0}{dc_2}\right)\frac{dc_2}{dr}\right]_{r=r^0} \tag{5.272}$$

$$\lambda'^0 = -\left(\frac{M_2}{M_1}\right)\left(\frac{c_2\alpha_{12}{}^0}{1+c_2\alpha_{22}{}^0}\right) \tag{5.273}*$$

The buoyancy condition becomes

$$\frac{1}{\rho^0(r^0)} = \bar{v}_s{}^0(r^0) = \left(\frac{\bar{v}_1{}^0 + \lambda'^0\bar{v}_2{}^0}{1+\lambda'^0}\right)_{r=r^0} \tag{5.274}$$

These equations would be a good approximation even for pressure-dependent systems, provided the values of $\bar{v}_1{}^0$, $\bar{v}_2{}^0$, and λ'^0 are replaced by those found at the band position.

D. DENSITY GRADIENT

The practical value of the theoretical formulations described above depends on the availability of methods which permit measurements of ρ^0 and its gradient $d\rho^0/dr$ at the band position r^0. This section concerns the problem relating to the latter quantity.

Since ρ^0 is a function of m_2, P, and T, we have at constant temperature

$$\frac{d\rho^0}{dr} = \left(\frac{\partial\rho^0}{\partial m_2}\right)_P \frac{dm_2}{dr} + \rho^0 K \frac{dP}{dr} \tag{5.275}$$

where K is the isothermal compressibility of the medium consisting of components 0 and 2. With substitution of equation 5.255 for the first term and of equation 5.5 for the second term, equation 5.275 becomes

$$\frac{d\rho^0}{dr} = \left[\frac{1}{\beta^0} + K(\rho^0)^2\right]\omega^2 r \tag{5.276}$$

where

$$\frac{1}{\beta^0} = \frac{M_2(1-\bar{v}_2{}^0\rho^0)}{RT[(1/m_2)+\beta_{22}{}^0]}\left(\frac{\partial\rho^0}{\partial m_2}\right)_P \tag{5.277}$$

The two terms in equation 5.276 are still functions of m_2 and P, but, as has been shown by Hearst et al.[76] for cesium chloride solutions, their pressure dependence may be ignored in many practical cases.

* See equation 5.96 for the definition of $\alpha_{22}{}^0$. $\alpha_{12}{}^0$ may be defined similarly.

The terms $(1/\beta^0)\omega^2 r$ and $K(\rho^0)^2\omega^2 r$ represent the density gradient produced by the redistribution of component 2 and the one by the mechanical compression of the fluid due to centrifugal force, respectively. Hence, they are often referred to as the *composition* density gradient and the *compression* density gradient.[74] Their sum is called the *physical* density gradient. The relative magnitudes of these two kinds of density gradient depend on the system as well as on the experimental conditions. For incompressible systems the compression density gradient vanishes. However, Hearst et al.[76] found for cesium chloride in water that it amounted to about 9% of the physical density gradient.

The quantities β^0 and $K(\rho^0)^2$ can be calculated if the following quantities are measured, at 1 atm, as functions of m_2 from appropriate thermodynamic experiments: K, $\bar{v}_2^0$, ρ^0, and γ_2^0 (the activity coefficient of component 2 on the molality scale). In this way, the data of β^0 have been prepared for a variety of salt solutions.[77-80] Under the neglect of pressure effects both β^0 and ρ^0 are functions of m_2 only, so that β^0 for a given mixture of components 0 and 2 is uniquely related to ρ^0. Figure 5.9, taken from the review article of Vinograd and Hearst,[72] illustrates this relationship for cesium chloride in water at 25°C. It is seen that the curve has a maximum at a density of 1.65 g/ml and that the values of β^0 remain substantially constant over a range about the maximum point.

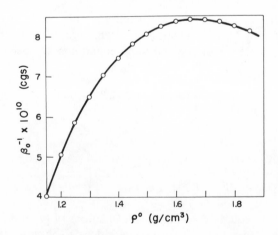

Fig. 5.9. Variation of $1/\beta_0$ with ρ^0 (density) for aqueous solution of CsCl at 25°C.[72] The curve has a maximum at a density of 1.65 g/ml.

E. BUOYANT DENSITY

The density ρ^0 at the band position is usually called the *buoyant density* of a macromolecular solute under study. It has a physical meaning as the reciprocal of a specific volume $\bar{v}_s^0$ defined by the third expression in equation 5.264. If, according to the usual convention, Γ'^0 is regarded as an amount (in grams) of component 2 bound on one gram of component 1, $\bar{v}_s^0$ may be taken to represent the specific volume of solvated component 1. Likewise, the quantity M_s^0 defined by equation 5.262 may be considered to be the molecular weight of solvated component 1. However, it should be noted that a solvated macromolecule is merely a conceptual one, although one often speaks of it as real entity.

Under the neglect of pressure effects, $\bar{v}_1^0$, $\bar{v}_2^0$, Γ'^0, and ρ^0 at a fixed temperature depend only on m_2, so that the first three quantities may be regarded as unique functions of the fourth one ρ^0. Therefore, in this case, equation 5.264 can be taken as a relation which determines the buoyant density. The density value thus determined is characteristic not of the macromolecular solute under consideration but of the given three-component system. It is designated by a notation ρ_b in the ensuing discussion.

The calculation of ρ_b by equation 5.264, however, is not feasible in practice, because no experimental means of estimating the binding coefficient is available. Direct measurement of ρ_b is obviously impractical. For these reasons we employ a convention in which the density of a solution measured pycnometrically at 1 atm is identified as ρ_b if the macromolecular solute in that solution bands just at the *isoconcentration point* r^c. Here r^c is the radial position at which the total solute concentration at sedimentation equilibrium coincides with that in the initial solution. If, as is usually the case in experiments with DNA, the macromolecular component is very dilute, the total solute concentration may be taken to be that of component 2. In this approximation, it can be shown that r^c is very accurately equated to the root-mean-square (rms) position $r^m = [(r_1^2 + r_2^2)/2]^{1/2}$ or the midpoint $r^M = (r_1 + r_2)/2$ of the solution column,* regardless of the rotor speed. Thus the buoyant density defined as above is essentially independent of rotor speed and may be regarded as characteristic of a given system.

When the density of a given solution, ρ_i^0, differs from the buoyant density, the macromolecular solute collects at a band position r^0 away

---

* In cesium chloride, at an initial density of 1.700 g/ml, the density at r^m in a 1.2-cm solution column at 44,770 rev/min is 1.706 g/ml, and that at r^M is 1.702 g/ml.[72]

from the isoconcentration point. In this case, ρ_b is related to ρ_i^0 by the equation

$$\rho_b = \rho_i^0 + \int_{r^c}^{r^0} \left(\frac{d\rho^0}{dr} \right) dr \tag{5.278}$$

Substitution of equation 5.276 for the integrand gives

$$\rho_b = \rho_i^0 + \int_{r^c}^{r^0} \left[\left(\frac{1}{\beta^0} \right) + K(\rho^0)^2 \right] \omega^2 r \, dr \tag{5.279}$$

A first approximation to the integral is to ignore the term of compression density gradient and to assume that β^0 is essentially constant over the solution column. Then

$$\rho_b = \rho_i^0 + \frac{\omega^2}{2\beta^0} \left[(r^0)^2 - (r^c)^2 \right] \tag{5.280}$$

This may be written

$$\rho_b = \rho_i^0 + \left(\frac{\omega^2 \bar{r}^0}{\beta^0} \right) \Delta r^0 \tag{5.281}$$

where

$$\bar{r}^0 = \frac{r^0 + r^c}{2}, \qquad \Delta r^0 = r^0 - r^c \tag{5.282}$$

In practice, the value of β^0 corresponding to the initial density ρ_i^0 may be used for the β^0 in equation 5.281, but a more appropriate β^0 would be the value for the mean density $(\rho_i^0 + \rho_b)/2$. Such a value may be determined by successive approximations. Obviously, it is required that a graph like Fig. 5.9 be known for the system under consideration in order to carry out these computations.

Often conducted for the analyses of DNAs is the experiment in which a new DNA and a marker DNA of known buoyant density are centrifuged together in a gradient of an appropriate high-density salt. If their band positions are denoted by $(r^0)_N$ and $(r^0)_M$, their buoyant-density difference $(\rho_b)_N - (\rho_b)_M$ is expressed by

$$(\rho_b)_N - (\rho_b)_M = \int_{(r^0)_M}^{(r^0)_N} \left(\frac{d\rho^0}{dr} \right) dr \tag{5.283}$$

which, in the same approximations as above, reduces to

$$(\rho_b)_N = (\rho_b)_M + \frac{\omega^2}{2\beta^0}[(r^0)_N{}^2 - (r^0)_M{}^2] \qquad (5.284)$$

This relation may be used to estimate $(\rho_b)_N$ from known values of $(\rho_b)_M$ and $(r^0)_M$.

F. THE EMPIRICAL METHOD OF SZYBALSKI

The treatments described above refer to the limit of zero concentration of the macromolecular component, and also to the simplest case in which the system contains no low-molecular-weight solutes other than a single high-density salt. Solutions treated in actual experiments contain not only a nonzero, though very small, amount of a macromolecular solute but also appropriate amounts of weak electrolytes added to buffer the solution. Even for such actual systems the above-mentioned convention for the buoyant density is still applicable, but, because equation 5.276 no longer holds, the density gradients must be estimated by some other means. The theoretical approach to this problem would involve a prohibitively difficult calculation. Recently, Szybalski[73] worked out a useful empirical method for the estimation of ρ_b and $(d\rho/dr)_{r=r^c}$ in actual solutions.

To explain his idea, consider a series of sedimentation equilibrium experiments in which a sample of DNA is banded at different positions in solutions of a high-density salt whose initial densities vary slightly over a range. It is assumed that all these experiments are conducted at the same rotor speed and with the same length of the solution column. The test solutions may contain, in addition to the DNA and the high-density salt, other simple electrolytes which buffer them. From the equilibrium schlieren patterns obtained for a series of different solutions two are arbitrarily picked up, and the band positions found for them are denoted by r' and r''. The initial densities of the chosen solutions are designated by ρ_i' and ρ_i''. The superscript 0 previously attached to these quantities has been omitted, because the solutions considered here are not always very dilute with respect to the macromolecular component. In principle, by suitable adjustment of the concentration of the high-density salt it is possible to obtain a pair of schlieren diagrams for which the condition $r' - (r^c)' = (r^c)'' - r''$ is satisfied. Here $(r^c)'$ and $(r^c)''$ denote the isoconcentration points for the solutions having initial densities of ρ_i' and ρ_i'', respectively. In what follows, the quantities for such a pair of schlieren diagrams are specified by a subscript s.

The basic idea of Szybalski is to assume that the isoconcentration point of a solution is not affected by its initial density. This would be a good approximation if, as is the case with actual experiments, the macromolecular component is sufficiently dilute. Then, referring to Fig. 5.10, it is seen that the buoyant density ρ_b of the DNA under study is given by

$$\rho_b = \frac{(\rho_i')_s + (\rho_i'')_s}{2} \tag{5.285}$$

and the *actual* density gradient $(d\rho/dr)_{r=r^c}$ at the isoconcentration point r^c for the solution having initial density ρ_b is represented by

$$\left(\frac{d\rho}{dr}\right)_{r=r^c} = \frac{(\rho_i'')_s - (\rho_i')_s}{(r')_s - (r'')_s} \tag{5.286}$$

if the distribution of densities in the cell is not too curved and the distance $(r')_s - (r'')_s$ is not too large, say less than one-third of the solution column.

When a pair of band positions symmetrically located about the isoconcentration point is not experimentally obtainable, the following procedure may be used. We choose an arbitrary pair of bands whose distance is relatively small. Then the density gradient at the midpoint of these bands, $(d\rho/dr)_{r=\bar{r}}$, where $\bar{r} = (r' + r'')/2$, may be represented approximately by

$$\left(\frac{d\rho}{dr}\right)_{r=\bar{r}} = \frac{\rho_i'' - \rho_i'}{r' - r''} \tag{5.287}$$

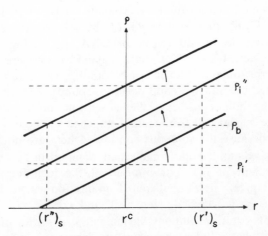

Fig. 5.10. Distributions of solution density ρ for different initial density ρ_i. r^c is the isoconcentration point. When ρ_i is equal to ρ_i' the macromolecular solute bands at $r = (r')_s$, where ρ coincides with ρ_b.

Division of equation 5.286 by equation 5.287 gives

$$\left(\frac{d\rho}{dr}\right)_{r=r^c}=\left(\frac{d\rho}{dr}\right)_{r=\bar{r}}(1+k) \tag{5.288}$$

where k is a correction factor given by

$$k=\frac{[(\rho_i'')_s-(\rho_i')_s]/[(r')_s-(r'')_s]}{(\rho_i''-\rho_i')/(r'-r'')}-1 \tag{5.289}$$

In a first approximation, k may be regarded as a function of $E=(r'+r'')/2-r^c$. This function for a given solution under given rotor speed and solution depth may be determined empirically from the measurements of the band position for a number of differing initial densities. Szybalski[73] determined k versus E curves for DNA in cesium chloride at $\omega=44,770$ rev/min and in cesium sulfate at $\omega=31,410$ rev/min. The results obtained are shown in Fig. 5.11 (in Szybalski's work, E was defined as the distance between $\bar{r}$ and the center of cell cavity, and 90–95% of the cell cavity was filled with the test solution in each run).

It is possible to write down the following approximate relations:

$$\rho_b-\rho_i'=\left(\frac{d\rho}{dr}\right)_{r.}\Delta r' \tag{5.290}$$

$$\rho_b-\rho_i''=\left(\frac{d\rho}{dr}\right)_{r=\bar{r}''}\Delta, \tag{5.291}$$

where

$$\Delta r'=r'-r^c, \qquad \Delta r''=r^c-r'' \tag{5.292}$$

$$\bar{r}'=r^c+\frac{\Delta r'}{2}, \qquad \bar{r}''=r^c-\frac{\Delta r''}{2}$$

By equation 5.288 we may transform equations 5.290 and 5.291 into

$$\rho_b=\rho_i'+\left(\frac{d\rho}{dr}\right)_{r=r^c}\frac{\Delta r'}{1+k'} \tag{5.293}$$

$$\rho_b=\rho_i''-\left(\frac{d\rho}{dr}\right)_{r=r^c}\frac{\Delta r''}{1+k''} \tag{5.294}$$

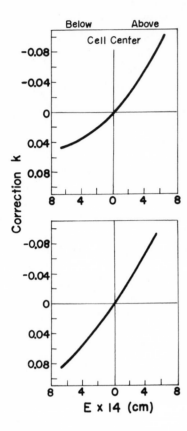

Fig. 5.11. Correction factors k as a function of $E = (r' + r'')/2 - r^c$ for CsCl (upper curve) and Cs_2SO_4 (lower curve).[73] For ways of using these graphs in actual analysis of DNA the reader should consult Szybalski's paper.[73]

where k' and k'' denote the values of k for $E = \Delta r'/2$ and $-\Delta r''/2$, respectively. Elimination of ρ_b from these two relations yields

$$\left(\frac{d\rho}{dr}\right)_{r=r^c} = (\rho_i'' - \rho_i')\Big/\left(\frac{\Delta r'}{1+k'} + \frac{\Delta r''}{1+k''}\right) \qquad (5.295)$$

which allows the evaluation of the desired $(d\rho/dr)_{r=r^c}$ from known values of ρ_i', ρ_i'', $\Delta r'$, $\Delta r''$, k', and k''. The buoyant density ρ_b can then be calculated from either equation 5.293 or 5.294.

These empirical procedures of Szybalski assume that the exact position of the isoconcentration point is known in advance. In practice, this assumption poses no serious problem, because the r^c position is replaced quite accurately by the rms position of the solution column. However, if three solutions of differing initial densities are subjected to experiment, it is possible to determine not only ρ_b and $(d\rho/dr)_{r=r^c}$ but also the exact location of the isoconcentration point from three relations, one for each solution.

Buoyant densities have been determined for a number of DNAs, RNAs, and synthetic polynucleotides in cesium chloride and in cesium sulfate. An extensive list of these results is available in Szybalski's article.[73]

G. HETEROGENEOUS MACROMOLECULAR SOLUTES

In general, macromolecular substances are mixtures of molecules which are heterogeneous not only in molecular weight but also in composition of constituting units, stereochemical configuration, spatial conformation, and so forth. These latter heterogeneities often result in a heterogeneity in partial specific volume or buoyant density. Macromolecular solutes having different buoyant densities collect at different positions when brought to sedimentation equilibrium in a density gradient. On the other hand, those having different molecular weights form bands with different spreads at the same band position. If there are no specific attractive interactions between any two molecules in a heterogeneous macromolecular solute, each component at density-gradient sedimentation equilibrium will distribute in Gaussian form. The total concentration profile is then given by the sum of Gaussian curves for all macromolecular components.

In order to express these situations in mathematical terms we employ equations based on the c-scale concentrations, together with the relevant assumptions. Then, by use of equations 5.270 and 5.6 it can be shown that the maximum concentration $c_i(r_i^0)$ of a particular solute component i is related to the initial concentration c_i^0 of the same component by

$$c_i(r_i^0) = \frac{c_i^0(r_2^2 - r_1^2)}{2r_i^0\sigma_i(2\pi)^{1/2}} \tag{5.296}$$

where r_i^0 and σ_i are the band position and standard deviation of the band formed by solute component i. If this is substituted back into equation 5.270 (with the subscript 1 replaced by i) and the result is summed over all macromolecular components (say q in total), we obtain for the total concentration of the macromolecular solutes, $c(r)$, at a radial position r

$$c(r) = \sum_{i=1}^{q} c_i^0 \left[\frac{(r_2^2 - r_1^2)}{2r_i^0\sigma_i(2\pi)^{1/2}} \right] \exp\left[-\frac{(r - r_i^0)^2}{2\sigma_i^2} \right] \tag{5.297}$$

This expression forms the basis for heterogeneity analysis by density-gradient sedimentation equilibrium experiments.

1. Solutes Heterogeneous in Molecular Weight

First, a macromolecular solute which is polydisperse only in molecular weight is considered. As has been mentioned repeatedly, such solutes are best exemplified by ordinary synthetic homopolymers. For this kind of solute equation 5.297 may be replaced by

$$c(r) = \frac{c_0(r_2^2 - r_1^2)}{2(2\pi)^{1/2}} \int_0^\infty \frac{g(M)\,dM}{r^0(M)\sigma(M)} \exp\left\{ -\frac{[r - r^0(M)]^2}{2[\sigma(M)]^2} \right\} \quad (5.298)$$

where c_0 is the solute concentration in the initial uniform solution, and $g(M)$ is the molecular weight distribution in the solute. It is to be noted that, as explicitly indicated, both r^0 and σ in this equation are generally dependent on M. To proceed further we introduce a restriction that the binding coefficient Γ'^0 may be treated as independent of M. Then it follows from equation 5.264 that all solutes have the same buoyant density so that r^0 becomes independent of M. The quantity σ may be expressed by (see equations 5.261 and 5.262)

$$\sigma^2 = (2\psi M)^{-1} \quad (5.299)$$

with a constant factor ψ defined by

$$\psi = [1 + \Gamma'^0(r^0)] \frac{\bar{v}_s^0(r^0)(d\rho/dr)_{r=r^0}^{\text{eff}} \omega^2 r^0}{2RT} \quad (5.300)$$

Equation 5.298 may then be written, with $x = r - r^0$,

$$\bar{c}(x) = \left(\frac{\psi}{\pi}\right)^{1/2} \int_{-\infty}^\infty M^{1/2} \exp(-\psi M x^2) g(M)\,dM \quad (5.301)$$

where $\bar{c}(x)$ is a reduced concentration defined by

$$\bar{c}(x) = \frac{2r^0 c(x)}{c_0(r_2^2 - r_1^2)} \quad (5.302)$$

Equation 5.301 may be solved for $g(M)$ by the method of Laplace transforms, or it may be treated by the method of moments as has been done by Hermans and Ende.[81]

As before, we designate the kth moment of $g(M)$ by ν_k, that is,

$$\nu_k = \int_0^\infty M^k g(M)\,dM \qquad (k = -1, 0, 1, \ldots) \qquad (5.303)$$

Then it can be shown[81] that equation 5.301 yields

$$\int_{-\infty}^\infty x^{-2k}\bar{c}(x)\,dx = \pi^{-1/2}\psi^k\Gamma(0.5 - k)\nu_k \qquad (k < \tfrac{1}{2}) \qquad (5.304)$$

$$\int_{-\infty}^\infty x^{1-2k}\frac{d\bar{c}(x)}{dx}\,dx = \pi^{-1/2}\psi^k(2k - 1)\Gamma(0.5 - k)\nu_k \qquad (k < \tfrac{3}{2}) \qquad (5.305)$$

where Γ denotes the gamma function. These relations give[23, 81]

$$\int_{-\infty}^\infty \bar{c}(x)\,dx = \nu_0 = 1 \qquad (5.306)$$

$$\int_{-\infty}^\infty x^2\bar{c}(x)\,dx = \tfrac{1}{2}\psi^{-1}\nu_{-1} = (2\psi M_n)^{-1} \qquad (5.307)$$

$$\int_{-\infty}^\infty x^{-1}\frac{d\bar{c}(x)}{dx}\,dx = -2\psi\nu_1 = -2\psi M_w \qquad (5.308)$$

where equation 5.209 has been referred to. According to 5.307, the number-average molecular weight of a polymer sample (polydisperse only in M) can be evaluated from the second moment of $\bar{c}(x)$ about the band position $x = 0$. It should be noted that the data at a single rotor speed are sufficient for this purpose. In this distinctive feature, the density-gradient sedimentation equilibrium experiment may be contrasted with the conventional one which does not allow an unambiguous determination of M_n from data at a single rotor speed unless the condition of meniscus depletion is employed.

2. Two Solutes Differing in Both Molecular Weight and Buoyant Density

Suppose that two different solutes, A and B, are brought to sedimentation equilibrium in a density gradient. Their buoyant densities are denoted by ρ_A and ρ_B, and their molecular weights are denoted by M_A and M_B. In general, $\rho_A \neq \rho_B$ and $M_A \neq M_B$, so that the two solutes may form bands of different widths at different band positions, r_A^0 and r_B^0. We ask what conditions must be satisfied in order that there may appear a bimodal concentration profile in which two peaks are distinctly observable. To answer this question Ifft et al.[78] introduced a resolution parameter δ defined by

$$\delta = \frac{r_A^0 - r_B^0}{\sigma_A + \sigma_B} \tag{5.309}$$

where σ_A and σ_B are the standard deviations of the bands formed by solutes A and B, respectively.

Substitution of equation 5.261 for the σ's, followed by assumption that the density gradient over the two bands is essentially constant and that no binding of the high-density salt on solutes A and B occurs, yields

$$\delta^2 \simeq \frac{(\tilde{\rho}_b)^2}{RT} \frac{M_A M_B}{(M_A^{1/2} + M_B^{1/2})^2} \left(\frac{\beta}{\rho^0} \right)_{r=\bar{r}} \tag{5.310}$$

where

$$\tilde{\rho}_b = \rho_A - \rho_B \tag{5.311}$$

$$\bar{r} = \frac{r_A^0 + r_B^0}{2} \tag{5.312}$$

and β is defined by

$$\beta^{-1} = (\beta^0)^{-1} + K(\rho^0)^2 \tag{5.313}$$

According to equation 5.310, the resolution of the two bands is improved as the two solutes have larger molecular weights and have a larger buoyant-density difference. The resolutiion also becomes better for a high-density salt which gives a larger value of β. For example, the β for cesium chloride in water is about twice as large as that for cesium sulfate in water. Thus it is more advantageous for a better resolution to use the former salt than the latter. Of great importance is the fact that the rotor speed has nothing to do with the resolution. This is because the factor β is little affected by pressure. It might be expected that the segregation of bands would be enhanced as the rotor speed is lowered, because the density gradient becomes smaller by this operation. However, this is not true, because, with decreasing the rotor speed, each band undergoes more spreading and, as a result, the extent of overlapping of the bands becomes greater.

If the two solutes have the same molecular weight M, equation 5.310 becomes

$$\delta^2 = \frac{(\tilde{\rho}_b)^2}{4RT}\left(\frac{M\beta}{\rho^0}\right)_{r=\bar{r}} \tag{5.314}$$

In this case, if the mixing ratio of A and B is $50:50$, the ratio, y, of the concentration at $r=\bar{r}$ to that at the band positoion of either solute is represented by[78]

$$y \simeq 2\exp\left(\frac{-\delta^2}{2}\right) \tag{5.315}$$

Therefore, if $\delta = 1$, only one peak appears; if $\delta = 2$, about 5% of the two solutes overlaps; and if $\delta = 3$, their bands are almost completely separated from one another.[78, 72] For typical DNA of molecular weight of about 10^7 in cesium chloride at 25°C, equation 5.314 gives[78]

$$\delta = 259\tilde{\rho}_b \tag{5.316}$$

Hence $\tilde{\rho}_b = 0.0077$ g/ml for $\delta = 2$, which means that as small a buoyant-density difference as 0.008 g/ml can be detected for giant DNA molecules in a cesium chloride density gradient. In their classic work, Meselson and Stahl[71] succeeded in demonstrating that there existed a density difference of 0.014 g/ml between DNAs containing ^{14}N and ^{15}N.

H. EFFECT OF DENSITY HETEROGENEITY ON MOLECULAR WEIGHT DETERMINATION

The high sensitivity of the band position to buoyant density suggests that even a slight heterogeneity in buoyant density would give rise to a band which spreads over a fairly wide range. This effect poses a serious problem when one wishes to determine the molecular weight of a macromolecular solute from the bandwidth in a density gradient. If one is ignorant of the heterogeneity, there is a great danger of committing a considerable underestimate of the molecular weight.

Baldwin[75] examined this problem theoretically with a sample which has a Gaussian distribution of buoyant densities, and showed that the band-width can be expressed as the sum of two contributions, one from Brownian motion (diffusion) and the other from density heterogeneity, and that the latter contribution is enhanced even by a very small density heterogeneity. For example, for the above-mentioned typical DNA the bandwidth is doubled (hence the molecular weight is underestimated by a factor of 2) if there is a density heterogeneity with a standard deviation of 0.003 g/ml, which corresponds to a deviation of only 0.2% from the average density of 1.70 g/ml. Heterogeneity in buoyant density of this degree would easily arise from chemical and conformational differences among the constituent molecules of a macromolecular substance. For example, denaturation of DNA was found to change the buoyant density in cesium chloride by about 0.016 g/ml (about 1% of the average density).[71] Differences in base composition of DNAs produce a buoyant-density difference (again in cesium chloride) of 0.001 g/ml for 1% variation of the (guanine + cytosine) content.[82-84] Similar results have been reported for synthetic polymers. Thus Buchdahl et al.[85] found a density difference of 0.032 g/ml (2.7% of the average 1.137 g/ml) between isotactic and atactic samples of polystyrene in a bromoform-benzene density gradient.

Various approaches to the problem of recognizing and separating weight and density heterogeneities have been proposed. To mention a few, Sueoka's method[86] requires separate sedimentation velocity measurements on the sample, from which the contribution of Brownian motion is estimated. Baldwin and Shooter[87] proposed varying density gradient by bringing DNA to equilibrium in a preformed, nonequilibrium density gradient of high-density salt. This method would require exceedingly long solution columns. Meselson and Nazarian[88] suggested that the second moment of the concentration profile during the approach to density-gradient sedimentation equilibrium may be used to obtain information about solute heterogeneity in buoyant density. Hermans[89] and Hermans and Ende[90] developed a theory of density-gradient sedimentation equilibrium for synthetic copolymers which are generally heterogeneous both in molecular weight and in composition. Despite these and other efforts of previous investigators, no satisfactory means of separating the effect of density heterogeneity from observed bands has been established to date.

Even if such a means becomes available, the molecular weights derived from corrected bandwidths refer to those of "solvated" macromolecules, M_s^0, defined by equation 5.262. Unfortunately, data for the binding coefficient needed to convert them to the true molecular weights are generally not available. Thus it is recognized that the density-gradient sedimentation equilibrium method is of very limited use for molecular weight determination. Its utmost merit is in the power of segregating, with a high sensitivity, macromolecules in accordance with buoyant density. Thus it has so far been used almost exclusively for "buoyant-density analysis," particularly in the studies of nucleic acids. It is beyond the scope of the present monograph to discuss the success and molecular biological significance of such analyses.

APPENDIX A

The method described below is due to Nazarian.[91] It derives from equation 5.40. According to Nazarian, we rewrite this equation in the form

$$c_1(q) = c_1(0) \exp(BM_1 q) \tag{A.1}$$

where

$$q = r^2 \tag{A.2}$$

$$B = \frac{\lambda}{r_2{}^2 - r_1{}^2} \tag{A.3}$$

and $c_1(0)$ is a constant independent of q. From equation A.1 it follows that

$$c_1(q + Q) - c_1(q) = [c_1(Q) - c_1(0)] \exp(BM_1 q) \tag{A.4}$$

For *fixed* Q, the concentration difference $c_1(q + Q) - c_1(q)$ varies exponentially with q just as does the absolute concentration $c_1(q)$. It should be noted that the quantity Q can be chosen arbitrarily. The absolute fringe number J is proportional to c_1. Hence, equation A.4 may be written

$$\Delta_Q J \equiv J(q + Q) - J(q) = K \exp(BM_1 q) \tag{A.5}$$

where K is a constant, and $\Delta_Q J$ is the number of fringes crossed in traversing the Rayleigh interference pattern from q to $q + Q$.
   Differentiation of equation A.5 gives

$$\frac{d \ln \Delta_Q J}{dq} = BM_1 \tag{A.6}$$

which indicates that the desired M_1 may be evaluated from the slope of a plot of $\ln \Delta_Q J$ versus q. For a construction of this plot no determination of absolute J is necessary. The only data needed, that is, $\Delta_Q J$ as a function of q, can be obtained directly and very accurately from observed fringe patterns. Thus, Nazarian's method, differing from the one based on equation 5.36 which may be rewritten

$$\frac{d \ln J}{dq} = BM_1 \tag{A.7}$$

is applicable to systems for which measurements of the initial solute concentration are not feasible. It is also useful in circumstances in which large aggregates of solutes are formed irreversibly and centrifuged down to the cell bottom before the system reaches sedimentation equilibrium.

This theory is restrictive to pseudoideal binary solutions. Nazarian[91] has described its extensions to nonideal binary solutions and pseudoideal paucidisperse solutions, but we do not enter into them here.

REFERENCES

1. R. J. Goldberg, *J. Phys. Chem.* **57**, 194 (1953).
2. A. Tiselius, *Z. Phys. Chem.* **124**, 449 (1926).
3. J. W. Williams, K. E. Van Holde, R. L. Baldwin, and H. Fujita, *Chem. Rev.* **58**, 715 (1958).
4. F. E. LaBar, *Proc. Natl. Acad. Sci., U.S.* **54**, 31 (1965).
5. H. Fujita, *Mathematical Theory of Sedimentation Analysis*, Academic Press, New York, 1962, Chap. V.
6. M. Wales, F. T. Adler, and K. E. Van Holde, *J. Phys. Colloid Chem.* **55**, 145 (1951).
7. K. E. Van Holde and R. L. Baldwin, *J. Phys. Chem.* **62**, 734 (1958).
8. J. M. Creeth and R. H. Pain, *Prog. Biophys. Mol. Biol.* **17**, 217 (1967).
9. O. Lamm, *Arkiv Mat. Astron. Fysik* **21B**, No. 2 (1929).
10. E. Marler, C. A. Nelson, and C. Tanford, *Biochemistry* **3**, 279 (1964).
11. L. W. Nichol, A. G. Ogston, and B. N. Preston, *Biochem. J.* **102**, 407 (1967).
12. W. E. Hill, G. P. Rossetti, and K. E. Van Holde, *J. Mol. Biol.* **44**, 263 (1969).
13. D. A. Yphantis, *Biochemistry* **3**, 297 (1964).
14. L. W. Nichol and A. B. Roy, *Biochemistry* **4**, 386 (1965).
15. T. L. Hill, *J. Am. Chem. Soc.* **79**, 4885 (1957); *J. Chem. Phys.* **30**, 93 (1959).
16. T. L. Hill, *An Introduction to Statistical Thermodynamics*, Addison-Wesley, Reading, Mass., 1960.
17. D. Stigter, *J. Phys. Chem.* **64**, 114 (1960).
18. K. O. Pedersen, *Z. Phys. Chem.* **A170**, 41 (1934).
19. C. Drucker, *Z. Phys. Chem.* **A180**, 359, 378 (1937).
20. T. F. Young, K. A. Kraus, and J. S. Johnson, *J. Chem. Phys.* **22**, 878 (1954); *J. Am. Chem. Soc.* **76**, 1436 (1954).
21. M. Wales. *J. Appl. Phys.* **22**, 735 (1951).
22. Th. G. Scholte, *J. Polymer Sci. A-2* **8**, 841 (1970).
23. M. Meselson, F. W. Stahl, and J. Vinograd, *Proc. Natl. Acad. Sci., U.S.* **43**, 581 (1957).
24. W. D. Lansing and E. O. Kraemer, *J. Am. Chem. Soc.* **58**, 1471 (1936).
25. O. Lamm, *Arkiv Kemi, Mineral., Geol.* **17A**, No. 25 (1944).
26. G. Scatchard, *J. Am. Chem. Soc.* **68**, 2315 (1946).
27. J. S. Johnson, K. A. Kraus, and G. Scatchard, *J. Phys. Chem.* **58**, 1034 (1954).
28. J. S. Johnson, G. Scatchard, and K. A. Kraus, *J. Phys. Chem.* **63**, 787 (1959).
29. G. Scatchard and J. Bregman, *J. Am. Chem. Soc.* **81**, 6095 (1959).
30. A. Vrij, "Light Scattering from Charged Colloidal Particles in Salt Solution," Thesis, Utrecht, 1959.
31. A. Vrij and J. Th. G. Overbeek, *J. Colloid Sci.* **17**, 570 (1962).
32. E. F. Casassa and H. Eisenberg, *J. Phys. Chem.* **64**, 753 (1960).

33. E. F. Casassa and H. Eisenberg, *J. Phys. Chem.* **65**, 427 (1961).

34. H. Eisenberg and E. F. Casassa, *J. Polymer Sci.* **47**, 29 (1960).

35. H. Eisenberg, *J. Chem. Phys.* **36**, 1837 (1962).

36. E. E. Casassa and H. Eisenberg, *Adv. Protein Chem.* **19**, 287 (1964).

37. E. T. Adams, Jr., in *Characterization of Macromolecular Structure* Natl. Acad. Sci., Washington, D.C., 1968, p. 84 *et seq.*

38. J. G. Kirkwood and R. J. Goldberg, *J. Chem. Phys.* **18**, 54 (1950).

39. W. H. Stockmayer, *J. Chem. Phys.* **18**, 58 (1950).

40. T. L. Hill, *Statistical Mechanics*, McGraw-Hill, New York, 1956.

41. I. H. Billick, *J. Phys. Chem.* **68**, 1798 (1964).

42. E. F. Casassa, *Polymer* **3**, 625 (1962).

43. H. Mosimann, *Helv. Chim. Acta* **26**, 369 (1943).

44. R. Signer and H. Gross, *Helv. Chim. Acta* **17**, 335 (1934).

45. R. Signer and P. von Travel, *Helv. Chim. Acta* **21**, 535 (1938).

46. G. V. Schulz, *Z. Phys. Chem. (Leipsig)* **A193**, 168 (1944).

47. M. Wales, M. Bender, J. W. Williams, and R. H. Ewart, *J. Chem. Phys.* **14**, 353 (1946).

48. M. Wales, *J. Phys. Colloid Chem.* **52**, 235 (1948).

49. M. Wales, *J. Phys. Colloid Chem.* **55**, 282 (1951).

50. H. Fujita, *J. Phys. Chem.* **63**, 1326 (1959).

51. H. Utiyama, N. Tagata, and M. Kurata, *J. Phys. Chem.* **73**, 1448 (1969).

52. H. Fujita, *J. Phys. Chem.* **73**, 1759 (1969).

53. R. C. Deonier and J. W. Williams, *Proc. Natl. Acad. Sci., U.S.* **64**, 828 (1969).

54. K. E. Van Holde and J. W. Williams, *J. Polymer Sci.* **11**, 243 (1953).

55. T. Svedberg and K. O. Pedersen, *The Ultracentrifuge*, Oxford University Press, London and New York (Johnson Reprint Corporation, New York), 1940, p. 237.

56. I. M. Mackie and J. J. Connell, *Biochem. Biophys. Acta* **93**, 544 (1964).

57. D. Yphantis, *Ann. N. Y. Acad. Sci.* **88**, 586 (1960).

58. H. Rinde, "The Distribution of the Sizes of Particles in Gold Sols Prepared According to the Nuclear Method," Thesis, Uppsala, 1928.

59. W. D. Lansing and E. O. Kraemer, *J. Am. Chem. Soc.* **57**, 1369 (1935).

60. G. Herdan, *J. Polymer Sci.* **10**, 1 (1953).

61. R. Koningsveld and C. A. F. Tuijnman, *J. Polymer Sci.* **39**, 445 (1959).

62. T. H. Donnelly, *J. Phys. Chem.* **70**, 1862 (1966).

63. S. W. Provencher, *J. Chem. Phys.* **46**, 3229 (1967).

64. D. A. Lee, *J. Polymer Sci. A-2* **8**, 1039 (1970).

65. M. Gehatia and D. R. Wiff, *J. Polymer Sci. A-2* **8**, 2039 (1970); *Eur. Polymer J.* **8**, 585 (1972).

66. H. Fujita and J. W. Williams, *J. Phys. Chem.* **70**, 309 (1966).

67. H. Fujita, *J. Chem. Phys.* **32**, 1739 (1960).

68. Th. G. Scholte, *J. Polymer Sci. A-2* **6**, 91 (1968).

69. H. W. Osterhoudt and J. W. Williams, *J. Phys. Chem.* **69**, 1050 (1965).

70. Th. G. Scholte, *J. Polymer Sci. A-2* **6**, 111 (1968); *Europ. Polym. J.* **6**, 51 (1970).

71. M. Meselson and F. W. Stahl, *Proc. Natl. Acad. Sci., U.S.* **44**, 671 (1958).

72. J. Vinograd and J. E. Hearst, *Fortschr. Chem. Org. Naturstoffe* **20**, 372 (1962).
73. W. Szybalski, in *Methods in Enzymology* (S. P. Colowick and N. O. Kaplan, Eds.), Academic Press, New York, 1968, Vol. 12, Part B.
74. J. E. Hearst and J. Vinograd, *Proc. Natl. Acad. Sci., U.S.* **47**, 999 (1961).
75. R. L. Baldwin, *Proc. Natl. Acad. Sci., U.S.* **45**, 939 (1959).
76. J. E. Hearst, J. B. Ifft, and J. Vinograd, *Proc. Natl. Acad Sci., U.S.* **47**, 1015 (1961).
77. R. Trautman, *Arch. Biochem. Biophys.* **87**, 289 (1960).
78. J. B. Ifft, D. H. Voet, and J. Vinograd, *J. Phys. Chem.* **65**, 1138 (1961).
79. D. B. Ludlum and R. C. Warner, *J. Biol. Chem.* **240**, 2961 (1965).
80. J. B. Ifft, W. R. Martin, III, and K. Kinzie, *Biopolymers* **9**, 597 (1970).
81. J. J. Hermans and H. A. Ende, *J. Polymer Sci.* **C1**, 161 (1963).
82. R. Rolfe and M. Meselson, *Proc. Natl. Acad. Sci., U.S.* **45**, 1039 (1959).
83. N. Sueoka, *J. Mol. Biol.* **3**, 31 (1961).
84. C. L. Schildkraut, J. Marmur, and P. Doty, *J. Mol. Biol.* **4**, 430 (1962).
85. R. Buchdahl, H. A. Ende, and L. H. Peebles, *J. Polymer Sci.* **C1**, 153 (1963).
86. N. Sueoka, *Proc. Natl. Acad. Sci., U.S.* **45**, 1480 (1959).
87. R. L. Baldwin and E. M. Shooter, in *Ultracentrifugal Analysis in Theory and Experiment* (J. W. Williams, Ed.), Academic Press, New York, 1963, p. 143.
88. M. Meselson and G. M. Nazarian, in *Ultracentrifugal Analysis in Theory and Experiment* (J. W. Williams, Ed.), Academic Press, New York, 1963, p. 131.
89. J. J. Hermans, *J. Chem. Phys.* **38**, 597 (1963); *J. Polymer Sci.* **C1**, 179 (1963).
90. J. J. Hermans and H. A. Ende, in *Newer Methods of Polymer Characterization* (B. Ke, Ed.), Wiley, New York, 1964, Chap. XIII.
91. G. M. Nazarian, *Anal. Chem.* **40**, 1766 (1968).

6

SEDIMENTATION EQUILIBRIUM IN CHEMICALLY REACTING SYSTEMS

A. INTRODUCTION

This chapter is a counterpart of Chapter 4. Here we discuss typical theories and methods developed for analyzing chemically reacting systems by sedimentation equilibrium experiments. This subject had been almost neglected until about a decade ago, but since then there has arisen much interest in it, with special reference to the investigations of self-associations of protein molecules in solution. As can be seen below, there still remains much to be desired in our current knowledge of the subject. Nevertheless, it is certain that the sedimentation equilibrium method is the most useful counterpart of the various nonequilibrium methods (such as sedimentation velocity, electrophoresis, chromatography, etc.) known at present for quantitative or qualitative characterization of chemically reacting systems.

A paper published by Tiselius[1] in 1926 marks the inception of studies in this field. In this classic paper, he formulated sedimentation equilibrium of self-associating solutes under the assumption that the solution is thermo-dynamically ideal (pseudoideal, more correctly). It is rather surprising that his theory had no followers until 1963 when Adams and Fujita[2] attempted, though restricting themselves to a monomer-dimer system, to correct it for thermodynamic nonideality. Rapid developments subsequently occurred, and there have become available a variety of methods by which useful information about relatively simple reacting systems can be deduced from sedimentation equilibrium measurements. This chapter purports to review certain typical examples of these methods, the ones considered to be fundamentally important, though we shall not necessarily follow the orig-inal derivations exactly.

Theoretically, sedimentation equilibrium of chemically reacting systems can be treated more easily than velocity sedimentation of the same systems, because the time variable disappears from theoretical formula-tions. Furthermore, we need no longer be troubled with hydrodynamic interactions between solutes, which lead to many complications, exempli-fied by the Johnston-Ogston effect, in the case of velocity sedimentation. As in Chapter 4, the theory for sedimentation equilibrium of chemically

377

reacting systems is available for both self-association and complex forma-tion. However, in this book, we are concerned chiefly with the former type of reaction, because work on complex-forming systems is still in a very early stage of development.

I. DISCRETE SELF-ASSOCIATION

A. THERMODYNAMIC RELATIONS

Discrete self-association is defined as a reaction in which a monomer, generally a macromolecule, associates with itself to form a finite number of higher aggregates (oligomers). In this section, we consider the situation in which these different solutes are at chemical equilibrium under given solvent conditions. To show thermodynamic relations descriptive of it, let us take a system involving monomer P_1, dimer P_2, and trimer P_3. The reactions which occur in this system are represented schematically by

$$2P \rightleftharpoons P_2, \qquad 3P_1 \rightleftharpoons P_3$$

It is important to note that, at equilibrium, only the monomer can be treated as a thermodynamic component, because the concentrations of P_2 and P_3 are related to that of P_1 by the conditions as explicitly shown below. Thus, in what follows, we refer to P_1, P_2, and P_3 as monomer species, dimer species, and trimer species, respectively.

According to thermodynamics, the equilibrium conditions for the above-mentioned reactions are given by

$$2\hat{\mu}_1 = \hat{\mu}_2$$
$$3\hat{\mu}_1 = \hat{\mu}_3$$

(6.1)

where $\hat{\mu}_i$ $(i = 1, 2, 3)$ is the chemical potential per mole of species i. As before, we write $\hat{\mu}_i$ in the form

$$\hat{\mu}_i = (\hat{\mu}_i^0)_c + RT \ln (y_i c_i)$$

(6.2)

where c_i is the c-scale concentration of species i, y_i is the practical activity coefficient of species i on the c-concentration scale, R is the gas constant, and T isd the absolute temperature. With equation 6.2, equation 6.1 yield the equilibrium relations

$$\frac{y_2 c_2}{(y_1 c_1)^2} = K_2 \tag{6.3}$$

$$\frac{y_3 c_3}{(y_1 c_1)^3} = K_3 \tag{6.4}$$

Here K_i ($i = 2, 3$), the equilibrium constant for monomer–i-mer association, is given by

$$K_i = \exp\left[\frac{i(\hat{\mu}_1^0)_c - (\hat{\mu}_i^0)_c}{RT}\right] \tag{6.5}$$

It is to be noted that K_i is a function of T and pressure P.

We now introduce a first assumption according to Adams and Fujita.[2] This is to take thermodynamic nonideality into account by imposing the following form on $\ln y_i$, that is,

$$\ln y_i = i M_1 B c \tag{6.6}$$

where M_1 is the molecular weight of the monomer, B is a thermodynamic nonideality parameter,* and c is the total solute concentration, that is,

$$c = c_1 + c_2 + c_3 \tag{6.7}$$

A more general expression for $\ln y_i$ is (see Chapter 5, Section IV.A)

$$\ln y_i = M_i \sum_{k=1}^{3} B_{ik} c_k + \text{higher terms in } c_k$$

*As in the preceding chapters, the solution in which $B = 0$ is referred to as pseudoideal (instead of ideal) in the ensuing discussion.

Equation 6.6 is an approximation to this, in which all B_{ik} are assumed to be the same and the higher terms in c_k are neglected.

From equation 6.6 it follows that

$$\frac{y_2}{y_1^2} = 1, \qquad \frac{y_3}{y_1^3} = 1 \tag{6.8}$$

Hence, the equilibrium conditions 6.3 and 6.4 are simplified to

$$c_2 = K_2 c_1^2 \tag{6.9}$$

$$c_3 = K_3 c_1^3 \tag{6.10}$$

Upon substitution of these relations into equation 6.7, there results

$$c = c_1 + K_2 c_1^2 + K_3 c_1^3 \tag{6.11}$$

which indicates that c_1 becomes a function of c only under the assumption made above. It follows then from equations 6.9 and 6.10 that c_2 and c_3 also depend only on c. If these concentrations of individual species, that is, the way in which the total concentration is shared among monomer, dimer, and trimer species, can be measured, the desired values for the equilibrium constants K_2 and K_3 are readily computed. Determinations of K_2 and K_3 then provide data from which thermodynamic information such as the entropy and enthalpy associated with the formation of a dimer or a trimer may be deduced.

The concentrations of individual species, however, are measurable only in special circumstances where each species can be assayed independently of others by virtue of its specific chemical, physical, or biological characteristics. Therefore, in general, one must resort to indirect experimental information for the evaluation of equilibrium constants. Such information is provided by several kinds of physical experiments. The sedimentation equilibrium experiment is presumably the most advantageous among them for reasons presented below.

1. Equations for Sedimentation Equilibrium

Consider an experiment in which a dilute solution containing monomer, dimer, and trimer species is brought to sedimentation equilibrium in an ultracentrifuge cell. It is assumed that the solution is incompressible and that the activity coefficients y_i of all species may be treated as independent of pressure. As discussed in Chapter 5, these assumptions are relevant as long as we are concerned with low-speed experiments. It is sufficient if we consider the equation descriptive of sedimentation equilibrium of the monomer species, because the concentrations of dimer and trimer species are calculable from equations 6.9 and 6.10 if c_1 is known. The desired equation can be derived from the condition that, at sedimentation equilibrium, the total potential per mole of monomer, $\hat{\mu}_1 - M_1(\omega^2 r^2/2)$, is independent of radial distance r, and we have

$$\left(\frac{\partial \hat{\mu}_1}{\partial c_1}\right)_T \frac{dc_1}{dr} = M_1(1 - \bar{v}_1 \rho)\omega^2 r \qquad (6.12)$$

Here $\bar{v}_1$ denotes the partial specific volume of the monomer species and all other symbols have the same meaning as in the preceding chapter.

Substituting for $\hat{\mu}_1$ from equation 6.2, with $\ln y_1$ given by equation 6.6, and considering equations 6.9, 6.10, and 6.11, we obtain

$$[1 + BM_1(c_1 + 2c_2 + 3c_3)]\frac{dc_1}{dr^2} = \frac{M_1(1 - \bar{v}_1 \rho)\omega^2}{2RT}c_1 \qquad (6.13)$$

or

$$[1 + B(M_1 c_1 + M_2 c_2 + M_3 c_3)]\frac{dc_1}{dr^2} = \frac{M_1(1 - \bar{v}_1 \rho)\omega^2}{2RT}c_1 \qquad (6.14)$$

where M_2 and M_3 are the molecular weights of dimer and trimer species, respectively. For sufficiently dilute solutions considered here, ρ may be replaced by the density ρ_0 of the pure solvent. Also, the assumption of the pressure independence of activity coefficients allows $\bar{v}_1$ ($\bar{v}_2$ and $\bar{v}_3$ as well) to be treated as independent of composition. With these approximations, equation 6.14, coupled with equations 6.9 and 6.10, may be integrated to obtain an expression from which c_1 may be calculated as a function of r. Such a result, however, is of no practical interest, because c_1, and the concentrations of other solute species as well, cannot be measured except in certain special circumstances.

The sedimentation equilibrium equations for dimer and trimer species can be derived in a similar manner, yielding

$$[1 + B(M_1 c_1 + M_2 c_2 + M_3 c_3)]\frac{dc_2}{dr^2} = \frac{M_2(1 - \bar{v}_2 \rho)\omega^2}{2RT}c_2 \qquad (6.15)$$

$$[1 + B(M_1 c_1 + M_2 c_2 + M_3 c_3)]\frac{dc_3}{dr^2} = \frac{M_3(1 - \bar{v}_3 \rho)\omega^2}{2RT}c_3 \qquad (6.16)$$

We add up equations 6.14, 6.15, and 6.16, and introduce an important assumption* that the three species have the same partial specific volume, say $\bar{v}$. Then, the result is greatly simplified to give, with ρ replaced by ρ_0,

$$(1 + BM_{wr}c)\frac{dc}{dr^2} = AM_{wr}c \tag{6.17}$$

where A is

$$A = \frac{(1 - \bar{v}\rho_0)\omega^2}{2RT} \tag{6.18}$$

and M_{wr} is defined as

$$M_{wr} = \frac{M_1c_1 + M_2c_2 + M_3c_3}{c} \tag{6.19}$$

The quantity M_{wr} represents the weight-average molecular weight of the solute in a local solution in which the total solute concentration is c. The letter r in the subscript indicates the radial position of such a solution. Equation 6.17 may be written

$$M_{wr}^{app} = \frac{M_{wr}}{1 + BM_{wr}c} \tag{6.20}$$

if one introduces a quantity M_{wr}^{app} by the relation

$$M_{wr}^{app} = \frac{1}{A}\frac{d\ln c}{dr^2} \tag{6.21}$$

It is convenient for the subsequent discussion to call M_{wr} the true *local* weight-average molecular weight and M_{wr}^{app} the apparent local weight-average molecular weight.

Since c_1, c_2, and c_3 are functions only of c under the assumptions made, it follows from equation 6.19 that M_{wr} depends only on c. Hence, from equation 6.20, this is also the case with M_{wr}^{app}. Thus, in what follows, we designate these local molecular weights as $M_{wr}(c)$ and $M_{wr}^{app}(c)$ when necessary for clarity.

The primary data provided by a single sedimentation equilibrium experiment are values of n_c, the refractive index of the solution over that of

*See Sections III.B and IV.B for implications of this assumption.

the solvent, as a function of r. As has been mentioned repeatedly in preceding chapters, it is necessary for conversion of n_c to c to assume that all solute components or species present in a given solution have the same specific refractive index increment.* This assumption is reasonable for self-associating systems.

Values of M_{wr}^{app} as a function of c can be calculated from the tangents to experimentally determined plots of $\ln c$ versus r^2 at various radial positions. The precision of the results depends primarily on the accuracy of raw data for $c(r)$ and secondarily on the method used for the determination of tangents. Usually, a best-fit smooth curve is drawn through data points by inspection, and its slopes are determined graphically. Alternatively, after a suitable polynomial describing the data has been determined with the aid of a least-squares method, it is differentiated analytically. The more recent trend is to carry out all operations on an electronic computer. A discussion of these operational problems is beyond the scope of this book.

A single sedimentation equilibrium experiment affords data for M_{wr}^{app} only over a limited interval of c which ranges from the value at the air-liquid meniscus of a solution column to that at the cell bottom. A series of such segments of curve may be obtained from experiments of differing initial concentration. If the assumptions made above are obeyed by all solute species present in the system and if the reactions between them are completely reversible, *these segments ought to superimpose on a single continuous curve, regardless of the type of self-association involved.* The range of concentration covered by this curve can be expanded by varying the initial concentration of individual experiments over a wide range.

It is to be recognized that a curve of M_{wr}^{app} versus c is the only data which are provided directly from sedimentation equilibrium experiments. The purpose of our theoretical investigations is to derive practical methods by which a maximum amount of information on a given self-associating system may be drawn out of this curve. During the last decade there has been much effort toward this goal. A predominant portion of the rest of this chapter is devoted to a detailed account of the typical methods as well as to illustrations of their applications to protein systems.

Probably, it is appropriate to remark here that data basic to the analysis of self-associations can be derived also from light-scattering or Archibald-type sedimentation experiments, although the proof is not presented in this book. When these methods are used, each experiment provides only a single value of M_{wr}^{app} corresponding to a particular concentration of the

---

*When the light-absorption method is used for the measurement, one must assume that all solute species have the same absorption coefficient for the wavelength of light used.

solution. Therefore, one is forced to repeat a great many experiments in order to construct a desired M_{wr}^{app} versus c curve over a wide range of concentration. The sedimentation equilibrium method is decidedly advantageous over these alternatives, because a single experiment allows a series of M_{wr}^{app} values to be determined over a certain finite range of concentration which encompasses the initial concentration of the test solution. This fact explains why, in recent years, there has been a preference to apply the sedimentation equilibrium method to studies of protein self-associations.

B.   THE THEORY OF ADAMS

The theory described below, due originally to Adams and Williams[3] and Adams,[4] evolved in an attempt to extend the earlier theory of Adams and Fujita[2] for monomer-dimer association to the case of monomer–dimer–m-mer association. Here m is an arbitrary positive integer. We outline it, taking the case of $m = 3$ as an example.

For use of the Adams theory, and all other similar theories as well, it is required that data for M_{wr}^{app} be obtained down to as low a concentration as possible so that they may be extrapolated to infinite dilution ($c = 0$) with accuracy. The reason is that the intercept at $c = 0$ is necessary for determination of the monomer molecular weight M_1. However, in practice, one often encounters situations in which the slope of M_{wr}^{app} versus c is so steep at low concentrations that the extrapolation cannot be attempted with any degree of accuracy. Thus, whenever possible, it is advisable either to determine M_1 from experiments with a solvent in which no association is supposed to take place or to derive its value from other sources, for example, the known amino acid composition in the case of proteins.

If one notes that $M_2 = 2M_1$ and $M_3 = 3M_1$, expresses c_2 and c_3 in terms of c_1 by use of equations 6.9 and 6.10, and considers equation 6.11, equation 6.19 for M_{wr} may be written

$$M_{wr} = M_1 \left(\frac{d \ln c}{d \ln c_1} \right) \tag{6.22}$$

Equation 6.20 may then be put in the form:

$$\left(\frac{M_1}{M_{wr}^{app}} - 1 \right) \frac{dc}{c} = d \ln \left(\frac{c_1}{c} \right) + B M_1 \, dc \tag{6.23}$$

which, upon integration, gives

$$\ln f_1^{app} = \ln f_1 + BM_1 c \tag{6.24}$$

where

$$f_1 = \frac{c_1}{c} \tag{6.25}$$

$$f_1^{app} = \exp\left[\int_0^c \left(\frac{M_1}{M_{wr}^{app}} - 1 \right) \frac{dc}{c} \right] \tag{6.26}$$

The integration constant has been determined from the condition that c_1 approaches c as the latter tends to zero. The quantity f_1 is the weight fraction of monomer species in a local solution where the total solute concentration is c. The quantity f_1^{app} then may be referred to as an apparent weight fraction of monomer species in the same solution. It coincides with f_1 when the solution is pseudoideal. For its determination as a function of c one must have data for M_{wr}^{app} from $c = 0$ to the desired values of c. The form of equation 6.26 immediately suggests the need for a greater accuracy of the data at lower concentrations.

Equation 6.23 may be rewritten

$$\left(\frac{M_1}{M_{wr}^{app}} \right) dc = \frac{c}{c_1} dc_1 + BM_1 c\, dc \tag{6.27}$$

which can be integrated to give

$$\frac{M_1}{M_{nr}^{app}} = \frac{1}{c} \int_0^{c_1} \left(\frac{c}{c_1} \right) dc_1 + \frac{BM_1}{2} c \tag{6.28}$$

where M_{nr}^{app} is a new quantity defined as

$$\frac{1}{M_{nr}^{app}} = \frac{1}{c} \int_0^c \frac{dc}{M_{wr}^{app}} \tag{6.29}$$

The M_{nr}^{app} is termed an apparent local number-average molecular weight.* It is seen that evaluation of this quantity too requires accurate M_{wr}^{app} data to be determined at low concentrations.

*Note that M_{nr}^{app} is not directly measurable from sedimentation experiment.

Substitution of equation 6.11 into equation 6.28, followed by integration, yields

$$\frac{M_1}{M_{nr}^{app}} = \frac{c_1}{c} + \frac{K_2}{2}c\left(\frac{c_1}{c}\right)^2 + \frac{K_3}{3}c^2\left(\frac{c_1}{c}\right)^3 + \frac{BM_1}{2}c \tag{6.30}$$

Again, using equation 6.11, this relation may be cast in the form

$$1 - \frac{M_1}{M_{nr}^{app}} = \frac{c}{2}\left[K_2\left(\frac{c_1}{c}\right)^2 - BM_1\right] + \frac{2K_3}{3}c^2\left(\frac{c_1}{c}\right)^3 \tag{6.31}$$

If one notes that c_1 tends to c as c approaches zero, it follows from this equation that the value of a quantity $H \equiv K_2 - BM_1$ may be determined from the initial slope of a plot of $1 - (M_1/M_{nr}^{app})$ versus $c/2$.

On the other hand, one obtains from equation 6.24

$$f_1^{app} = \left[1 - K_2c\left(\frac{c_1}{c}\right)^2 - K_3c^2\left(\frac{c_1}{c}\right)^3\right]\exp(BM_1c) \tag{6.32}$$

Expansion of the right-hand side in powers of c gives

$$f_1^{app} = 1 + c\left[BM_1 - K_2\left(\frac{c_1}{c}\right)^2\right] + O(c^2) \tag{6.33}$$

This indicates that H may also be evaluated from the initial slope of a plot of f_1^{app} versus c. The plot should have an ordinate intercept of unity, which may be a useful guide in determining the initial slope of the plot.

One can eliminate the term $K_3c_1{}^3$ from equations 6.30 and 6.11 to obtain the relation

$$\frac{3M_1}{M_{nr}^{app}} - 1 = 2\left(\frac{c_1}{c}\right) + \frac{K_2}{2}c\left(\frac{c_1}{c}\right)^2 + \frac{3BM_1}{2}c \tag{6.34}$$

Equation 6.20 may be rewritten, with substitution of equations 6.9, 6.10, and 6.19,

$$\frac{1}{(M_1/M_{wr}^{app}) - BM_1c} = \frac{c_1}{c} + 2K_2c\left(\frac{c_1}{c}\right)^2 + 3K_3c^2\left(\frac{c_1}{c}\right)^3 \tag{6.35}$$

Elimination of the term $K_3 c_1{}^3$ with the aid of equation 6.11 yields

$$3 - \frac{1}{(M_1/M_{wr}^{app}) - BM_1 c} = 2\left(\frac{c_1}{c}\right) + K_2 c \left(\frac{c_1}{c}\right)^2 \qquad (6.36)$$

One multiplies equation 6.34 by 2 and then subtracts equation 6.36 from the product. This operation yields

$$\frac{6M_1}{M_{nr}^{app}} + \frac{1}{(M_1/M_{wr}^{app}) - BM_1 c} - 5 = 2\left(\frac{c_1}{c}\right) + 3BM_1 c \qquad (6.37)$$

The term c_1/c can be eliminated by use of equation 6.24, and one obtains the final relation

$$\frac{6M_1}{M_{nr}^{app}} + \frac{1}{(M_1/M_{wr}^{app}) - BM_1 c} - 5 = 2f_1^{app} \exp\left(-BM_1 c\right) + 3BM_1 c \qquad (6.38)$$

In this equation, all quantities but the nonideality parameter B can be evaluated from experimentally obtained $M_{wr}^{app}(c)$ data. If the system contains no other species than monomer, dimer, and trimer and if the assumptions made above are rigorously obeyed, there should exist, at any concentration chosen in the range studied, a value of B which satisfies equation 6.38. The values of B thus obtained at different concentrations should be constant within the limits of accuracy expected for primary experimental data.

Once the value of B is known in this way, substitution of it, together with the known values of H and M_1, into $H = K_2 - BM_1$ permits the equilibrium constant K_2 to be calculated. This procedure, however, is not recommendable, because the determination of H by either procedure mentioned above may suffer a considerable degree of uncertainty. Instead, one calculates f_1 or c_1 as a function of c from equation 6.24 with introduction of the data for f_1^{app} and the known values of B and M_1, and then inserts the resulting values into equation 6.34 or 6.36 (the latter equation is preferable) to compute K_2. Finally, substitution of these values of c_1 and K_2 into equation 6.11 allows the computation of K_3 for chosen c values. Obviously, the values of K_2 and K_3 thus obtained must be independent of c within experimental errors.

A theory similar to that described above can be developed for monomer–dimer–m-mer associations. The reader should refer to Adams and Williams[3] and also Adams[4] for its details.

1. Cases of Monomer-Dimer and Monomer-Trimer Association

The monomer-dimer system is a special case, in which K_3 is zero, of the system treated in the preceding section. Likewise, the monomer-trimer system corresponds to the special case in which K_2 is zero. For these systems, too, one may apply equation 6.38 to evaluate B, because no division by K_2 and K_3 was made in the process of deriving this relation. With known values of B and M_1, one can calculate c_1 as a function of c from equation 6.24. For the monomer-dimer system the equilibrium constant K_2 is then obtained by using the relation

$$K_2 = \frac{c - c_1}{c_1^2} \tag{6.39}$$

In the case of monomer-trimer association, the corresponding equation for the equilibrium constant K_3 is

$$K_3 = \frac{c - c_1}{c_1^3} \tag{6.40}$$

Probably, it is relevant to remark here that, in their formulation of monomer-dimer associating systems, Adams and Fujita[2] derived a relation of the form

$$2\left(\frac{M_1}{M_{wr}^{app}}\right) - 1 = (1 + 4K_2c)^{-1/2} + 2BM_1c \tag{6.41}$$

It is a simple matter to verify that this equation is equivalent to equation 6.35 for $K_3 = 0$, and can be solved for K_2 to give[5]

$$\frac{(M_{wr}^{app}/M_1)^2}{[2(1 - BM_{wr}^{app}c) - (M_{wr}^{app}/M_1)]^2} - 1 = 4K_2c \tag{6.42}$$

According to this equation, the equilibrium constant K_2 can be calculated pointwise for an arbitrarily chosen value of B, and by successive approximations of B the best choice for K_2 will be a least-squares fit to a straight line with zero slope.

C. THE CHUN-TANG THEORY

It can be verified that, as long as equation 6.6 holds, equation 6.23 applies for any self-associating system. The same is also the case with equation 6.28. Elimination of BM_1 from these two equations yields

$$U(c) \equiv 2\left(\frac{M_1}{M_{nr}^{app}}\right) - \frac{M_1}{M_{wr}^{app}}$$

$$= \frac{2}{c}\int_0^{c_1}\left(\frac{c}{c_1}\right)dc_1 - \left(\frac{c}{c_1}\right)\frac{dc_1}{dc} \tag{6.43}$$

This idea of eliminating the nonideality parameter is due to Chun et al.[6] and independently to Roark and Yphantis.[7] The latter authors designated $U(c)$ as $1/M_{y1}(c)$.

We now consider a monomer–m-mer associating system which obeys all of the assumptions introduced in the preceding sections. Then we have

$$c = c_1 + c_m \tag{6.44}$$

with

$$c_m = K_m c_1^{\,m} \tag{6.45}$$

Substitution of equation 6.44 into equation 6.43 yields

$$U(c) = \frac{2}{c}\left(c_1 + \frac{c - c_1}{m}\right) - \frac{c}{c_1 + m(c - c_1)} \tag{6.46}$$

With $f_1 = c_1/c$ (the local weight fraction of monomer), this equation may be written

$$U(c) = 2(1 - m^{-1})f_1 + 2m^{-1} - \frac{1}{f_1 + m(1 - f_1)} \tag{6.47}$$

It is to be noted that this is quadratic in f_1, yielding as an appropriate root

$$f_1 = \frac{U + 2 - (2/m) - \left\{[U + 2 - (2/m)]^2 - 8(Um - 1)/m\right\}^{1/2}}{4(m-1)/m} \tag{6.48}$$

Since $U(c)$ is obtainable from experimental data, it is possible to calculate f_1 as a function of c for any assigned value of m. This important conclusion was discovered by Tang et al.[8]

Division of equation 6.44 by c_1, followed by rearrangement, leads to

$$\frac{1 - f_1}{f_1} = K_m(cf_1)^{m-1} \tag{6.49}$$

Thus, if the assigned value of m is correct, a plot of $(1-f_1)/f_1$ (calculated from equation 6.48) versus $(cf_1)^{m-1}$ will give a straight line which passes through the origin. The desired value of K_m can then be determined from the slope. In practice, the process will be such that one tries this plot for a variety of m values until an m is found which gives a most satisfactory straight line passing through the origin.

By substituting equation 6.44 into equation 6.23, another important relation[6,8] is obtained, that is,

$$\frac{M_1}{M_{wr}^{app}} - \frac{1}{m-(m-1)f_1} = BM_1 c \tag{6.50}$$

Hence, BM_1 can be evaluated from the slope of a plot of the quantity on the left-hand side against c.

D.   EXAMPLES OF SYNTHETIC DATA

For a comprehension of the methods described above it will be of help to have an idea of what the graphs of $M_{wr}^{app}(c)$ and $M_{nr}^{app}(c)$ for typical systems look like. For this purpose we quote the synthetic examples prepared by Adams and Filmer.[9]

Figure 6.1 shows plots of M_1/M_{wr}^{app} versus c for a hypothetical monomer-dimer-trimer system characterized by $K_2=0.65$ and $K_3=0.5$, with BM_1 as a parameter. It is seen that if BM_1 is positive, as will be the case in general, then the curve shows a minimum; whereas if BM_1 is zero or negative, the curve is monotonically decreasing with increase in c. Relatively steep rises of the curves in the vicinity of $c=0$ suggest that if the data for M_{wr}^{app} are restricted to relatively high concentrations, there will be considerable uncertainty in estimate of the monomer molecular weight by extrapolation to infinite dilution. Figure 6.2 shows similar plots for a hypothetical monomer-trimer system in which $K_3=0.5$. An interesting feature is that whenever BM_1 is positive, there appears a maximum in the curve at a position near but not at $c=0$. Further, when BM_1 is positive, the curve exhibits a minimum at a higher value of c. For BM_1 zero or negative the curves are monotonically decreasing, as in Fig. 6.1. Especially, for $BM_1=0$, the curve has zero limiting slope at $c=0$. Comparison of Figs. 6.1 and 6.2 indicates how appreciably the behavior of $1/M_{wr}^{app}$ versus c of a monomer-dimer-trimer system in the region near $c=0$ is affected by the presence or absence of the dimer species. This theoretical consequence again stresses the importance of extending the measurement down to lower concentrations.

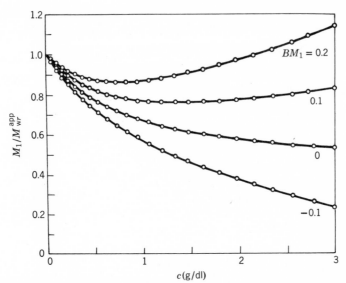

Fig. 6.1. Effects of the second virial coefficient B on the shape of M_1/M_{wr}^{app} for a hypothetical monomer-dimer-trimer association as a function of the total solute concentration c. The equilibrium constants chosen for the computation are $K_2 = 0.65$ (dl/g) and $K_3 = 0.50$ (dl^2/g^2).

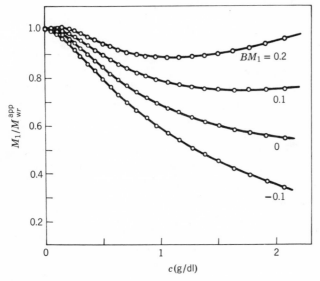

Fig. 6.2. Effects of the second virial coefficient B on the shape of M_1/M_{wr}^{app} for a hypothetical monomer-trimer association as a function of the total solute concentration c. The same K_3 value as in Fig. 6.1 is used for the computation.

391

Figure 6.3 displays the plots of M_1/M_{nr}^{app} versus c and of cf_1^{app} versus c calculated from the curve for $BM_1 = 0.2$ in Fig. 6.1. The curve for M_1/M_{nr}^{app} exhibits a minimum as does that for M_1/M_{wr}^{app}, but the effect is much less in Fig. 6.3 than in Fig. 6.1.

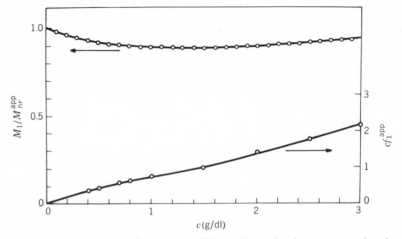

Fig. 6.3. Dependence of M_1/M_{nr}^{app} and cf_1^{app} on the total solute concentration for a hypothetical monomer-dimer-trimer associating system in which $K_2 = 0.65$, $K_3 = 0.50$, and $BM_1 = 0.20$ (dl/g).

The effect of K_2 on the plot of M_1/M_{wr}^{app} versus c is illustrated in Fig. 6.4 for a monomer-dimer system with $BM_1 = 0.10$. As the value of K_2 is increased, that is, the reaction prefers formation of dimer species, the plot exhibits a more pronounced minimum and becomes steeper in the region near $c = 0$. If the plotted points in the graph are actual experimental values, there will be no possibility of extrapolating them for $K_2 = 6.5$ and 65 to infinite dilution with any degree of certainty. For such large equilibrium constants, therefore, the availability of a technique which allows accurate measurements of M_{wr}^{app} at very low concentrations becomes imperative. Whenever applicable, the photoelectric scanner developed by Schachman and colleagues (see Chapter 3 for its references) may be of great help in such circumstances.

E.   EXAMPLES OF ACTUAL DATA

For illustration of the M_{wr}^{app} data for actual systems we choose the results reported recently by Visser et al.,[10] who studied, with great care, the

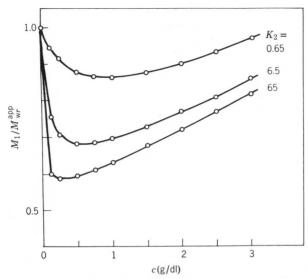

Fig. 6.4. Effects of the dimerization constant K_2 on the shape of M_1/M_{wr}^{app} as a function of the total solute concentration c for a hypothetical monomer-dimer associating system where $BM_1 = 0.1$ (dl/g).

sedimentation equilibrium of β-lactoglobulin B in pH 2.64 NaCl–glycine buffer of ionic strength 0.16 at various temperatures between 5 and 35.5°C. There are presented in Fig. 6.5 the experimental values of M_{wr}^{app} at 15 and 25°C as a function of protein concentration (expressed here in terms of the fringe number). The various groups of symbols denote data derived from single experiments at different initial protein concentrations. It is seen that the sets of data points from individual experiments at either 15 or 25°C superimpose fairly well to form a single curve over the entire range of concentration examined, in agreement with the theoretical prediction. Visser et al. state that this fact indicates that the protein was pure, that pressure effects were not significant,* and that the association reaction was reversible.

The solid curves in Fig. 6.5 rise very sharply, pass through a broad maximum, and then gradually decline as the concentration is increased. The trend of the data at low concentrations suffices to discourage us from attempting extrapolation to infinite dilution. Thus, Visser et al. took the limiting value of M_{wr}^{app} at $c = 0$, that is, the molecular weight of β-

*See Section IV.B for pressure effects.

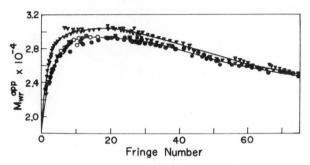

Fig. 6.5. Apparent local weight-average molecular weights, M_{wr}^{app}, of β-lactoglobulin B at 15°C (upper curve) and at 25°C (lower curve) as a function of the protein concentration in fringes.[10] Both curves are for pH 2.64 and ionic strength 0.16, and include data for different initial concentrations.

lactoglobulin B monomer, to be 18,333 on the basis of the amino acid composition reported by Frank and Braunitzer,[11] and used this M_1 value in their data analysis.

Figure 6.6 shows another example of $M_{wr}^{app}(c)$ data, taken from a paper by Hancock and Williams,[12] on a preparation of chymotrypsinogen A in pH 7.90 veronal buffer of ionic strength 0.03 at 25°C. Again, the data points derived from experiments at different initial protein concentrations merge quite satisfactorily to form a single continuous curve, as required by theory. They can be extrapolated with no great difficulty to obtain a monomer molecular weight which agrees with 25,600 known from amino acid analyses, a value confirmed by the sedimentation equilibrium studies of LaBar.[13] The general trend of the data points in Fig. 6.6 is quite different from that in Fig. 6.5. It is monotonically increasing with a small downward curvature, and exhibits no maximum in the range of concentration examined.

It is to be noted that until recently there has been a serious problem concerning the theoretically required merger of plots of M_{wr}^{app} versus c from experiments at differing initial concentrations. As can be seen in several places in the literature, most of the earlier experiments failed to obtain as good a merger of data points as observed in Figs. 6.5 and 6.6. An example showing the nonoverlapping of data points is presented in Fig. 6.7. Various causes for this phenomenon were suggested and examined experimentally.[14-16] With Hancock and Williams,[12] the author feels that the conformity of the more recent data to the theoretical requirement of the formation of a single curve is mainly due to an enhanced precison of the experimental results.

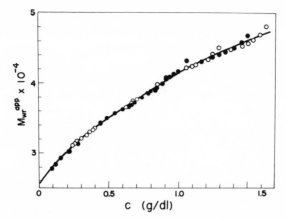

Fig. 6.6. Data of $M_{wr}^{app}(c)$ for chymotrypsinogen in buffer at pH 7.9 and ionic strength 0.03.[12] Different marks are for differing initial concentrations.

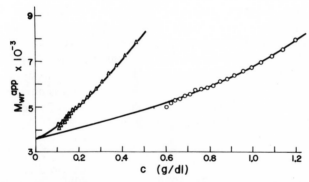

Fig. 6.7. The experimental data of Squire and Li for M_{wr}^{app} as a function of the total solute concentration c obtained with a sample of α_s-adrenocorticotropin at pH 1.30 and ionic strength 0.200 [P. G. Squire and C. H. Li, *J. Am. Chem. Soc.* **83**, 3521 (1961)]. $\bigcirc$, for c_0 (initial protein concentration) $= 0.869$ g/dl and $\omega = 20{,}410$ rev/min; $\triangle$, for $c_0 = 0.248$ g/dl and $\omega = 3{,}410$ rev/min.

F. DATA ANALYSIS

The initial, and doubtlessly the very crucial, step in an analysis of given M_{wr}^{app} data is the choice of a reaction model which permits a simple and accurate description of the data. In the absence of independent information from other sources, this is essentially a guesswork, consisting of a trial-and-error process. In some instances, the general trend of the data points will afford some clues by which inadequate models may be ruled out. The analysis to be made is, in essence, a kind of curve-fitting work. Thus, even if a model, with appropriate assignment of values to the equilibrium constants and nonideality parameter, is found to describe the experimental results with accuracy, one cannot claim its correspondence to the reactions which must be taking place in the system. In general, there will be more than one model which fits the experimental data equally well or even with a higher accuracy. The question arises as to what then should be done to establish the reaction model that can be considered to be the most realistic. We defer a discussion on this "uniqueness" problem to Section II.D.

1. Results on β-Lactoglobulins

Visser et al.[10] used the theories described in the preceding sections in an attempt to analyze their β-lactoglobulin B data. The reaction was assumed to be monomer-dimer association.

Figure 6.8 shows comparison of equation 6.49 for $m = 2$ with the 15°C data. The values of f_1 were calculated from equation 6.48, the Tang equation, after the M_{wr}^{app} data over the range of c from 0 to 3 in fringe numbers had been suitably adjusted. Though not strictly, the plotted points follow a straight line passing through the origin, as required by theory. The K_2 value estimated from the slope of the indicated line is about 35 dl/g. Visser et al.[10] found that the upturn of the points at higher values of cf_1 disappears, that is, an almost strictly linear plot is obtained over the whole range of the abscissa, if f_1 is reevaluated by use of an equation given by Tang and Adams[17] who assumed for $\ln y_i$

$$\ln y_i = iM_1 B_1 c + iM_1 B_2 c^2 \qquad (6.51)$$

a form introduced earlier by Adams and Williams[3] and Adams[4] as an empirical extension of equation 6.6. It must be pointed out, however, that, theoretically, a more correct coefficient for the c^2 term in equation 6.51 should be $(iM_1)^2$ and that, with the assignment of this coefficient, the analysis of self-associations becomes difficult or virtually impossible, because the equilibrium relation of the form $K_i = c_i/c_1^i$ then no longer holds. Thus, the Adams-Tang equation as well as Visser's finding on the basis of it are of doubtful validity, at least from the theoretical point of view. The upturn of the plots in Fig. 6.8, therefore, must be attributed to either errors in the data or incorrect choice of the model.

As another test of equation 6.49, let us quote a study on β-lactoglobulin A in 0.2 M glycine buffer (pH 2.46, ionic strength 0.10) at 11°C from Tang's thesis.[8] Figure 6.9 shows plots of $(1 - f_1)/f_1$ versus $(cf_1)^{m-1}$ for $m = 2$, 3, and 4. It is seen that the monomer-dimer association model is the

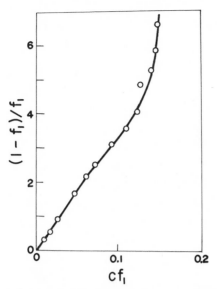

Fig. 6.8. Plots of $(1-f_1)/f_1$ versus cf_1 for β-lactoglobulin B in buffer of pH 2.64 and ionic strength 0.16 at 15°C.[10]

most appropriate among the three choices, even though the data points for $m=2$ do not perfectly obey the theoretical requirement. A point to note is the fact that this type of plot is very sensitive to the choice of m value. Test of equation 6.50 with Tang's data is displayed in Fig. 6.10. Again, it may be concluded that the monomer-dimer model is the most relevant, because only the points for $m=2$ are fitted, though not exactly, by a straight line passing through the origin, as required by theory. A reasonably accurate value of BM_1 for the system is then obtained as the slope of this line.

If equations 6.49 and 6.50 were accurately obeyed by given experimental data, it could be concluded that the reaction model with the chosen m value was relevant and the values of K_m and BM_1 obtained from the resulting straight lines were desired ones for the system. However, in cases when this is not true, as in the graphs given here, it is desirable to determine the optimum values of the parameters by use of a least-squares procedure.[10]

This consists of calculating $M_{wr}^{app}(c)$ pointwise for approximate values of the parameters selected on a primary analysis by a suitable theory, and substituting the results into a measure of the "goodness of fit," χ^2, defined as

$$\chi^2 = \sum_k \frac{[y_k - y(x_k)]^2}{\sigma_k} \qquad (6.52)$$

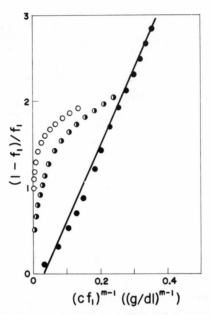

Fig. 6.9. Plots of $(1-f_1)/f_1$ versus $(cf_1)^{m-1}$ for β-lactoglobulin A in buffer of pH 2.46 and ionic strength 0.10 at 11°C.[8] ●, for $m=2$; ◑, for $m=3$; ○, for $m=4$. The line is drawn to fit the data points for $m=2$.

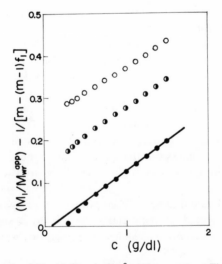

Fig. 6.10. Test of equation 6.50 with Tang's data[8] used to construct Fig. 6.9. ●, for $m=2$; ◑, for $m=3$; ○, for $m=4$.

398

Here y_k is the calculated value of M_{wr}^{app} at concentration x_k, $y(x_k)$ is the value of this quantity taken at x_k from an interpolation curve to the observed data, and σ_k is the uncertainty in the data point at x_k. The desired optimum values of the parameters are obtained as ones which minimize χ^2 with respect to each of the parameters simultaneously. The chosen model may permit different combinations of parameters, each yielding the calculated M_{wr}^{app} values which coincide with the interpolation curve better than the experimental uncertainty. The final choice of the parameters then becomes a difficult, largely subjective, matter.

Visser et al.,[10] applying the least-squares method (with σ_k being taken as 350 g/mole independent of x_k), obtained the numerical results summarized in the left-hand column of Table 6.1. They also made a similar analysis with the Tang-Adams modified equations which contain two nonideality coefficients, but the results are not shown here.

TABLE 6.1. Equilibrium Constants K_2 at Different Temperatures and Corresponding Best BM_1 for β-Lactoglobulin B in pH 2.64 NaCl–Glycine Buffer of Ionic Strength 0.16[10]

Temperature (°C)	χ^2 Test		Adams-Fujita Equation	
	BM_1^a	K_2^b	BM_1	K_2
5	0.055	50	0.055	50.4 ± 5.5
10	0.105	52	0.105	53.5 ± 15.1
15	0.109_5	44	0.109	38.9 ± 10.1
25	0.108	22.0	0.110	23.1 ± 2.5
35.5	0.111	10.2	0.111	10.4 ± 0.8

^{a}In dl/g.
^{b}In dl/g.

A test of the Adams-Fujita equation 6.42 with Visser's β-lactoglobulin B data at 25°C is illustrated in Fig. 6.11, where the values of K_2 calculated for $BM_1 = 0.110$ dl/g are plotted against protein concentration. Visser et al. state that the deviations from the indicated horizontal line are taken to be reflections of errors in the data, rather than of incorrect model choice. Values of K_2 and BM_1 that have been obtained in this way are listed in the right-hand column of Table 6.1. A similar evaluation of parameters with the Adams-Fujita equation has been carried out by Denoier and Williams[5] for lysozyme dissolved in pH 7.0 sodium cacodylate buffer of ionic strength 0.20 at 25°C. These authros provided still another equation which may be of use for analyzing monomer-dimer associations.

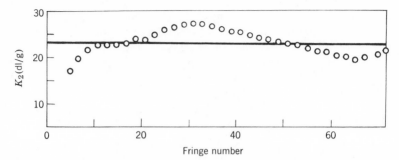

Fig. 6.11. The equilibrium constant K_2 as a function of concentration in fringes, calculated from the lower curve in Fig. 6.5 by use of equation 6.42 with $BM_1 = 0.110$ dl/g.

G. THE SVRY THEORY

The idea underlying the theory described in this section was first reported by Sophianopoulos and Van Holde,[18] and a few years later extended to a more general form by Roark and Yphantis.[7] So the designation SVRY is given to it.

The SVRY theory is concerned with monomer–m-mer–n-mer associations in both pseudoideal and nonideal solutions. Here, for simplicity, we confine our description to pseudoideal systems.

Let us first consider the case of monomer–m-mer association. The local solute concentration c is represented by

$$c = c_1 + K_m c_1{}^m \tag{6.53}$$

where use has been made of the equilibrium relation

$$c_m = K_m c_1{}^m \tag{6.54}$$

It follows from equation 6.20 that if the solution is pseudoideal, the apparent local weight-average molecular weight M_{wr}^{app} coincides with the true local weight-average molecular weight M_{wr}. Hence, equation 6.22 may be written

$$M_{wr}^{app} = M_1 \left(\frac{d\ln c}{d\ln c_1} \right) \tag{6.55}$$

Note that, as can be easily verified, equation 6.22 is applicable for any self-associating system. Substitution of equation 6.53 yields

$$M_{wr}^{app} = \frac{M_1(c_1 + K_m m c_1^m)}{c}$$

which may be rewritten

$$M_{wr}^{app} = M_1\left(\frac{c_1}{c}\right) + M_m\left(\frac{c_m}{c}\right) \tag{6.56}$$

Next, equation 6.55 is inserted into equation 6.29 and equation 6.53 is used. Then, after integration,

$$\frac{1}{M_{nr}^{app}} = \left(\frac{1}{M_1}\right)\left(\frac{c_1}{c}\right) + \left(\frac{1}{M_m}\right)\left(\frac{c_m}{c}\right) \tag{6.57}$$

It will be recognized that the right-hand side represents the reciprocal of the true local number-average molecular weight of the solute at total solute concentration c.

The quantities c_1/c and c_m/c can be eliminated from equations 6.56 and 6.57 to yield an interesting relation

$$M_{wr}^{app}(c) = M_1 + M_m - \frac{M_1 M_m}{M_{nr}^{app}(c)} \tag{6.58}$$

Thus, if only two-self associating species are present and, moreover, the solution is pseudoideal, a plot of $M_{wr}^{app}(c)$ against $1/M_{nr}^{app}(c)$ gives a straight line, and its slope and ordinate intercept allow the computation of M_1 and M_m to be made. Once M_1 and M_m are known in this way, c_1 and c_m can be calculated as functions of c from equations 6.56 and 6.57, and the results are used to compute the equilibrium constant K_m pointwise from equation 6.54. Also M_m/M_1 gives the value of another unknown m.

It is a simple task to prove that analogous relations hold between M_{zr}^{app} and M_{wr}^{app}, between $M_{z+1,r}^{app}$ and M_{zr}^{app}, and so forth.[18] For example,

$$M_{zr}^{app}(c) = M_1 + M_m - \frac{M_1 M_m}{M_{wr}^{app}(c)} \tag{6.59}$$

where M_{zr}^{app}, an apparent local z-average molecular weight, is defined by

$$M_{zr}^{app} = \frac{d^2 c/d(r^2)^2}{A[dc/d(r^2)]} \tag{6.60}$$

with A represented by equation 6.18. In practice, it is advantageous to evaluate M_{zr}^{app} by means of the relation

$$M_{zr}^{app} = \frac{d(cM_{wr}^{app})}{dc} \tag{6.61}$$

or[7]

$$M_{zr}^{app} = \frac{(M_{wr}^{app})^2 d(1/M_{wr}^{app}c)}{d(1/c)} \tag{6.62}$$

Mathematical equivalence of these relations to equations 6.60 may be verified easily. Roark and Yphantis[7] called the graph of $M_{k+1,r}^{app}$ versus $1/M_{k,r}^{app}$ a *two-species plot*. Here k stands for n, w, z, ... as $k = 0, 1, 2, \ldots$.

Prior to Roark and Yphantis, Sophianopoulos and Van Holde[18] had derived for pseudoideal monomer–m-mer associating systems

$$M_z^{app} = M_1 + M_m - \frac{M_1 M_m}{M_w^{app}} \tag{6.63}$$

a relation similar in form to equation 6.59. Here M_w^{app} and M_z^{app} are the apparent weight-average and apparent z-average molecular weights of the solute over the entire solution column, defined, respectively, by,

$$M_w^{app} = \frac{(c)_{r=r_2} - (c)_{r=r_1}}{c_0 \lambda} \tag{6.64}*$$

and

$$M_z^{app} = \left[\left(\frac{dc}{dr^2}\right)_{r=r_2} - \left(\frac{dc}{dr^2}\right)_{r=r_1} \right] \frac{(r_2^2 - r_1^2)}{M_w^{app} c_0 \lambda^2} \tag{6.65}*$$

where

$$\lambda = A(r_2^2 - r_1^2) = \frac{(1 - \bar{v}\rho_0)(r_2^2 - r_1^2)\omega^2}{2RT} \tag{6.66}$$

The quantity c_0 denotes the total solute concentration of the solution

---

*See equations 5.200 and 5.201.

before centrifugation. In the case of reacting systems, both M_w^{app} and M_z^{app} depend on c_0 even if the systems are thermodynamically pseudoideal. Equation 6.63 can be derived as an integral form of equation 6.59.*

If a nonlinear two-species plot is obtained, one may suspect either the presence of three or more reacting species or a departure of the solution from pseudoideality. In such a case, it is advisable to try a *three-species plot*,[7] provided the data are so accurate and extensive that the derivatives presented below may be estimated with precision.

For a pseudoideal monomer–m-mer–n-mer associating system the following relations may be derived easily[7]:

$$\frac{1}{M_{nr}^{app}} = \frac{1}{M_1}\left[1 + (m^{-1}-1)\frac{c_m}{c} + (n^{-1}-1)\frac{c_n}{c}\right] \qquad (6.67)$$

$$M_{wr}^{app} = M_1\left[1 + (m-1)\frac{c_m}{c} + (n-1)\frac{c_n}{c}\right] \qquad (6.68)$$

$$M_{wr}^{app}M_{zr}^{app} = M_1^{\,2}\left[1 + (m^2-1)\frac{c_m}{c} + (n^2-1)\frac{c_n}{c}\right] \qquad (6.69)$$

From these equations the variables (c_m/c) and (c_n/c) can be eliminated to obtain

$$M_{nr}^{app}M_{wr}^{app}M_{zr}^{app} - \alpha M_{nr}^{app}M_{wr}^{app} - \beta M_{nr}^{app} + \gamma = 0 \qquad (6.70)$$

where α, β, and γ are constants. Differentiation of this relation with respect to M_{nr}^{app} yields

$$\frac{d(M_{nr}^{app}M_{wr}^{app}M_{zr}^{app})}{dM_{nr}^{app}} = \alpha\frac{d(M_{nr}^{app}M_{wr}^{app})}{dM_{nr}^{app}} + \beta \qquad (6.71)$$

where

$$\alpha = M_1 + M_m + M_n, \qquad \beta = -(M_1M_m + M_1M_n + M_mM_n) \qquad (6.72)$$

Thus, if three and only three associating species are present and if the solution is pseudoideal, a plot of the quantity on the left-hand side of equation 6.71 against $d(M_{nr}^{app}M_{wr}^{app})/dM_{nr}^{app}$ should be linear, and one may

*Appendix A.

evaluate from the slope and ordinate intercept of the line any two of M_1, M_m, and M_n when the rest is known from another source. This type of plot has been referred to as a three-species plot by Roark and Yphantis.[7] If there are only two species present, the derivatives in equation 6.71 become constants, as can be easily verified.* Therefore, the system containing only a monomer and an oligomer defines a point on the three-species plot.

The three-species plot method, though interesting theoretically, will have to meet a formidable difficulty in its practical applications. Roark and Yphantis[7] state "it has not yet proven feasible to calculate reliable values for the derivatives used in the three-species plot for all points throughout the (concentration) distribution (at sedimentation equilibrium)" and, moreover, "with present experimental precision it appears unlikely that these derivatives can be determined well, except near the middle of a solution column."

1. Illustrations with Actual Data

A two-species plot of the Roark-Yphantis type is illustrated in Fig. 6.12, in which the data points refer to a preparation of β-lactoglobulin A in pH 4.55 buffer of ionic strength 0.1 at 4.6°C.[7] Plots lie close to a monomer-trimer line with a monomer molecular weight M_1 of about 36,000. This M_1 value is twice the molecular weight known for the unassociated β-lactoglobulin A protein, indicating that the smallest species present under the solvent conditions studied was a dimer. Thus, the higher species was actually a hexamer. These results are at variance with those deduced from light-scattering measurements under similar conditions by Townend and Timasheff,[19] who showed a dimer (36,000 molecular weight species)-octamer association to be adequate for a description of their data.

As an example of the two-species plot of the Sophianopoulos-Van Holde type, data[18] for a lysozyme preparation in 0.15 M KCl at 20°C are displayed in Fig. 6.13. The graph includes all values in the range of pH from 5 to 9 and at protein concentrations c_0 from 0.6 to 1.4 g/dl. The line shows monomer-dimer association with a monomer molecular weight of 13,900. Sophianopoulos and Van Holde[18] showed that polymers higher than dimer are required to explain their data at pH 11.

$$d(M_{nr}^{app}M_{wr}^{app})/ dM_{nr}^{app}= M_1+ M_m$$

$$d(M_{nr}^{app}M_{wr}^{app}M_{zr}^{app})/ dM_{nr}^{app}= M_1^2+ M_1M_m+ M_m^2$$

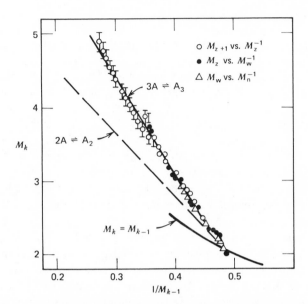

Fig. 6.12. "Two-species plot" for the trailing-edge fraction of β-lactoglobulin A in pH 4.55 buffer of ionic strength 0.1 at 4.6°C.[7] Intersection of the two-species line with the hyperbola (solid curve) yields a monomer molecular weight of about 36,000. The abscissa and ordinate are expressed in somewhat unusual units. See the original paper of Roark and Yphantis[7] for this material.

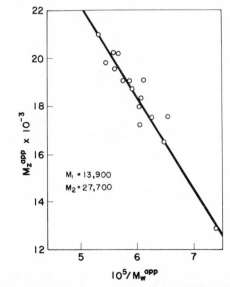

Fig. 6.13. "Two-species plot" for a lysozyme preparation in 0.15 M KCl at 20°C.[18]

H.   REMARKS

The degree of agreement of a theory with experiment can, in general, be improved by increasing the number of adjustable parameters. Thus, in the case of self-associations, given experimental data for M_{wr}^{app} may be described more accurately by a model in which there are more oligomers of higher order, hence there are more equilibrium constants that can be adjusted. However, in practice, it seems that there will be a maximum number of reacting species for which an analysis of data may be carried out with a fair degree of accuracy. Adams[20] proposed a theory for self-associating systems which contain four reacting species, but it is doubtful as to whether the limited accuracy of current sedimentation equilibrium measurements warrants the use of his method of analysis.

There have been attempts[3,4,10,21] in which, instead of increasing the number of reacting species, a term in c^2 was added to the right-hand side of equation 6.6 for $\ln y_i$. However, as noted in Section I.F.1, this way of increasing an adjustable parameter leads to a theory virtually impossible to use for actual analysis, unless such a theoretically unacceptable expression as equation 6.51 is assumed. In the author's opinion, inclusion of a so-called third virial coefficient in the expression for $\ln y_i$ seems to be too artificial a procedure for relatively small macromolecular solutes such as typical globular proteins.

Undoubtedly, thermodynamic nonideality is one of the major factors which make the analysis of sedimentation equilibrium data for self-associating systems a difficult task. In fact, if it can be ignored, as was considered reasonable in the early stages of development, it is possible to evaluate in a more straightforward way the equilibrium constants of the systems from data for $M_{wr}^{app}(c)$.[22]

For instance, when $B = 0$, f_1^{app} reduces to f_1. Hence, equation 6.26 becomes

$$f_1 = \exp\left[\int_0^c \left(\frac{M_1}{M_{wr}^{app}} - 1 \right) \frac{dc}{c} \right] \tag{6.73}$$

which is usually referred to as Steiner's equation. This equation applies for any pseudoideal self-associating system. On the other hand, we have the relation

$$M_{wr} = \frac{M_1 c_1 + M_2 c_2 + \cdots + M_m c_m}{c}$$

$$= M_1 f_1 \left[1 + 2K_2(cf_1) + \cdots + mK_m(cf_1)^{m-1} \right] \tag{6.74}$$

which may be transformed to

$$X(c) \equiv \frac{[M_{wr}^{app}/(M_1 f_1)] - 1}{cf_1} = 2K_2 + 3K_3(cf_1) + \cdots + mK_m(cf_1)^{m-2} \quad (6.75)$$

where the fact that $M_{wr}^{app} = M_{wr}$ for $B = 0$ has been considered.

By combination of equations 6.73 and 6.75 it is possible, in principle, to determine all equilibrium constants up through K_m when data for $M_{wr}^{app}(c)$ are obtained with high accuracy. The actual process for this purpose is essentially a curve fitting to a plot of $X(c)$ versus cf_1 calculated from the experimental data. If the plot is linear, it may be taken to mean that the system is a monomer-dimer-trimer one, and the intercept at $cf_1 = 0$ and the slope of the plot may be equated to $2K_2$ and $3K_3$, respectively. Rao and Kegeles[23] applied this method to α-chymotrypsin in a phosphate buffer and found that the data for M_{wr}^{app} (obtained from Archibald-type sedimentation experiments) were fitted approximately by a monomer-dimer-trimer self-association model.

Presently, the well-established notion is that even dilute solutions of relatively small proteins are, in general, thermodynamically nonideal and that thermodynamic analyses of these solutions will commit gross errors unless nonideality effects are taken into proper account. Thus, it must be concluded that the method of data analysis based on Steiner's equation is of limited value at the present stage, except for very dilute solutions as can be investigated by high-speed sedimentation equilibrium experiments.

As shown in Appendix B, a plot of M_{wr}^{app} versus c for any pseudoideal self-associating system increases monotonically with increase in c. Thus, if one observes a curve for $M_{wr}^{app}(c)$ which exhibits a maximum, it may be taken as an indication that a significant nonideality effect is operating in the system. However, one should note that a positive B value does not always give rise to a maximum in the curve. The detailed shape of the curve is governed by relative magnitudes of equilibrium constants K_i and nonideality parameter B.

II. INDEFINITE SELF-ASSOCIATION

A. EQUILIBRIUM RELATIONS

When the number of reacting species taking part in a discrete self-association increases indefinitely, the system is said to undergo an indefinite self-association. This type of self-association is represented by the following set reactions:

$$2P_1 \rightleftarrows P_2, \qquad 3P_1 \rightleftarrows P_3, \quad \ldots, \quad iP_1 \rightleftarrows P_i, \quad \ldots \quad (6.76)$$

where, as before, P_i denotes an i-mer of a monomer P_1. An alternative representation of these reactions is

$$P_1 + P_1 \rightleftarrows P_2, \qquad P_2 + P_1 \rightleftarrows P_3, \qquad \ldots, \qquad P_{i-1} + P_1 \rightleftarrows P_i, \qquad \ldots \quad (6.77)$$

Let us denote the molar equilibrium constants for the successive reactions in this latter set by $K_{12}, K_{23}, \ldots, K_{i-1,i}, \ldots$. Then it can be shown that if the activity coefficient y_i of i-mer species is expressed by equation 6.6, the following set of equilibrium relations holds:

$$K_{12} = \frac{C_2}{C_1^{\,2}}, \qquad K_{23} = \frac{C_3}{C_1 C_2}, \qquad \ldots \qquad (6.78)$$

Here C_i is the molar concentration of i-mer species, that is, the number of moles of i-mer species per liter of solution. In deriving equations 6.78, it should be noted that the same activity coefficient y_i can be used for both c-scale concentration and molarity concentration.

Uncompromisingly, systems undergoing general indefinite self-association are not an inviting object of either theoretical or experimental investigation because of the infinite number of parameters with which one has to be concerned. Probably, the only case of practical interest is the system in which an essentially constant change in free energy accompanies the addition of 1 mole of monomer to any i-mer. In this case, all molar equilibrium constants may be assumed to be equal, that is,

$$K_{12} = K_{23} = K_{34} = \cdots = K_{i-1,i} = \cdots \equiv K \qquad (6.79)$$

An indefinite self-association obeying this condition is called *isodesmic*.[5,24] With equation 6.79, equations 6.78 yield

$$C_2 = K(C_1)^2, \qquad C_3 = K^2(C_1)^3, \ldots, C_i = K^{i-1}(C_1)^i, \qquad \ldots \quad (6.80)$$

If the molar concentrations are converted to the c-concentrations, then we have

$$c_2 = 2kc_1^{\,2}, \qquad c_3 = 3k^2 c_1^{\,3}, \qquad \ldots, \qquad c_i = ik^{i-1}c_1^{\,i}, \qquad \ldots \quad (6.81)$$

where the quantity k, defined by

$$k = \frac{1000K}{M_1} \qquad (6.82)$$

is often termed the intrinsic equilibrium constant. It is seen from equations

6.81 that, for this special type of indefinite self-association, the previously used equilibrium constants K_2, K_3, ..., are expressed in terms of k as

$$K_2 = 2k, \qquad K_3 = 3k^2, \quad ..., \quad K_i = ik^{i-1}, \qquad ... \qquad (6.83)$$

B. THE THEORY OF VAN HOLDE AND ROSSETTI

In relation to their study of the stacking of purine in aqueous solution, Van Holde and Rossetti[25] developed a method for analyzing $M_{wr}^{app}(c)$ data with the assumption of an indefinite isodesmic association. The essential point of their development is as follows.

The total solute concentration c at a particular radial position in the cell is represented as

$$c = c_1 + c_2 + \cdots + c_i + \cdots \qquad (6.84)$$

Substitution of equations 6.81 for c_i's allows this infinite series to be summed up if the condition $kc_1 < 1$ is obeyed. The result is

$$c = \frac{c_1}{(1 - kc_1)^2} \qquad (6.85)$$

We substitute this into equation 6.22 (note that equation 6.22 is applicable to any self-associating system, as was noted in Section I.G) to obtain a relation

$$\left(\frac{M_{wr}}{M_1} \right)^2 - 1 = 4kc \qquad (6.86)$$

It can be shown that equation 6.20 also applies for any self-associating system, as long as equation 6.6 holds. Combination of this equation with equation 6.86 yields

$$\frac{M_{wr}^{app} / M_1}{1 - BM_{wr}^{app}c} = (1 + 4kc)^{1/2} \qquad (6.87)$$

or

$$k = (4c)^{-1} \left\{ \frac{(M_{wr}^{app} / M_1)^2}{[1 - BM_{wr}^{app}c]^2} - 1 \right\} \qquad (6.88)$$

These are due originally to Van Holde and Rossetti.[25] Thus, if the proper value of B is chosen, the quantity on the right-hand side of equation 6.88 should become constant and yield the desired value for the intrinsic equilibrium constant k. The actual procedure to be used is then such that one calculates the right-hand side of equation 6.88 from experimentally given data for M_{wr}^{app} and M_1, assuming various values for B, and seeks the value which yields a constant k. When the calculated values of k for different c exhibit statistical scatter around a certain value, it is preferable to determine the optimum values of the parameters (k and B) by applying a χ^2 test as described before.

1. Application to Association of Purine in Water

Van Holde and associates,[24, 25] performing careful sedimentation equilibrium experiments for the purpose of investigating the association of purine and of cytidine in water, showed that the data obtained for both substances could be fitted very well by the theory described in the preceding section. Here, the data on purine are presented, together with the anaylsis by Van Holde and Rossetti.[25] Before moving ahead to this end, we emphasize the importance and advantages of the sedimentation equilibrium study with well-defined low-molecualr-weight materials, in accordance with Van Holde et al.,[24] who write:

1.   The monomer weight is always known with accuracy.
2.   Problems of purity and stability are much less serious (than in high-molecular-weight substances). The stability of the substances allows very long experiments, in which the temperature is varied in a stepwise fashion, to be entirely practical.
3.   Since the diffusion coefficients are large, equilibrium is attained in very short times with short columns. Alternatively, long columns can be employed to increase resolution.

Figure 6.14 shows observed values of M_{wr}^{app} for purine in water at 24.9°C as a function of the molar concentration of monomeric purine. Differently filled circles indicate single experiments at different initial concentrations. It is seen that, as required by theory, the data points merge well to form a single continuous curve over the entire range of concentration investigated. The curve exhibits a broad maximum at a higher concentration, which implies that a significant nonideality effect is operating in the system. It is to be noted that the sedimentation equilibrium method can be employed with accuracy for investigation of such a small molecule as purine whose molecular weight is 120.

Tests of equation 6.88 with the data of Fig. 6.14 are illustrated in Fig. 6.15. One point from each experiment (marked with a vertical bar in Fig.

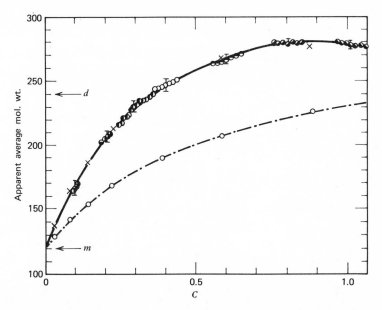

Fig. 6.14. Apparent local average molecular weights (M_{wr}^{app} and M_{nr}^{app}) as functions of molar concentration (C) for purine in water at 24.9°C.[25] Solid line, drawn to fit M_{wr}^{app} data for initial concentrations of 0.1 (◑), 0.2 (◒), (0.3) (●), 0.4 (◓), 0.6 (◐), and 0.8 (◕). The vertical bars show estimated maximum error at each initial concentration. The crosses (x) are M_{wr}^{app} values calculated from the deduced equilibrium constant and second virial coefficiennt. Chain line, M_{nr}^{app} calculated from the solid line for M_{wr}^{app}. The open circles (○) are M_{nr}^{app} values computed from the deduced equilibrium constant and second virial coefficient. The arrows indicate theoretical molecular weights for monomer and dimer, respectively.

6.14) has been used. Van Holde and Rosetti[25] state that since the precision within an experiment is better than that between experiments, inclusion of more points would not really yield more information. From the graph one finds that the value of $BM_1 = 1.2$ ml/g leads to k values essentially independent of concentration, and also that the results are fairly sensitive to the choice of B value. The k values corresponding to $BM_1 = 1.2$ are 23.3 ± 0.5 ml/g, which give $K = 2.80 \pm 0.06$ 1/mole, in substantial agreement with the value found by Gill et al.[26] by an independent procedure.

Solving equation 6.88 for M_{wr}^{app} and substituting the result into equation 6.29 (which is also valid for any self-associating system), we obtain, after integration,

$$\frac{M_1}{M_{nr}^{app}} = \frac{1}{2kc}\left[(1+4kc)^{1/2}-1\right] + \frac{BM_1}{2}c \qquad (6.89)$$

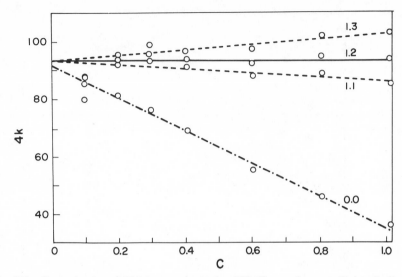

Fig. 6.15. Determination of BM_1 by use of equation 6.88. The numbers attached to the lines are the assumed values of BM_1 in cubic centimeters per gram. The data are for purine in water at 24.9°C.[25]

The values of M_{nr}^{app} computed pointwise from this equation with the above-mentioned k and BM_1 values are plotted by open circles in Fig. 6.14. They stand in good agreement with the chain line obtained by substitution of the observed M_{wr}^{app} into equation 6.29. It is dubious, however, whether this comparison may be used as a semiindependent check of the k and BM_1 values deduced from M_{wr}^{app} data (see Section II.D).

There are several published examples showing the relevance of the indefinite isodesmic association model to observed data for high-molecular-weight substances. For example, the reader should consult the paper by Adams and Lewis[15] on β-lactoglobulin A in solution of pH 4.61 and ionic strength 0.2 at 16°C, and that by Hancock and Williams[12] on chymotrypsinogen A in solution of pH 7.90 and ionic strength 0.03 at 25°C.

C. OTHER METHODS OF ANALYSIS

The method of Van Holde and Rossetti described above is somewhat inconvenient because it requires a trial-and-error process. More direct methods for evaluating k and BM_1 have been worked out by several authors. Some of these methods are briefly described below.

It is a simple matter to derive, by eliminating either k or BM_1 from equations 6.88 and 6.89, a set of equations

$$2V - (\tfrac{1}{2})\left\{3 + U - \left[(3+U)^2 - 16U\right]^{\frac{1}{2}}\right\} = BM_1c$$

$$\times \left\{\frac{3 - U + \left[(3-U)^2 - 4U\right]^{\frac{1}{2}}}{2U}\right\}^2 - 1 = 4kc$$

(6.90)

where U is the Chun-Tang function defined by equation 6.43, and V is $M_1/M_{nr}^{app}(c)$. Equations 6.90 allow BM_1 and k to be computed pointwise for different values of c at which the experimental data for $M_{wr}^{app}(c)$ and $M_{nr}^{app}(c)$, and hence U and V, have been obtained. The first equation of (6.90) is attributable to Adams and Lewis.[15]

Next, if BM_1 is eliminated from equations 6.24 and 6.28 and equation 6.85 is substituted into the resulting relation, we obtain

$$2V - \ln f_1^{app} = 2(f_1)^{\frac{1}{2}} - \ln f_1$$

(6.91)

which allows f_1 to be evaluated as a function of c, since the quantities on the left-hand side are determinable from experimental $M_{wr}^{app}(c)$ data. With these $f_1(c)$ we can prepare two kinds of plot: $1 - (f_1)^{\frac{1}{2}}$ versus $f_1 c$ and $V - (f_1)^{\frac{1}{2}}$ versus $c/2$. They should be linear, giving k from the slope of the former plot and BM_1 from the slope of the latter. This is because equations 6.85 and 6.89 yield a set of relations

$$1 - (f_1)^{\frac{1}{2}} = k f_1 c$$

$$V - (f_1)^{\frac{1}{2}} = \frac{BM_1}{2} c$$

(6.92)

This method of estimating k and BM_1 is from Chun et al.[6] These authors also showed that plots of $U - (f_1)^{\frac{1}{2}}/[2 - (f_1)^{\frac{1}{2}}]$ versus c ought to follow a straight line whose slope is BM_1.

D. THE UNIQUENESS PROBLEM

The most perplexing problem for those who are concerned with sedimentation equilibrium analysis of self-associating systems is the lack of criteria by which the reaction model that must correspond to the actual situation can be sorted out when there are two or more models which describe the given data equally well. Now that much improvement of the data themselves has been attained, this problem is becoming increasingly serious.

Figure 6.16, a reproduction from the paper by Hancock and Williams[12] on chymotrypsinogen A, compares experimental values of $M_{wr}^{app}(c)$ (solid line) with the calculated ones for a monomer-dimer-trimer association model with $K_2 = 0.908$, $K_3 = 0.844$, and $BM_1 = -0.01$ (open circles) and those for an indefinite isodesmic association model with $k = 49.6$ and $BM_1 = 4.0$ (filled triangles). Both models fit the observed data very well. It may seem to be unrealistic to be forced to utilize a negative value for the nonideality parameter B, but even so no one will be able to claim, with this result alone, a preference for one reaction mechanism over the other. An essentially similar situation was encountered by Deonier and Williams[5] when they dealt with their lysozyme data in terms of monomer-dimer association and indefinite isodesmic association. These authors discussed very thoroughly the reasons for the indistinguishability of the two models and the related problems.

Another important reference to the uniqueness problem is a recent paper

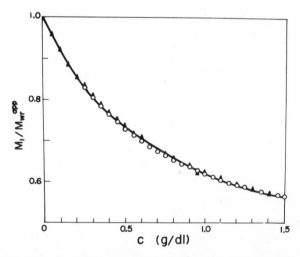

Fig. 6.16. Comparison of experimental and calculated values of M_1/M_{wr}^{app} for chymotrypsinogen in solution at pH 7.9 and ionic strength 0.03.[12] Solid line, drawn to fit experimental data (not shown). ▲, calculated for isodesmic association with $k = 49.6$ and $BM_1 = 4.0$ (dl/g). ○, calculated for monomer-dimer-trimer association with $K_2 = 0.908$, $K_3 = 0.844$, and $BM_1 = -0.01$ (dl/g).

written by Van Holde et al.,[24] who attempted analyzing their cytidine data in terms of various self-association mechanisms including the indefinite isodesmic one. Table 6.2 summarizes the results of their analyses. Here, ΔM_{wr}^{app}, a measure of the "goodness of fit," denotes the root-mean-square deviation of calculated M_{wr}^{app} from observed values, in atomic mass units. The best-fit curve for each of the chosen reaction models was determined with the aid of a digital computer, in which the procedure used was analogous in some respects to those employed earlier by Reinhardt and Squire.[27] It is concluded from the table that the inclusion of thermodynamic nonideality is almost imperative and that the indefinite isodesmic, the monomer-dimer-trimer-tetramer, and the monomer-dimer-trimer models with suitable nonideality terms provide the best descriptions

TABLE 6.2. Models for Cytidine Association in Water at $25°C$[24]

Model[a]	ΔM_{wr}^{app}	BM_1	Remarks
M-D (I)	14.9[b]	0.0	Rejected on ΔM[c]
M-D (N)	3.4	−0.664	ΔM slightly too large, negative B
M-Tr (I)	15.0	0.0	Rejected on ΔM
M-Tr (N)	6.9	0.614	Rejected on ΔM
M-Te (I)	31.8	0.0	Rejected on ΔM
M-Te (N)	17.4	1.792	Rejected on ΔM
M-D-Tr (I)	1.4	0.0	Good on M_{wr}^{app}, M_{nr}^{app}. M_{zr}^{app} pred. too low
M-D-Tr (N)	1.3	0.0814	Good on M_{wr}^{app}, M_{nr}^{app}; M_{zr}^{app} pred. too low
M-D-Te (I)	3.1	0.0	ΔM slightly too large; M_{zr}^{app} good
M-D-Te (N)	1.1	0.543	Good on M_{wr}^{app}, M_{nr}^{app}; M_{zr}^{app} large
M-D-Tr-Te (I)	1.3	0.0	Rejected: negative amount of tetramer; M_{zr}^{app} too low
M-D-Tr-Te (N)	1.1	0.427	Very good on M_{wr}^{app}, M_{nr}^{app} but pred. M_{zr}^{app} somewhat high
Isodesmic (N)	1.9	0.35	Quite good on M_{wr}^{app} M_{nr}^{app}; M_{zr}^{app} is fair

[a] Code: M = monomer, D = dimer, Tr = trimer, Te = tetramer; I = pseudoideal, N = nonideal, pred = predicted.
[b] Root-mean-square deviations from observed M_{wr}^{app} values, in atomic mass units.
[c] Abbreviation for ΔM_{wr}^{app}.

of the data. At present, one cannot proceed beyond this point, the three models being left as equally plausible assignments to the actual reactions which must be taking place in the cytidine solutions studied. The results of a similar analysis of the Van Holde-Rossetti data on purine (see Fig. 6.14) were very much alike.[25] Again, the reaction was very accurately described by an indefinite isodesmic association mechanism as well as by discrete association schemes.

The general conclusion that follows from these recent studies is discouraging: The sedimentation equilibrium in its present form does not allow unique establishment of a reaction mechanism to be made when applied to self-associating systems. This does not mean, however, the uselessness of the method, because, though remarkably accurate experimental data are required in general, it provides information on the basis of which one will be able to narrow the latitude of reaction models assignable to a given system. It is highly desirable that further researches, both in theory and in experiment, are undertaken so that the maximum capability of the sedimentation equilibrium method for self-associating systems may be elucidated.

Van Holde et al.[24] emphasize that even the degree of surety of the conclusions they obtained on the association of purine and of cytidine is possible only with an accurate method capable of *measuring* several average molecular weights. However, this emphasis seems to be irrelevant, because no other average molecular weights than M_{wr}^{app} can be measured directly. For example, M_{nr}^{app} and M_{zr}^{app} as functions of concentration are obtainable only by the calculations in which observed values of M_{wr}^{app} are substituted into equations 6.29 and 6.61, respectively. It is to be noted that these equations are not anything but the defining expressions of M_{nr}^{app} and M_{zr}^{app} in terms of M_{wr}^{app}, the only quantity that can be determined by experiment. Thus, for instance, the comparison of the two sets of M_{nr}^{app} values, one from the directly obtained M_{wr}^{app} values and the other from the equation for a chosen reaction model, as shown in Fig. 6.14, does not seem to have much theoretical significance.

III. COMPLEX FORMATION

A. INTRODUCTION

In solutions of macromolecules of biological interest, we often encounter reactions of the following types:

$$A + B \rightleftarrows AB \tag{6.93}$$

$$mA + nB \rightleftarrows A_m B_n \qquad (m, n = 1, 2, \ldots) \tag{6.94}$$

The self-associations discussed in the preceding sections are included as the special cases of reaction 6.94. Osmometry and light scattering of the systems involving these complex-forming reactions have been studied in detail both theoretically and experimentally by Steiner,[22,28-30] under the assumption that the solutions are pseudoideal. Recently, Adams[31], Nichol and Ogston[32], and Chun and Kim[33] describe methods for analyzing sedimentation equilibrium experiments on these systems.

Nichol and Ogston assumed that the partial specific volumes and specific refractive index increments of all reacting species are equal. While this assumption makes the theoretical treatment easier, it is by no means mandatory, as has been shown by Adams.[31] In what follows, we first give an account of Adams' theory, though some modifications of his original derivation are made. The central problem here is to devise a procedure which allows the equilibrium constant for reaction 6.93 to be determined from experimental data when the molecular weights of reacting species, A, B, and AB, are known in advance.

B.   THE ADAMS THEORY FOR THE REACTION $A + B \rightleftarrows AB$

On a molar basis, the equilibrium condition for this type of reaction is given by

$$\hat{\mu}_A + \hat{\mu}_B = \hat{\mu}_{AB} \qquad (6.95)$$

where $\hat{\mu}_i$ $(i = A, B, AB)$ is the chemical potential per mole of reacting species i. The reactants A and B are thermodynamic components, and hence their chemical potentials are well-defined quantities, because the amounts of A and B can be varied independently. On the other hand, the complex AB cannot be taken as a thermodynamic component, because its concentration is automatically fixed by virtue of equation 6.95 if those of A and B are given. Equation 6.95 should be regarded as the defining equations for the chemical potential of AB. As in self-associating systems, A, B, and AB are referred to as solute species in the subsequent presentation.

The condition of pseudoideality allows $\hat{\mu}_i$ to be expressed by

$$\hat{\mu}_i = \left(\hat{\mu}_i^0 \right)_c + RT \ln c_i \qquad (i = A, B, AB) \qquad (6.96)$$

Substitution into equation 6.95 gives the familiar equilibrium relation:

$$\frac{c_{AB}}{c_A c_B} = K \qquad (6.97)$$

where K, the equilibrium constant for the reaction considered, is given by

$$K = \exp\left\{ -\frac{(\hat{\mu}_{AB}{}^0)_c - (\hat{\mu}_A{}^0)_c - (\hat{\mu}_B{}^0)_c}{RT} \right\} \qquad (6.98)$$

Because of equation 6.95 it is sufficient if one is concerned with the sedimentation equilibrium equations for species A and B. With the condition that the solution is pseudoideal, incompressible, and dilute, these equations are shown to be given by

$$\frac{dc_A}{dr^2} = L_A M_A c_A \qquad (6.99)$$

$$\frac{dc_B}{dr^2} = L_B M_B c_B \qquad (6.100)$$

Here M_A and M_B are the molecular weights of species A and B, and L_i $(i = A, B)$ is defined by

$$L_i = \frac{(1 - \bar{v}_i \rho_0)\omega^2}{2RT} \qquad (6.101)$$

with $\bar{v}_i$ being the partial specific volume of species i.

The total refractive index of the solution over that of the solvent is denoted by n_c. Then it follows from equation 3.19 that

$$n_c = R_A c_A + R_B c_B + R_{AB} c_{AB} \qquad (6.102)$$

where R_i $(i = A, B, AB)$ is the specific refractive index increment of species i in the solvent considered. Using equation 6.97, this expression for n_c may be written

$$n_c = R_A c_A + R_B c_B + K R_{AB} c_A c_B \qquad (6.103)$$

Differentiation with respect to r^2, followed by substitution of equations 6.99 and 6.100, yields

$$\frac{dn_c}{dr^2} = L_A R_A M_A c_A + L_B R_B M_B c_B + K R_{AB} Q c_A c_B \qquad (6.104)$$

in which

$$Q = L_A M_A + L_B M_B \qquad (6.105)$$

Now, by integration, we obtain from equations 6.99 and 6.100

$$c_A(r) = c_{A1} \exp[\phi_A(r)] \tag{6.106}$$

$$c_B(r) = c_{B1} \exp[\phi_B(r)] \tag{6.107}$$

where

$$\phi_A(r) = L_A M_A (r^2 - r_1^2), \qquad \phi_B(r) = L_B M_B (r^2 - r_1^2) \tag{6.108}$$

and c_{A1} and c_{B1} are the concentrations of species A and B at the air-liquid meniscus of the solution column, $r = r_1$. The equilibrium constant K may be eliminated from equations 6.103 and 6.104 to yield*

$$\frac{dn_c}{dr^2} - Qn_c = -L_B M_B R_A c_{A1} \exp[\phi_A(r)] - L_A M_A R_B c_{B1} \exp[\phi_B(r)] \tag{6.109}$$

This relation indicates that a plot of quantity $Z(r)$ defined by

$$Z(r) = \left(Qn_c - \frac{dn_c}{dr^2} \right) \exp[-\phi_A(r)] \tag{6.110}$$

against $\exp[\phi_B(r) - \phi_A(r)]$ should give a straight line, whose intercept I, at $r = r_1$, where the abscissa takes the value of unity, and whose slope S are represented by

$$I = L_B M_B R_A c_{A1} + L_A M_A R_B c_{B1} \tag{6.111}$$

$$S = L_A M_A R_B c_{B1} \tag{6.112}$$

Both dn_c/dr^2 and n_c as functions of r may be determined experimentally by use of Rayleigh interference optics or its combined use with schlieren optics. Of the auxiliary quantities M_A, M_B, R_A, R_B, $\bar{v}_A$, and $\bar{v}_B$ necessary to prepare Z plots the data for the last four are obtainable only from measurements in which the test solutions contain either A or B alone as a macromolecular component. Once all these quantities are determined, one may evaluate the unknown concentrations c_{A1} and c_{B1} by substitution of I and S obtained from the Z plot into equations 6.111 and 6.112.

---

*Adams[31] eliminated the c_A terms on the right-hand sides of equations 6.103 and 6.104 by making a quantity $(dn_c/dr^2) - n_c$.

If the value of n_c at the air-liquid meniscus is denoted by n_{c1}, it follows from equation 6.103 that

$$n_{c1} = R_A c_{A1} + R_B c_{B1} + K R_{AB} c_{A1} c_{B1} \qquad (6.113)$$

Since n_{c1} may be estimated by suitable extrapolation of the experimentally determined n_c versus r curve and since the numerical values of R_{A1}, R_{B1}, c_{A1}, and c_{B1} are now at hand, the combined quantity $K R_{AB}$ can be calculated from equation 6.113. Obviously, the quantity R_{AB} cannot be measured by experiment, because it is impossible to prepare a solution which contains species AB only. The solution being treated here is always a mixture of species A, B, and AB that are at chemical equilibrium. The best one can do here is to follow a convention introduced by Svedberg and Pedersen,[34] who computed R_{AB} by the relation

$$M_{AB} R_{AB} = M_A R_A + M_B R_B \qquad (6.114)$$

where

$$M_{AB} = M_A + M_B \qquad (6.115)$$

The partial specific volume of species AB, $\bar{v}_{AB}$, is also not measurable experimentally. Again, one may use for its evaluation another convention of Svedberg and Pedersen,[34] which reads

$$M_{AB} \bar{v}_{AB} = M_A \bar{v}_A + M_B \bar{v}_B \qquad (6.116)*$$

If, in actual applications of the method described above, the Z plots are found to exhibit a curvature, it is attributable to a thermodynamic nonideality of the solution or to the presence of other complexes such as A_2B or AB_2 or to their combined effect. Even if a linear Z plot is obtained from a single experiment, it is advisable to perform experiments at different initial concentrations made up by changing composition of A and B and check whether equally good linear plots are obtained at other concentrations and also whether the K values derived from them remain constant within experimental error. Incorporation of thermodynamic nonideality into the Adams theory is not a mathematically difficult task, but the results, containing three more adjustable parameters, will be too complicated to be of practical use.

---

*The assumption used in the discussion of self-associating systems that $\bar{v} = \bar{v}_2 = \cdots = \bar{v}_m$ $= \bar{v}$ is equivalent to this Svedberg-Pedersen convention.

C. THE NICHOL-OGSTON THEORY FOR THE REACTION
$mA + nB \rightleftharpoons A_m B_n$

In essence, this is a formal extension of the Adams theory, so that its description here is limited to a minimum. The basic assumptions employed are that the solution is pseudoideal, incompressible, dilute, and all reacting species, A, B, C ($= A_m B_n$), have the same partial specific volume ($\bar{v}$) and the same specific refractive index increment. The last assumption is unrealistic, and if desired, it can be deleted.

The equilibrium condition for the reaction is given by

$$K c_A^m c_B^n = c_C \qquad (6.117)$$

where c_i ($i = A, B, C$) is the c-scale concentration of solute species i. Under the assumptions made the sedimentation equilibrium for the three species are as follows:

$$\frac{dc_A}{dr^2} = LM_A c_A \qquad (6.118)$$

$$\frac{dc_B}{dr^2} = LM_B c_B \qquad (6.119)$$

$$\frac{dc_C}{dr^2} = L(mM_A + nM_B)c_C \qquad (6.120)$$

where

$$L = \frac{(1 - \bar{v}\rho_0)\omega^2}{2RT} \qquad (6.121)$$

Addition of equations 6.118 through 6.120 gives

$$\frac{d\ln c}{dr^2} = LM_{wr} \qquad (6.122)$$

where

$$c = c_A + c_B + c_C \qquad (6.123)$$

and

$$M_{wr} = \frac{M_A c_A + M_B c_B + M_C c_C}{c} \qquad (6.124)$$

with

$$M_C = mM_A + nM_B \qquad (6.125)$$

As before, M_{wr} denotes the weight-average molecualr weight of the solute at a local position r, and c is the total solute concentration at the same position. From equations 6.117, 6.124, and 6.125 it immediately follows that M_{wr} is not a function of a single variable c but depends on both c_A and c_B. This is a distinctly different feature from the self-associating systems discussed in Sections II and III.

Elimination of c_C from equations 6.123 and 6.124 gives

$$c_A(M_A - M_C) + c_B(M_B - M_C) = c(M_{wr} - M_C) \qquad (6.126)$$

On the other hand, integration of equations 6.118 and 6.119 yields

$$c_A = c_{A1} \exp[\phi_A(r)] \qquad (6.127)$$

$$c_B = c_{B1} \exp[\phi_B(r)] \qquad (6.128)$$

where

$$\phi_A(r) = LM_A(r^2 - r_1^2), \qquad \phi_B(r) = LM_B(r^2 - r_1^2) \qquad (6.129)$$

and c_{A1} and c_{B1} are the concentrations of species A and B at the air-liquid meniscus. Introduction of equations 6.127 and 6.128 into equation 6.126, followed by rearrangement, leads to

$$c(r)(M_{wr} - M_C)\exp[-\phi_A(r)]$$
$$= c_{A1}(M_A - M_C) + c_{B1}(M_B - M_C)\exp[\phi_B(r) - \phi_A(r)] \qquad (6.130)$$

This relation gives a statement that plots of a quantity $Y(r)$ defined by

$$Y(r) = c(r)(M_{wr} - M_C)\exp[-\phi_A(r)] \qquad (6.131)$$

against $\exp[\phi_A(r) - \phi_B(r)]$ follow a straight line whose intercept I at $r = r_1$ and whose slope S are given by

$$I = c_{A1}(M_A - M_C) + c_{B1}(M_B - M_C) \qquad (6.132)$$

$$S = c_{B1}(M_B - M_C) \qquad (6.133)$$

Therefore, once I and S are determined from a Y plot, the two unknown quantities c_{A1} and c_{B1} may be evaluated by solving equations 6.132 and 6.133 simultaneously. Obviously, this procedure can be put to use only in circumstances in which the values of M_A, M_B, m, n, and all quantities contained in the factor L can be estimated in advance from separate experiments or from other sources of information. With c_{A1} and c_{B1} known, the values of c_A and c_B at any radial position in the cell can be computed from equations 6.127 and 6.128, and then the results, in conjunction with the experimental data for c, allow the evaluation of c_C at the same position. These values of the concentrations of the three species are inserted into equation 6.117 to calculate the desired K value. It is important to make similar calculations at various positions in order to check whether the resulting K values do not show any systematic variation with position.

In the general case, where m and n are unknown in advance, two more independent relations are required. The reader is referred to the paper by Nichol and Ogston[32] for a treatment of such a case.

IV. OTHER TOPICS

A. THE CONSERVATION OF MASS FOR REACTING SYSTEMS

One may inquire why the conservation-of-mass law, which has been so important in the theoretical formulation of sedimentation equilibrium phenomena of nonreacting solutes, does not appear in the treatment of chemically reacting systems. This problem has been considered by Adams.[35]

It is obvious that the total mass of all solute species in a chemically reacting system must be conserved whatever reactions may take place between the species, because the ultracentrifuge cell is a closed system. This condition is expressed by

$$\frac{c_0}{2}\left(r_2{}^2 - r_1{}^2\right) = \int_{r_1}^{r_2} c(r) r \, dr \qquad (6.134)$$

where the notation has the same meaning as defined previously. In nonreacting systems, this type of relation holds for each solute component too (see equation 5.6). But this is no longer the case for any reacting species, because the amount of it increases or decreases upon its conversion from or to other species coexisting in the solution. A mathematical proof of this almost self-evident fact is given below, with a monomer-dimer associating system taken as an example.

Let us assume that the conservation-of-mass law applies for each of the two species. Then

$$c_1{}^0 = \frac{2}{r_2{}^2 - r_1{}^2} \int_{r_1}^{r_2} c_1(r) r \, dr \qquad (6.135)$$

$$c_2{}^0 = \frac{2}{r_2{}^2 - r_1{}^2} \int_{r_1}^{r_2} c_2(r) r \, dr \qquad (6.136)$$

Under the assumption of pseudoideality and $\bar{v}_1 = \bar{v}_2 = \bar{v}$, we have from equations 6.14 and 6.15

$$\frac{dc_1}{dr^2} = AM_1 c_1 \tag{6.137}$$

$$\frac{dc_2}{dr^2} = 2AM_1 c_2 \tag{6.138}$$

where A is a constant defined by equation 6.18. If the solutions to these differential equations, after the integration constants have been determined by equations 6.135 and 6.136, are introduced into the equilibrium condition

$$c_2 = K_2 c_1^{\,2}$$

we obtain

$$2(e^x - 1) = x(e^x + 1) \tag{6.139}$$

where

$$x = AM_1(r_2^{\,2} - r_1^{\,2}) \tag{6.140}$$

Obviously, equation 6.139 does not hold except for zero of x, whereas x is an experimental variable. This implies that equations 6.135 and 6.136, from which equation 6.139 was derived, are in error.

B.   PRESSURE EFFECTS

Of the various assumptions that have been incorporated into the theoretical developments described above, the neglect of the pressure dependence of partial specific volumes and activity coefficients of individual solute species is probably good enough for usual low-speed sedimentation equilibrium experiments with a short solution column. In the treatments of chemically reacting systems, however, we must consider another pressure effect. It is concerned with equilibrium constants. This problem has been investigated by Kegeles et al.,[36] TenEyck and Kauzmann,[37] Josephs and Harrington,[38] and Howlett et al.[39,40]

Taking, for example, the case of self-association, the equilibrium constant for m-merization, K_m, is given by (see equation 6.5)

$$K_m = \exp\left[\frac{m(\hat{\mu}_1^{\,0})_c - (\hat{\mu}_m^{\,0})_c}{RT} \right] \tag{6.141}$$

Logarithmic differentiation with respect to pressure P gives

$$\left(\frac{\partial \ln K_m}{\partial P}\right)_T = \frac{mM_1(\bar{v}_1{}^0 - \bar{v}_m{}^0)}{RT} \tag{6.142}$$

where the thermodynamic relation $(\partial(\hat{\mu}_i{}^0)_c/\partial P)_T = iM_i\bar{v}_i{}^0$ has been considered. The superscript 0 refers to infinite dilution of the solution. From equation 6.142 we find that K_m is independent of pressure if $\bar{v}_1{}^0 = \bar{v}_m{}^0$. Under the neglect of the pressure dependence of activity coefficients this condition may be replaced by $\bar{v}_1 = \bar{v}_m$, which is just the assumption we have incorporated into the theories for self-associating systems. As has been noted in Section III.B, this equality of partial specific volumes is an example of the Svedberg-Pedersen convention, and, physically, it is equivalent to demanding that no volume change occurs upon the formation of an m-mer from m monomers.

We now integrate equation 6.142 from the air-liquid meniscus to a position r of a solution column in the ultracentrifuge cell. Then

$$K_m(P) = K_m(P_1)\exp\left[\frac{-\Delta V^0(P - P_1)}{RT}\right] \tag{6.143}$$

where P is the hydrostatic pressure at the position r, P_1 is its value at the air-liquid meniscus, and ΔV^0 is the change in molar volume accompanying the association of m monomers.* For very dilute solutions equation 5.5 gives $P - P_1 = \frac{1}{2}\rho_0\omega^2(r^2 - r_1^2)$, where ρ_0 is the solvent density. Thus, equation 6.143 may be replaced by

$$K_m(r) = K_m(r_1)\exp\left[\frac{-\Delta V^0\rho_0\omega^2(r^2 - r_1^2)}{2RT}\right] \tag{6.144}$$

Equation 6.144 indicates that if the volume change ΔV^0 is not zero, the equilibrium constant varies over the solution column, in a manner dependent on the rotor speed. For a fixed ΔV^0 the effect is more pronounced as the rotor speed is higher. As can be seen from the definition of ΔV^0, the magnitude of ΔV^0 may be larger for higher molecular weight solutes. Thus, in the situations encountered in sedimentation velocity studies of macromolecular reactions, both ΔV^0 and ω are so large that there may occur an enormous variation of the equilibrium constant with radial

---

*$\Delta V^0 = \bar{V}_m{}^0 - m\bar{V}_1{}^0 = mM_1(\bar{v}_m{}^0 - \bar{v}_1{}^0)$.

distance.[36,38] The effect will be appreciable also in the case of density-gradient sedimentation equilibrium experiments,[37] since the rotor speeds used are generally quite high. On the other hand, it may not be too much or even may be negligibly small for usual sedimentation equilibrium experiments, because not only are the rotor speeds relatively low but also a short solution column is employed (i.e., both ω and $r^2 - r_1^2$ are small). To examine this prediction let us take a numerical example: $m = 2$, $M_1 = 50,000$, $\bar{v}_2^0 - \bar{v}_1^0 = -0.015$ ml/g, $r_1 = 6.0$ cm, $r_2 = 6.3$ cm, $\omega = 10,000$ rev/min, $\rho_0 = 1.00$ g/ml, and $T = 298°K$. Then

$$\frac{K_2(r_2)}{K_2(r_1)} = \exp(4.08 \times 10^{-2}) = 1.04$$

which means that the effect is small, though not entirely negligible.

If a solution is run at two different rotor speeds, and the values of M_{wr}^{app} at points of equal concentration are found to differ, a volume change in the reaction may be suspected. Furthermore, in such a case, the data obtained from experiments of differing initial concentration will fail to merge, because points of equal concentration will correspond to different radial distances, and thus different pressures and different equilibrium constants. Exact correction for such effects will not be easy, since if there is indeed a volume change, the assumption based on the Svedberg-Pedersen convention becomes invalid. For efforts toward overcoming this difficulty, the reader is referred to a recent paper by Howlett et al.,[40] in which nonoverlapping of M_{wr}^{app} data for lysozyme in pH 8.0 diethylbarbiturate buffer of ionic strength 0.15 at 15°C was attributed to a negative volume change accompanying dimerization of the protein. However, as these authors also point out, the observed phenomenon may have been due to charge effects and specific ion binding effects.

C. CHARGE EFFECTS

So far in this chapter we have treated the solvent as a single thermodynamic component even though, as in actual experiments with proteins, it is a multicomponent system of low-molecular-weight substances. This approximation will break down for reacting solutes which have appreciable interactions with added low-molecular-weight substances, and the experimentally determined values of M_{wr}^{app} then have to be corrected for the interactions. Such corrections will become particularly important in solutions of highly charged macromolecular solutes, such as proteins far removed from their isoelectric points. In the usual experimental situations for studies of polyelectrolytes and proteins, where a sufficient amount of supporting electrolytes is added to the solution in order to repress electrostatic interactions between polyions, the problem is chiefly concerned with correcting observed data for the residual (or secondary) charge effect discussed in Chapter 5, Section III.A.2. This effect, which persists at infinite dilution of the macromolecular component, may be restated as an effect due to specific ion binding onto the polyion, and, according to the Casassa-Eisenberg theory (see Chapter 5, Section III.B and following), it could be minimized if data are taken with solutions which have been dialyzed against their solvent (containing supporting electrolytes). In fact, this predialysis procedure is routinely employed by current biophysical chemists. Finally, the reader is referred to a paper by Roark and Yphantis[41] for a more specific discussion of the charge effects upon sedimentation equilibrium of self-associating solutes.

APPENDIX A

Substitution of equations 6.21 and 6.61 into equation 6.59, followed by integration with respect to c between the limits of the solution column, yields

$$\frac{1}{A}\left[\left(\frac{dc}{dr^2}\right)_{r_2} - \left(\frac{dc}{dr^2}\right)_{r_1}\right]$$
$$= (M_1 + M_m)[(c)_{r_2} - (c)_{r_1}] - AM_1M_m\int_{r_1}^{r_2} c\, dr^2 \qquad \text{(A.1)}$$

By the conservation-of-mass law (see equation 6.134) the integral on the right-hand side of this equation is equated to $c_0(r_2^2 - r_1^2)$, where c_0 is the initial concentration of the solution. Hence, referring to equations 6.64 through 6.66, equation A.1 may be written

$$M_w^{app}M_z^{app} = (M_1 + M_m)M_w^{app} - M_1M_m \qquad \text{(A.2)}$$

which is equivalent to the desired Sophianopoulos-Van Holde equation 6.63.

APPENDIX B

Let us consider the case of monomer-dimer association. Then we have

$$M_{wr} = \frac{M_1(c_1 + 2K_2c_1^2)}{c} \qquad \text{(B.1)}$$

$$c = c_1 + K_2c_1^2 \qquad \text{(B.2)}$$

From these it follows that

$$c\frac{dM_{wr}}{dc} = M_1\frac{1+4K_2c_1}{1+2K_2c_1} - M_1\frac{1+2K_2c_1}{1+K_2c_1}$$

$$= \frac{M_1K_2c_1}{(1+2K_2c_1)(1+K_2c_1)} \geqslant 0 \qquad (B.3)$$

where the equality sign holds only for $K_2 = 0$. Thus, the M_{wr} for pseudoideal monomer-dimer associating systems are monotonically increasing functions of c. This conclusion can be verified for any pseudoideal self-associating system by a similar procedure, though the algebra becomes somewhat more complex.

REFERENCES

1. A. Tiselius, *Z. Phys. Chem.* (*Leipzig*) **124**, 449 (1926).
2. E. T. Adams, Jr., and H. Fujita, in *Ultracentrifugal Analysis in Theory and Experiment* (J. W. Williams, Ed.), Academic Press, New York, 1963, p. 119.
3. E. T. Adams, Jr., and J. W. Williams, *J. Am. Chem. Soc.* **86**, 3454 (1964).
4. E. T. Adams, Jr., *Biochemistry* **4**, 1646 (1965).
5. R. C. Denoier and J. W. Williams, *Biochemistry* **9**, 4260 (1970).
6. P. W. Chun, S. J. Kim, J. D. Williams, W. T. Cope, L.-H. Tang, and E. T. Adams, Jr., *Biopolymers* **11**, 197 (1972).
7. D. E. Roark and D. A. Yphantis, *Ann. N. Y. Acad. Sci.* **164**, 245 (1969).
8. L.-H. Tang, "Sedimentation Equilibrium Studies on β-Lactoglobulin A at Acid pH's, " Thesis, Illinois Institute of Technology, Chicago, 1971; L.-H. Tang and E. T. Adams, Jr., *Arch. Biochem. Biophys.* **157**, 520 (1973).
9. E. T. Adams, Jr., and D. L. Filmer, *Biochemistry* **5**, 2971 (1966).
10. J. Visser, R. C. Denoier, E. T. Adams, Jr., and J. W. Williams, *Biochemistry* **11**, 2634 (1972).
11. G. Frank and G. Braunitzer, *J. Physiol. Chem.* **348**, 1691 (1967).
12. D. K. Hancock and J. W. Williams, *Biochemistry* **8**, 2598 (1969).
13. F. E. LaBar, *Proc. Natl. Acad. Sci., U.S.* **54**, 31 (1965).
14. J. C. Nichol, *J. Biol. Chem.* **243**, 4065 (1968).
15. E. T. Adams, Jr., and M. S. Lewis, *Biochemistry* **7**, 1044 (1968).
16. P. D. Jeffrey and J. H. Coates, *Biochemistry* **5**, 489 (1966).
17. L.-H. Tang and E. T. Adams, Jr., *Fed. Proc., Fed. Am. Soc. Exp. Biol.* **30**, 1303 (1971).
18. A. J. Sophianopoulos and K. E. Van Holde, *J. Biol. Chem.* **239**, 2516 (1964).
19. R. Townend and S. N. Timasheff, *J. Am. Chem. Soc.* **82**, 3168 (1960); see also T. F. Kumosinski and S. N. Timasheff, *ibid.* **88**, 5635 (1966).
20. E. T. Adams, Jr., *Biochemistry* **6**, 1864 (1967).
21. D. A. Albright and J. W. Williams, *Biochemistry* **7**, 67 (1968).
22. R. F. Steiner, *Arch. Biochem. Biophys.* **39**, 333 (1952).

23. M. S. N. Rao and G. Kegeles, *J. Am. Chem. Soc.* **80**, 5724 (1958).

24. K. E. Van Holde, G. P. Rossetti, and R. D. Dyson, *Ann. N. Y. Acad. Sci.* **164**, 279 (1969).

25. K. E. Van Holde and G. P. Rossetti, *Biochemistry* **6**, 2189 (1967).

26. S. J. Gill, M. Downing, and G. F. Sheats, *Biochemistry* **6**, 272 (1967).

27. W. P. Reinhardt and P. G. Squire, *Biochim. Biophys. Acta* **94**, 566 (1965).

28. R. F. Steiner, *Arch. Biochem. Biophys.* **47**, 56 (1953).

29. R. F. Steiner, *Arch. Biochem. Biophys.* **49**, 71 (1954).

30. R. F. Steiner, *Arch. Biochem. Biophys.* **49**, 400 (1954).

31. E. T. Adams, Jr., in *Characterization of Macromolecular Structure* (D. M. McIntyre, Ed.), National Academy of Sciences, Washington, D. C., 1968, p. 84.

32. L. W. Nichol and A. G. Ogston, *J. Phys. Chem.* **69**, 4365 (1965).

33. P. W. Chun and S. J. Kim, *J. Phys. Chem.* **74**, 899 (1970).

34. T. Svedberg and K. O. Pedersen, *The Ultracentrifuge*, Oxford University Press, London and New York (Johnson Reprint Corporation, New York), 1940, p. 53.

35. E. T. Adams, Jr., A paper presented at a New York Academy of Sciences conference on Advances in Ultracentrifugal Analysis, New York, N. Y., February 1968.

36. G. Kegeles, L. Rhodes, and J. L. Bethune, *Proc. Natl. Acad. Sci., U. S.* **58**, 45 (1967).

37. L. F. TenEyck and W. Kauzmann, *Proc. Natl. Acad. Sci., U. S.* **58**, 888 (1967).

38. R. Josephs and W. F. Harrington, *Proc. Natl. Acad. Sci., U. S.* **58**, 1587 (1967).

39. G. J. Howlett, P. D. Jeffrey, and L. W. Nichol, *J. Phys. Chem.* **74**, 3607 (1970).

40. G. J. Howlett, P. D. Jeffrey, and L. W. Nichol, *J. Phys. Chem.* **76**, 777 (1972).

41. D. E. Roark and D. A. Yphantis, *Biochemistry* **10**, 3241 (1971).

APPROACH TO SEDIMENTATION EQUILIBRIUM

A. INTRODUCTION

Mathematically, it takes an infinitely long time for a solution in the ultracentrifuge to reach the state of sedimentation equilibrium. However, in practice, after a certain finite length of time the solution is brought to a state which is indistinguishable from the equilibrium state within the precision of the measurement. It is for this time that the experimental worker wants to have a theoretical prediction, because it should enable him to estimate the necessary duration of centrifugation for a successful sedimentation equilibrium experiment and also to select conditions which allow a quicker attainment of equilibrium. This chapter reviews some typical theories proposed so far for the increase in the rate of approach to sedimentation equilibrium.

B. THE THEORY OF WEAVER

As early as 1926, in a paper dealing with the settling of particles in a liquid under gravity, Weaver[1] calculated the time at which the transient terms in the Mason-Weaver solution[2] for this phenomenon become effectively zero. Since for the reason mentioned in Chapter 2, Section IV.B.1 the Mason-Weaver solution can be adapted to the sedimentation in the ultracentrifuge if the terms in it are properly redefined, Weaver's conclusion is also transcribable to the ultracentrifuge case. Thus, if the parameter α defined by equation 2.244 is much smaller than unity, the transient terms in equation 2.241 become negligible for values of time greater than

$$t_L = \frac{2(r_2 - r_1)}{s\omega^2 \bar{r}} \tag{7.1}$$

where $\bar{r}$ may be taken as the mean of r_1 and r_2. This expression indicates that t_L becomes smaller, that is, sedimentation equilibrium will be attained more quickly as the rotor speed becomes higher, a conclusion which one might find intuitively obvious.

The expression for α may be rewritten

$$\alpha = \frac{\epsilon'}{(r_2/r_1)^2 - 1} \tag{7.2}$$

where, as before, ϵ' denotes $2D/(s\omega^2 r_1^2)$. Hence the condition $\alpha \ll 1$, on which equation 7.1 is based, is actually equivalent to $\epsilon' \ll 1$. As has been mentioned frequently in preceding chapters, the latter condition is usually met in conventional sedimentation velocity experiments, but not in sedimentation equilibrium experiments except for the case in which the meniscus-depletion method is used. Thus for usual low-speed equilibrium runs Weaver's equation 7.1 may not be applied, but we must seek another criterion.

Elimination of s from equations 7.1 and 7.2 leads to

$$t_L = \frac{2\alpha(r_2 - r_1)^2}{D} \tag{7.3}$$

This may be put in the form

$$t_L = \frac{2(r_2 - r_1)^2}{D \ln[(c)_{r=r_2}/(c)_{r=r_1}]} \tag{7.4}$$

because it follows from the first term in equation 2.241 that the equilibrium concentrations at the ends of the solution column, $(c)_{r=r_1}$ and $(c)_{r=r_2}$, are related to α by

$$\frac{(c)_{r=r_2}}{(c)_{r=r_1}} = \exp\left(\frac{1}{\alpha}\right) \tag{7.5}$$

Equation 7.3 or 7.4, due originally to Svedberg and Pedersen,[3] indicates the inverse proportionality of t_L to D, which implies that sedimentation equilibrium will be attained more quickly by lower molecular weight solutes if compared at the same values of $(c)_{r=r_2}/(c)_{r=r_1}$ and $r_2 - r_1$. What is more important is the proportionality between t_L and $(r_2 - r_1)^2$. This predicts that the approach of a given system to sedimentation equilibrium will be markedly accelerated by reducing the length of solution column or the volume of solution placed in the cell cavity if the rotor speed is chosen so that the value of $(c)_{r=r_2}/(c)_{r=r_1}$ may not come close to unity upon the reduction of the solution column. Presumably based on this prediction, the early equilibrium studies at Uppsala[4] adjusted solution columns to about 0.5 cm in length, nearly one-half the values usually employed in sedimentation velocity experiments. However, we must note that Uppsala's procedure actually had no theoretical justification, because Svedberg and Pedersen's equation should have been inapplicable to conventional low-speed equilibrium runs.

C.   THE THEORY OF VAN HOLDE AND BALDWIN

In a very important paper published in 1958, Van Holde and Baldwin[5] derived for the first time a relation which permits prediction of the rate of approach to equilibrium under the conditions of low-speed sedimentation experiments. They defined a dimensionless parameter δ by

$$\delta = \frac{\Delta c_{eq} - \Delta c_t}{\Delta c_{eq}} \tag{7.6}$$

to measure the departure of a transient state from sedimentation equilibrium. Here Δc_t denotes the difference in concentration between the cell bottom and the meniscus at time t from the initiation of centrifugation, and Δc_{eq} is its value at equilibrium ($t = \infty$). If equation 2.241 is used to express Δc_t, and the result is substituted into equation 7.6 together with

$$\Delta c_{eq} = \left(\frac{1}{\alpha}\right) c_0 \tag{7.7*}$$

we obtain

$$\delta = 64\pi^2 \alpha^4 \exp\left(\frac{-t}{4\alpha p}\right)$$

$$\times \sum_{m=1}^{\infty} \frac{m^2 [1 - (-1)^m \cosh(1/2\alpha)] \exp[-(\alpha m^2 \pi^2 t/p)]}{(1 + 4\pi^2 \alpha^2 m^2)^2} \tag{7.8}$$

where

$$p = \frac{2(r_2 - r_1)}{s\omega^2(r_1 + r_2)} \tag{7.9}$$

It should be noted that p is just one-half the time t_L introduced by Weaver.

The series 7.8 converges rapidly for large values of α and t. Even for α as small as 0.1 and for $(t/p) = 1$, the ratio of the second to the first term is about 0.9×10^{-2}. Thus for large values of t and for α not too small it is sufficient if one retains only the first term. Then

$$\delta = \frac{4 \exp[-\pi^2 t U(\alpha)(\alpha/p)]}{\pi^2 [U(\alpha)]^2} \left[1 + \cosh\left(\frac{1}{2\alpha}\right)\right] \tag{7.10}$$

where

$$U(\alpha) = 1 + \frac{1}{4\pi^2 \alpha^2} \tag{7.11}$$

*This relation readily follows from equation 2.241 with $t = \infty$.

Equation 7.10 may be rearranged to yield the value of t required to reduce the departure from equilibrium to a given value of δ. Designating this time by t_δ, we obtain the Van Holde-Baldwin equation:

$$t_\delta = \frac{(r_2 - r_1)^2}{D} F(\alpha) \qquad (7.12)$$

where

$$F(\alpha) = -\frac{1}{\pi^2 U(\alpha)} \ln\left(\frac{\pi^2 [U(\alpha)]^2 \delta}{4[1 + \cosh(1/2\alpha)]} \right) \qquad (7.13)$$

The graph of $F(\alpha)$ for $\delta = 1 \times 10^{-3}$ is shown in Fig. 7.1. It is seen that for α larger than 0.6 the values of $F(\alpha)$ vary very slowly with α. This feature is maintained for other values of δ, say 0.01. As can be deduced from equation 7.5, low-speed equilibrium experiments are featured by values of α which are about unity. Therefore, for this kind of experiment the following expression is sufficient to estimate t_δ:

$$t_\delta = \frac{C_\delta (r_2 - r_1)^2}{D} \qquad (7.14)$$

Here C_δ is a constant dependent on the chosen value of δ. For example,

$$C_{0.01} = 0.47, \qquad C_{0.001} = 0.67 \qquad (7.15)$$

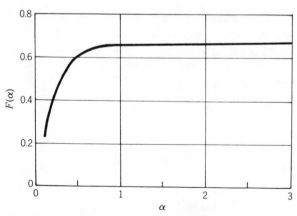

Fig. 7.1. Variation of the function $F(\alpha)$ with α for the case when $\delta = 0.001$. Note that $F(\alpha)$ is essentially constant for α above 0.8.

Equation 7.14 leads to the important conclusion that as long as we choose the experimental conditions such that α becomes greater than about 0.6, the rate of approach to sedimentation equilibrium can be increased by reducing the length of solution column. Since the square of $r_2 - r_1$ appears in the equation, the rate is increased nine times if the solution column is reduced to one-third. Thus for a quick attainment of equilibrium we should use as short a solution column as possible, although, in actuality, we will have to compromise at some stage, because the reduction of the solution column is attended by an increased loss of precision in the optical measurements of concentration or concentration gradient distributions. At present, it is the usual practice to adjust the solution column to 2 to 3 mm. Then, except for very high-molecular-weight solutes for which the diffusion coefficient D is quite small, the sedimentation equilibrium state is obtained within one to three days, and sometimes within hours. For example, in experiments with ribonuclease in 3-mm solution columns, Van Holde and Baldwin[5] found equilibrium, within experimental error, after 14 hours and obtained excellent values for the molecular weight of that protein.

For actual use of equation 7.14, we must know the value of D for a given system. This is often estimated by inserting a separately determined s value and a suitably inferred molecular weight into the Svedberg relation. Accurate data are not always needed for these quantities. In practice, some safety factor (1.2–1.5) is multiplied by the t_δ value computed with these data in determining the total length of time for a run to be performed. In this case, the criterion $\delta = 0.001$ would be too stringent, and $\delta = 0.01$ would be sufficient. It is also necessary to check in advance whether the chosen values of ω and $r_2 - r_1$ satisfy the condition $\alpha > 0.6$ required for equation 7.14 to be valid.

Equation 7.12 indicates that t_δ also may be reduced by decreasing $F(\alpha)$. Figure 7.1 shows that we must reduce α below 0.6 before $F(\alpha)$ decreases appreciably with decreasing α. Since α is inversely proportional to ω^2 (see equation 2.244), the desired reduction of t should be obtained by working at very high rotor speeds. Van Holde and Baldwin[5] have shown that, in the range of α near zero, t_δ is approximately represented by

$$t_\delta = \frac{4(r_2 - r_1)}{s\omega^2(r_1 + r_2)} \tag{7.16}$$

which is equivalent to Weaver's equation 7.1 for t_L. This agreement confirms the statement made in the preceding section that Weaver's equation can be of effective use for experiments at very high speeds.

D. THE THEORY OF HEXNER ET AL. FOR THE OVERSPEEDING TECHNIQUE

Although the use of short columns is very effective for reducing the centrifuging time required to reach sedimentation equilibrium, it suffers from a loss of the accuracy of measurement. In view of this fact, Hexner et al.[6] proposed another method, now often termed the *overspeeding technique*.[7] This method consists of centrifuging given solution at an angular speed ω' for a certain interval of time, t', until the concentration distribution in the cell closely approximates the equilibrium distribution which would be obtained if the rotor were spun at a lower speed ω. Then the rotor speed is decreased from ω' to ω, and the centrifugation is continued until equilibrium is reached.

In order to estimate ω' and t', Hexner et al. used, as in the Van Holde-Baldwin theory, the rectangular-cell approximation to the Lamm equation and, after solving the equation so as to satisfy the conditions required by the above-mentioned choice of rotor speed, arrived at the conclusion: Equilibrium will be reached more quickly than in the case when the rotor is spun at a constant speed ω from the beginning if ω' and t' are selected so that the following relation is satisfied:

$$\frac{(1-2\alpha Q)(1+e^{Q})}{(\alpha' e^{1/\alpha'}-1)(Q^2+\pi^2)}$$

$$= -8\alpha'^2 \sum_{m=1}^{\infty} F_m \frac{me^{-P'_m t''}[1-(-1)^m e^{-1/2\alpha'}][1+(-1)^m e^R]}{(1+4\alpha'^2 m^2\pi^2)^2} \tag{7.17}$$

where

$$F_m = \frac{2\pi^2(1+m)(\alpha m+\alpha')+R(1-4\pi^2 m\alpha\alpha')}{R^2+(m+1)^2\pi^2}$$

$$-\frac{2\pi^2(m-1)(\alpha m-\alpha')+R(1+4\pi^2 m\alpha\alpha')}{R^2+(m-1)^2\pi^2} \tag{7.18}$$

$$Q=(\alpha')^{-1}-(2\alpha)^{-1} \tag{7.19}$$

$$R=(2\alpha')^{-1}-(2\alpha)^{-1} \tag{7.20}$$

$$P'_m=\alpha' m^2\pi^2+(4\alpha')^{-1} \tag{7.21}$$

$$t''=\frac{t'}{\beta'} \tag{7.22}$$

with

$$\alpha = \frac{D}{s\omega^2\bar{r}(r_2 - r_1)}\ , \qquad \alpha' = \frac{D}{s(\omega')^2\bar{r}(r_2 - r_1)} \qquad (7.23)$$

$$\beta' = \frac{\alpha'(r_2 - r_1)^2}{D}$$

Equation 7.17 may be solved numerically for t'' as a function of α' for different fixed values of α. To this end, Hexner et al. used only the first term in the infinite series, with the result reproduced in the graph in Fig. 7.2. The values of α chosen here roughly correspond to the conditions for low-speed equilibrium experiments. In a recent paper, Teller et al.,[8] taking more terms in the infinite series, made a thorough theoretical investigation of the overspeeding method relevant to the conditions for high-speed experiments of the meniscus-depletion type, but we do not enter into it here.

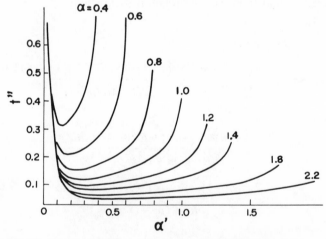

Fig. 7.2.   Relations between t'' and α' for various fixed values of α.

In looking at the family of curves shown in Fig. 7.2, we should note that smaller values of α and α' correspond to larger values of ω and ω', respectively, if the length of solution column is fixed. It is also to be noted that we have the relation

$$\alpha' = \left(\frac{\omega}{\omega'}\right)^2 \alpha \qquad (7.24)$$

The theory described above may be put into practical use in the following way. First, we choose ω and $r_2 - r_1$ suitably so that a value of about unity may be obtained for α. For this purpose it is necessary that approximate values of s and D or of molecular weight M be known or they may be inferred from appropriate sources; with an M known the ratio s/D may be calculated by use of the Svedberg relation. For the assigned value of α we select a pair of α' and t'' from Fig. 7.2. There exists a broad range of choices of these variables. Hexner et al. proposed the selection of a smallest experimentally convenient value for α' and then the corresponding t''. However, this proposal has no theoretical bases. It is desirable to work out a theory by which the selection of a best pair of α' and t'' can be made. From the chosen α', together with the assigned α and ω, we calculate ω' by equation 7.24. Also, substitution of the chosen α' and the known values of D and $r_2 - r_1$ into equation 7.23 allows the determination of β'. Then the value of t' can be obtained from equation 7.22. In this way, all data required for an overspeeding run become available.

The effectiveness of the overspeed method was demonstrated by Hexner et al.[6] with experiments on typical proteins. Their results are summarized below.

Substance	Ribonuclease	Insulin	Lysozyme
$r_2 - r_1$ (mm)	3	8	3
ω' (rev/sec)	350	320	250
t' (hr)	1.1	15.2	4.1
ω (rev/sec)	220	240	216
t (hr)	2.2	16	4.8
t_e (hr)	14	76	17
M_w	13,690	5800	14,750

Here t is the total time for the experiment, and t_e is the time normally required for an experiment at a constant rotor speed. It is seen that by the overspeeding technique we can acquire a several-fold reduction in the time required to reach equilibrium. We may question, however, how long the ultracentrifuge should be run after the prescribed period of overspeeding run. This problem was investigated in detail by Teller et al.[8] for the case of high-speed experiments, but the corresponding study for low-speed experiments is not as yet available in the literature. Thus there remains much to be explored in the theory of the overspeeding method, although, at present, this method is in routine use for accelerating the attainment of sedimentation equilibrium. Probably, if the short column technique of Van

Holde and Baldwin and the overspeeding technique of Hexner et al. had not been invented, the sedimentation equilibrium experiment would not have established its present-day status as an almost standard and routine means for molecular weight determination of macromolecular substances, especially of proteins and other macromolecules of biological interest.

E. APPLICATION OF SYNTHETIC BOUNDARY CELLS

The theories described above are concerned with experiments which use conventional cells, that is, which are initiated with a solution of uniform concentration. Pasternak et al.[9] have shown that the rate of approach to sedimentation equilibrium can be increased if the experiment is initiated with a suitable single-step concentration distribution set up in the synthetic boundary cell. The underlying idea is that such a distribution is closer to that at sedimentation equilibrium than is the uniform distribution.

If the rectangular-cell approximation is used, the Lamm equation gives as a general solution satisfying the boundary conditions at the cell ends

$$c(y,\tau) = A_0 \exp\left(\frac{y}{\alpha}\right) + \exp\left[\left(\frac{y}{2\alpha}\right) - \left(\frac{\gamma\tau}{4\alpha}\right)\right]$$

$$\times \sum_{m=1}^{\infty} A_m \left[\cos(m\pi y) + (2\alpha m\pi)^{-1}\sin(m\pi y)\right]\exp(-m^2\pi^2\alpha\gamma\tau) \quad (7.25)$$

where $A_0, A_1, \ldots$ are unknown constants to be determined from initial conditions. The meaning of all other symbols is the same as defined in Chapter 2, Section IV.B.1. Because the time exponential in equation 7.25 involves m^2, the transient term represented by the infinite series will decay rapidly if it is possible to choose initial conditions so that A_1 vanishes. Pasternak et al. showed that for the initial condition represented by

$$c(y,0) = \begin{cases} c_1 & (0 < y < y_0) \\ c_2 & (y_0 < y < 1) \end{cases} \quad (7.26)$$

the constant A_1 can be made to vanish if c_2/c_1 and r_0 are chosen to satisfy the relation

$$\frac{c_2}{c_1} = \frac{\cos(\pi y_0) - [(4\pi^2\alpha^2 - 1)/(4\pi\alpha)]\sin(\pi y_0) - \exp[y/(2\alpha)]}{\cos(\pi y_0) - [(4\pi^2\alpha^2 - 1)/(4\pi\alpha)]\sin(\pi y_0) + \exp[(y_0 - 1)/(2\alpha)]} \quad (7.27)$$

Here y_0 stands for

$$y_0 = \frac{r_0 - r_1}{r_2 - r_1} \quad (7.28)$$

It is possible to calculate from equation 7.27 curves of c_2/c_1 versus y_0 for various values of α. Figure 7.3 illustrates one for $\alpha = 1/1.2$. The curve indicates that there exists a wide range of choice for the location of the initial boundary r_0. The pure solvent ($c_1 = 0$) may be used as the medium for the region $r_1 < r < r_0$ if the initial boundary is placed at the position corresponding to $y_0 \approx 0.2$. Also it is seen that in the vicinity of $y_0 = 0.55$ the curve has a broad minimum, and the system is insensitive to errors which may occur while filling the cell.

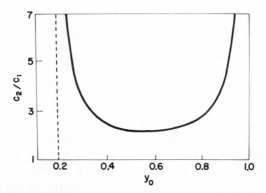

Fig. 7.3. The initial boundary $y_0 = (r_0 - r_1)/(r_2 - r_1)$ and the corresponding relative concentration c_2/c_1 which satisfy equation 7.27 for the case when $\alpha = 1/1.2$.

Pasternak et al. computed from equation 7.25 (after all the A_k have been determined in accordance with the initial condition 7.26) the time required to reach the state in which the concentration distribution does not deviate more than 1% from that at equilibrium. The reduced time, τ^*, corresponding to this time is given as a function of α, y_0, and c_2/c_1. In Fig. 7.4 is illustrated the variation of τ^* with $1/(2\alpha)$ for a special case in which $y_0 = 0.5$ and $c_2/c_1 = 2.17$. The curve has a sharp minimum at $1/(2\alpha) = 0.6$ or $\alpha = 1/1.2$. This fact implies that for a single-step initial concentration distribution with $y_0 = 0.5$ and $c_2/c_1 = 2.17$ there would be obtained a maximum reduction in time if the rotor speed and the length of solution column were chosen so as to yield $\alpha = 1/1.2$.

For comparison, Pasternak et al.[9] computed τ^* for the experiment initiated with a uniform concentration distribution. Its ratio to τ^* for a single-step initial concentration distribution is plotted against $1/(2\alpha)$ by a dashed line in Fig. 7.4 for the case where $y_0 = 0.5$ and $c_2/c_1 = 2.17$. It is observed that there is an eight fold reduction in time compared with the

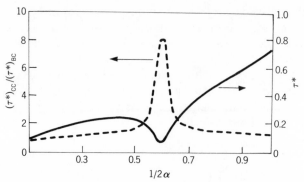

Fig. 7.4. Solid line, dependence of τ^* on α when the experiment is initiated with $y_0 = 0.5$ and $c_2/c_1 = 2.17$. Dashed line, ratio of times required to reach sedimentation equilibrium in two cells, conventional cell (CC) and synthetic boundary cell (BC). In BC, $y_0 = 0.5$ and $c_2/c_1 = 2.17$.

case of uniform initial concentration if α is chosen exactly equal to $1/1.2$. However, the graph shows that if the assumed value of α differs from $1/1.2$ by 15%, one acquires only a twofold reduction. Unfortunately, it is frequently impossible to select the rotor speed and the length of solution column so that the corresponding α satisfies equation 7.27 for a given initial concentration distribution, because either the values of s and D or that of molecular weight M of the substance under test may not be given exactly prior to the experiment. Thus it must be concluded that the method of Pasternak et al. is of less actual value than the method described in the preceding sections.

Charlwood[10] extended the theory of Van Holde and Baldwin[5] to the experiment in which the initial condition is given by equation 7.26 with $c_1 = 0$. Figure 7.5 shows his calculated results for $\delta = 0.005$ in the form of $F(\alpha, y_0)/y_0$ plotted against α, with y_0 as a parameter. Here $F(\alpha, y_0)$ is a dimensionless quantity defined by the relation $t_\delta = (r_2 - r_1)^2 F(\alpha, y_0)/D$, where, as before, t_δ denotes the time required to reduce the departure from sedimentation equilibrium to a fraction δ defined by equation 7.6. Thus, $F(\alpha, y_0)$ is equivalent to the function $F(\alpha)$ in the Van Holde-Baldwin theory. It is seen from Fig. 7.5 that the curves for $y_0 = 0.2$ and 0.3 exhibit minima. In particular, the minimum for $y_0 = 0.2$ appears in the range of α that is normally of interest from the experimental point of view; that is, 0.6–1.5. However, these minima are so sharp that a precise predetermination of α becomes imperative for actual utilization of the theoretical features displayed in the graph. Unfortunately, as has been noted above in relation to the theory of Pasternak et al., this requirement is not easily met in practice. Excepting the narrow regions near the minima, the curves in

Fig. 7.5 yield values of $F(\alpha, y_0)$ ranging from 0.4 to 0.6, regardless of α and y_0. These values do not differ greatly from those of $F(\alpha)$ for $\delta = 0.005$ in the Van Holde-Baldwin theory. Accordingly, unless α can be predetermined accurately, the use of a synthetic boundary cell is not very advantageous for the increase in the rate of approach to sedimentation equilibrium.

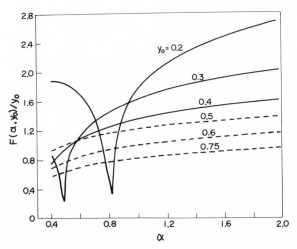

Fig. 7.5. Relations between $F(\alpha, y_0)$ for $\delta = 0.005$ and α. y_0 is the relative position of the initial sharp boundary between solution and solvent, with $y_0 = 0$ and $y_0 = 1$ corresponding to the air-liquid meniscus and the cell bottom, respectively.

1. Multiple-Step Concentration Distribution

If we were able to set up an initial concentration distribution which more closely approximates the distribution to be obtained at equilibrium, there should be available a further reduction in the time required to reach equilibrium.This idea has been used by Griffith,[11] who designed a special type of synthetic boundary cell which accommodates a four-step concentration distribution.

First, we assume an appropriate value M' for the molecular weight M of the substance to be examined. The success of Griffith's method primarily depends on how closely we can choose M' to M. Then we compute a parameter σ, called "effective reduced molecular weight" by Yphantis,[12] from

$$\sigma = \frac{(1 - \bar{v}\rho_0)\omega^2 M'}{RT} \tag{7.29}$$

where the meaning of the symbols is the same as defined in the preceding chapters. We also calculate the quantity $\Delta(r^2)$ defined by

$$\Delta(r^2) = r_2^2 - r_1^2 \tag{7.30}$$

By successive dilution of a given test solution we prepare four solutions whose concentrations c_I, c_{II}, c_{III}, and c_{IV} satisfy the condition

$$c_I : c_{II} : c_{III} : c_{IV} = k : k^3 : k^5 : k^7 \tag{7.31}$$

where

$$k = \exp\left[\frac{\sigma}{2} \frac{\Delta(r^2)}{8} \right] \tag{7.32}$$

Equal volumes of these solutions are placed in Griffith's special cell in such a way that the concentration distribution, as schematically shown in Fig. 7.6, may be set up. Then centrifugation is commenced at the given rotor speed ω.

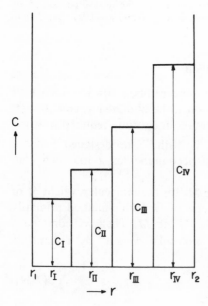

Fig. 7.6. An example of the four-step concentration distribution to be set up at the start of an experiment with Griffith's cell.

If the solution is binary and if nonideality effects need not be considered, the equilibrium concentration is given by equation 5.40. Computations of the concentrations at the midpoints of four equal volumes, r_I, r_{II}, r_{III}, and r_{IV} in Fig. 7.6, yield quite accurately the same relation as equation 7.31. Thus a four-step concentration distribution made up in accordance with equation 7.31 should represent a close approximation to the final equilibrium distribution if the assumed M' were fairly correct (and the system were less heterogeneous and less nonideal).

When this method was applied to ribonuclease A in 0.025 M H_3PO_4 buffer, pH 7.12, Griffith[11] found that equilibrium attained only 48 min after the operational speed of rotation had been reached. It is interesting to compare this value with 14 hr in the Van Holde-Baldwin work for the same protein and with 2.2 hr reported by Hexner et al. who applied the overspeeding technique for this protein.

F. MEASUREMENT OF THE DIFFUSION COEFFICIENT FROM THE RATE OF APPROACH TO EQUILIBRIUM

Equation 7.10 may be rewritten in the form

$$\ln \delta = \ln \left\{ \frac{4[1 + \cosh(1/2\alpha)]}{\pi^2 [U(\alpha)]^2} \right\} - \frac{D\pi^2 U(\alpha)}{(r_2 - r_1)^2} t \qquad (7.33)$$

This equation should be asymptotically correct (i.e., more valid as t is increased more), because, in its derivation, the second and higher terms in the series 7.8 were neglected. From Rayleigh fringe patterns at various transient times and also at equilibrium we can evaluate $\delta = (\Delta c_{eq} - \Delta c_t)/\Delta c_{eq}$ as a function of t. According to equation 7.33, there should be obtained a straight line at large values of t when these data are plotted in the form of $\ln \delta$ versus t, and the slope of the line should be equal to $- D\pi^2 U(\alpha)/(r_2 - r_1)^2$. Since α, hence $U(\alpha)$, can be evaluated by inserting the observed value of Δc_{eq} into equation 7.7, we here find a means of measuring the diffusion coefficient D. As an example, the plot of $\ln \delta$ versus t for a bovine ribonuclease obtained by Van Holde and Baldwin[5] is shown in Fig. 7.7. It is seen that the plotted points follow a straight line almost from the beginning of the experiment. Van Holde and Baldwin[5] found for sucrose and ribonuclease that the values of D determined by this procedure agreed with results from the literature, on the average, within 1.5%. They have developed a similar treatment for multicomponent solutions in which all solute interactions are negligible.[5]

Based on his series solution to the Lamm equation described in Chapter 2, Section IV.B.2, Nazarian[13] has shown that a semilogarithmic plot of

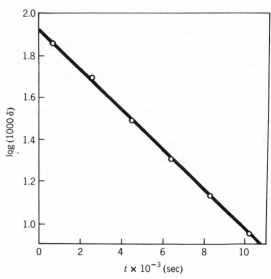

Fig. 7.7. Plots of ln δ versus t from an experiment on a sample of bovine ribonuclease.[5]

$-dc/dt$ against t for any given position in the vicintiy of $r = \frac{1}{4}(3r_1 + r_2)$ $= r_1 + \frac{1}{4}(r_2 - r_1)$ should tend quite rapidly toward a straight line. The slope of the straight line is equal to

$$-\frac{D\pi^2 U(\alpha)}{(r_2 - r_1)^2}$$

provided terms of order $(r_2 - r_1)^2 / (r_1 + r_2)^2$ may be ignored in comparison to unity. As before, the constant α may be determined from the observed value of Δc_{eq}. Thus we have another means of determining D. Incidentally, the expression given above for the slope is identical to the coefficient for t in equation 7.33, which implies that both δ and $-dc/dt$ at a position near $r = \frac{1}{4}(3r_1 + r_2)$ tend to zero at the same rate for large values of t. Nazarian's approach, though interesting from the theoretical point of view, has a practical disadvantage, because it resorts to data of dc/dt which, in actuality, may not be obtained with precision from measured concentration distributions. This disadvantage, however, can be circumvented, since it readily follows from Nazarian's formalism that a plot of $\ln(c_{eq} - c_t)$ against t at any position near $r = \frac{1}{4}(3r_1 + r_2)$ also approaches a straight line having the same slope as does a semilogarithmic plot of $-dc/dt$ against t at the same position. Here c_t is the concentration at time t, and c_{eq} its value at sedimentation equilibrium. Since c_t and c_{eq} can be determined directly from primary fringe data, it may be expected that the new plot will allow a more accurate determination of D.

G. EFFECTS OF CONCENTRATION DEPENDENCE OF s AND D ON THE APPROACH TO EQUILIBRIUM

The Van Holde-Baldwin theory described in Chapter 5, Section I.C deals with the case in which both s and D are constant. However, since these coefficients for macromolecular solutes are generally dependent on solute concentration c, it is of both theoretical and practical significance to investigate how the approach to equilibrium is affected by their concentration dependence. This problem was first treated by Weiss and Yphantis[14] by using their analytic solution to the Lamm equation in which D is held constant but s is allowed to vary with c in the following manner:

$$s = s_0(1 - k_s c) \tag{7.34}$$

where s_0 is the value of s at infinite dilution, and k_s is a positive constant. Recently, Dishon et al.[15] made a more detailed study of the approach to equilibrium by solving the Lamm equation numerically under conditions corresponding to equilibrium runs for the following three cases:

CASE 1. s is given by equation 7.34 and D is constant.

CASE 2. s has a concentration dependence of the form

$$s = \frac{s_0}{1 + k_s c} \tag{7.35}$$

and D is constant.

CASE 3. s is held constant and D is given by

$$D = D_0(1 + k_D c) \tag{7.36}$$

The values of the parameters chosen for the calculations were as follows: $r_1 = 6.4$ cm, $r_2 = 6.7$ cm; $\sigma_0 \equiv \omega^2 s_0 / D_0 = 0.509$ and 5.09 cm^{-2}. The former value of σ_0 corresponds to conventional low-speed equilibrium runs, and the latter to high-speed runs which are used for the meniscus-depletion method.

Dishon et al. considered two parameters δ and δ' for the description of the approach to sedimentation equilibrium. The δ is the same as that of Van Holde and Baldwin,[5] that is,

$$\delta = \frac{\Delta c_{eq} - \Delta c_t}{\Delta c_{eq}} \tag{7.37}$$

and the δ' is defined by

$$\delta' = \frac{\Delta c_{eq} - \Delta c_t}{c_0} \tag{7.38}$$

Apparently, δ and δ' are proportional. Dishon et al. state that δ is most useful for estimating the time required so that an experimentally determined weight-average molecular weight (which is proportional to Δc_t) be within a fraction δ of its equilibrium value, while δ' is most useful in determining the time required so that the directly observed quantity Δc_t be within a given experimental error δ' of its final value.

In all the cases examined, both δ and δ' for large values of a reduced time $\tau = 2s_0\omega^2 t$ had the limiting forms

$$\delta = Ae^{-\Lambda\tau}, \qquad \delta' = Be^{-\Lambda\tau} \tag{7.39}$$

where A, B, and Λ are independent of τ. Of theoretical importance is how the factor Λ is affected by the concentration dependence of s and D. The following is a summary of the answers to this problem from the numerical work by Dishon et al.

CASE 1. Weiss and Yphantis[14] showed, based on their analytic solution to the Lamm equation, that Λ is not affected by the magnitude of $k_s c_0$, where c_0 is the initial concentration of the solution. Thus

$$\Lambda = \frac{\pi^2(r_1 + r_2)\alpha_0}{4(r_2 - r_1)}\left(1 + \frac{1}{4\pi^2\alpha_0^2}\right) \tag{7.40}$$

where α_0 is defined by

$$\alpha_0 = \frac{2D}{s_0\omega^2(r_2^2 - r_1^2)} \tag{7.41}$$

Therefore, the diffusion coefficient D can be evaluated from the slope of a $\ln\delta$ versus t plot in exactly the same manner as explained in Section F, regardless of the magnitude of $k_s c_0$. This theoretical consequence has been confirmed by Dishon et al. in terms of their numerical solutions. Both A and B depended on $k_s c_0$.

CASE 2. For $\sigma_0 = 0.509$ cm^{-2} the values of Λ for different $k_s c_0$ are very nearly the same as that for $k_s = 0$. This implies that D may be evaluated the same way as in Section F, regardless of the magnitude of $k_s c_0$. For $\sigma_0 = 5.09$ cm^{-2} the values of Λ differ for different $k_s c_0$, being smaller for smaller $k_s c_0$ and tending to a constant value for larger $k_s c_0$. Dishon et al. predicted that if σ_0 is of the order of unity as encountered in conventional low-speed equilibrium experiments, the factor Λ is essentially independent of the form and degree of the dependence of s on c.

From the numerical data for A, B, and Λ, Dishon et al. computed the times required so that δ and δ' may reach 0.001. Figure 7.8a shows these times, relative to those for $k_s c_0 = 0$, as functions of $s(c_0)/s_0$ for the two types of $s(c)$ and a value of $\sigma_0 = 0.509$ cm^{-2}. Figure 7.8b shows the corresponding functions for $\sigma_0 = 5.09$ cm^{-2}. The upper panel indicates that for this low σ_0 both forms of $s(c)$ give equilibrium times which are identical and substantially independent of the degree of c-dependence when $\delta = 0.001$ is utilized as a criterion for the attainment of equilibrium, while the choice of $\delta' = 0.001$ as such a criterion leads to equilibrium times which, though slightly different for the two forms of $s(c)$, decrease appreciably with increasing degree of c-dependence. The behavior for $\sigma_0 = 5.09$ cm^{-2} is striking: Both the δ and δ' criteria predict large increases in equilibrium times for the nonlinear c-dependence of s $[s = s_0/(1 + k_s c)]$, while the linear c-dependence of s $[s = s_0(1 - k_s c)]$ leads to significant decreases. Examination of the effects of small c-dependence reveals, however, that the two forms of $s(c)$ yield curves which are tangent with each other at the limit of vanishing c-dependence. With increasing nonlinear c-dependence the equilibrium times for both $\delta = 0.001$ and $\delta' = 0.001$ exhibit a shallow minimum and then increase.

CASE 3. For both $\sigma_0 = 0.509$ and 5.09 cm^{-2} the values of Λ increase with increasing $k_D c_0$, and the times required to reach $\delta = \delta' = 0.001$ decrease monotonically as $k_D c_0$ increases. Thus the attainment of equilibrium is accelerated as there is more marked increase of D with c.

H. OTHER TOPICS

The magnetically suspended ultracentrifuge designed by Beams et al.[16] undergoes rotor deceleration due to pressure and eddy current effects, although its degree is slight, usually of the order of 1%/day or less. Billick et al.[17–19] have attempted a theoretical analysis of the effects of nonconstant rotor speed on sedimentation equilibrium experiments, hoping that it can serve as a check on Beams' assertation that rotor slowing might be used to speed up the attainment of equilibrium.[16] Their mathematical development is too complex to be reproduced here, but the results clearly

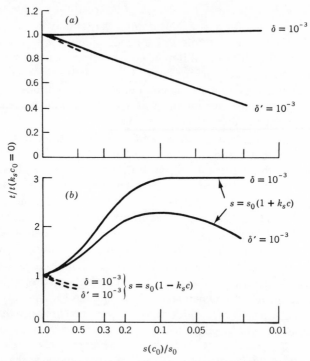

Fig. 7.8. Effects of concentration dependence of s on the time required to attain sedimentation equilibrium, when $\delta = \delta' = 0.001$. Solid line, for $s = s_0/(1 + k_s c)$; dashed line, for $s = s_0(1 - k_s c)$. (a) $\sigma_0 \equiv \omega^2 s_0^2 / D = 0.509$ cm^{-2}, typical of conventional low-speed experiments; (b) $\sigma_0 = 5.09$ cm^{-2}, typical of high-speed experiments.

demonstrate that a considerable saving in time over the current techniques (using constant rotor speed) is obtained by introducing rotor deceleration.

Klenin et al.[20] considered the case in which the rotor is run at constant speed till the experiment comes to equilibrium, followed by instantaneous change to a different but constant rotor speed. Equations were derived which permit the computation of the time for equilibrium to be reestablished, and the calculated values were found to be in good agreement with data for an oligostyrene in a theta solvent.

REFERENCES

1.   W. Weaver, *Phys. Rev.* **27**, 499 (1926).
2.   M. Mason and W. Weaver, *Phys. Rev.* **23**, 412 (1924).

3. T. Svedberg and K. O. Pedersen, *The Ultracentrifuge*, Oxford University Press, London and New York (Johnson Reprint Corporation, New York), 1940, p. 57.

4. See Ref. 3, p. 305.

5. K. E. Van Holde and R. L. Baldwin, *J. Phys. Chem.* **62**, 734 (1958).

6. P. E. Hexner, L. E. Radford, and J. W. Beams, *Proc. Natl. Acad. Sci., U.S.* **47**, 1848 (1961).

7. E. G. Richards, D. C. Teller, and H. K. Schachman, *Biochemistry* **7**, 1054 (1968).

8. D. C. Teller, T. A. Horbett, E. G. Richards, and H. K. Schachman, *Ann. N.Y. Acad. Sci.* **164**, 166 (1969).

9. R. A. Pasternak, G. M. Nazarian, and J. R. Vinograd, *Nature* **179**, 92 (1957).

10. P. A. Charlwood, *Biopolymers* **5**, 663 (1967).

11. O. M. Griffith, *Anal. Biochem.* **19**, 243 (1967).

12. D. A. Yphantis, *Biochemistry* **3**, 297 (1964).

13. G. M. Nazarian, *J. Phys. Chem.* **62**, 1607 (1958).

14. G. H. Weiss and D. A. Yphantis, *J. Chem. Phys.* **42**, 2117 (1965).

15. M. Dishon, G. H. Weiss, and D. A. Yphantis, *Biopolymers* **4**, 457 (1966).

16. J. W. Beams, D. M. Spitzer, and J. P. Wade, *Rev. Sci. Instrum.* **33**, 151 (1962).

17. I. H. Billick, M. Dishon, M. Schulz, G. H. Weiss, and D. A. Yphantis, *Proc. Natl. Acad. Sci., U.S.* **56**, 399 (1966).

18. I. H. Billick, M. Schulz, and G. H. Weiss, *J. Phys. Chem.* **71**, 2496 (1967).

19. I. H. Billick, M. Dishon, G. H. Weiss, and D. A. Yphantis, *Biopolymers* **5**, 1021 (1967).

20. S. I. Klenin, H. Fujita, and D. A. Albright, *J. Phys. Chem.* **70**, 946 (1966).

INDEX